AAPG REPRINT SERIES NO. 17

Exploration Methods and Concepts II

Selected Papers Reprinted from

AAPG BULLETIN

Compiled by

J. KASPAR ARBENZ

• Published by The American Association of Petroleum Geologists
Tulsa, Oklahoma, U.S.A., 1976

Exploration Methods and Concepts II: Preface

This volume of reprints represents the continuation of Reprint Series 16 and contains papers dealing with *trap prediction* and *evaluation and prediction of hydrocarbon charge*. Papers reprinted in Reprint Series 16 concern *evaluation of risk* and *prediction of reservoir rocks*.

The reader is referred to the preface of Reprint Series 16 for a brief discussion of the purpose and contents of these two reprint volumes and for a list of additional papers concerning methods of exploration that have been published by AAPG over the past 20 years.

J. KASPAR ARBENZ
Shell Oil Company
Houston, Texas
February 19, 1976

Reprinted from:
BULLETIN OF THE AMERICAN ASSOCIATION OF PETROLEUM GEOLOGISTS
VOL. 42, NO. 3 (MARCH, 1958), PP. 561-587, 7 FIGS.

CHIEF TOOL OF THE PETROLEUM EXPLORATION GEOLOGIST: THE SUBSURFACE STRUCTURAL MAP[1]

LOUIE SEBRING, JR.[2]
Lafayette, Louisiana

ABSTRACT

This paper, prepared primarily for the inexperienced geologist, describes the subsurface structural map with stratigraphic additions and modifications, its preparation, and its uses in the search for petroleum. An accurate, relatively uncluttered base map with the wells spotted correctly on it, is a prerequisite. Well logs provide the bulk of the basic data required, and the most useful of these logs is the electric log. Electric-log correlation is based on electrical characteristics of the datum and its position in a sequence of correlative events. Correlation difficulty is a function of lateral stratigraphic variation and of structural complication.

The mapping datum should reflect the structure of prospective producing zones, should occur over a wide area, should be encountered by most wells, and the geologist must be able to recognize it and correlate it.

Mapping should begin in areas of greatest control, and should reveal producing characteristics and trapping elements of the fields already productive. The most reliable geological interpretations are based on production information, as well as on the correlation of well logs. When production information is sketchy and the geology complicated, mapping on multiple horizons is a useful technique to determine the correct geological solution.

The contour interval selected for regional mapping will be based on studies of fields that will reveal closure necessary for accumulation.

In exploratory mapping, the geologist will direct his primary search for traps already known in the area, but he should not overlook the possibility that traps not previously recognized may occur in the area studied.

The occurrence of oil or gas shows in a non-productive well is direct evidence of a trap, and their presence is an important aid in interpretation.

Fault-throw computations can give additional control on the opposite side of the fault plane, but as they are subject to many errors, they should be used with caution. Determination of the dip and strike of faults is best determined by contouring on the fault plane.

The reflection seismograph is the geologist's most important source of additional information, but its limitations should be recognized.

The modification of the basic subsurface structural map by the addition of boundary lines of permeability of the various prospective zones, will make it useful in exploration for combination structural and stratigraphic traps.

Current maintenance of exploratory maps, made necessary by intense competition, will also enable the geologist to evaluate best his previous mapping efforts and those of his predecessors.

The exploration geologist's proper attitude is described as reasonable optimism. This leads to the conclusion that exploration for large reserves in the older producing districts should be concentrated in areas where they are known to exist. In areas previously explored unsuccessfully, new basic approaches and techniques would seem most likely of success.

The geologist must actively "sell" his maps and its prospects to his management, since all of his previous efforts are useless unless his work is used in acquiring land and drilling wells.

Barring the discovery of a direct method of finding oil, the various subsurface methods based on well control will eventually tend to supplant all other methods in the search for petroleum.

INTRODUCTION

This paper was prepared mainly for the inexperienced geologist who is just beginning his exploratory work. The ideas expressed are based on ten years of

[1] Manuscript received, June 17, 1957. Presented before the Gulf Coast Association of Geological Societies at New Orleans, Louisiana, November 8, 1957. Published by permission of Champlin Oil & Refining Co.

[2] Geologist, Champlin Oil & Refining Co. Many of the ideas expressed in this paper are not originally the writer's, but were conveyed to him by geologists employed by Champlin Oil & Refining Co. and The Standard Oil Company of Texas. The assistance of Bernard Bonnecaze in drafting the figures and Mrs. Leo Franques in typing the manuscript is gratefully acknowledged.

561

3

exploratory work in Southwest Texas, South Louisiana, and the Denver-Jules-burg basin. The writer believes that the principles of exploratory mapping that apply in these areas will also apply in other areas where sandstone production predominates.

Subsurface geology may be defined as the study of the rocks and rock forma-tions below the surface of the earth. A subsurface map is constructed from data supplied by wells penetrating these formations and from geophysical surveys that measure significant physical properties of the subsurface rocks.

Wells are now being drilled at a constantly increasing rate. More than 58,000 wells were completed in the United States in 1956. These wells provide an increas-ing wealth of subsurface information.

Many types and kinds of maps may be prepared from the available data. These include the subsurface structural map, the isopachous map, the various types of facies maps, paleogeologic maps, geophysical maps and many others. The maps most commonly used by the exploration geologist in his search for petroleum are the subsurface structural map, with some basic stratigraphic addi-tions and modifications, and the various geophysical maps, most of which are also structural maps prepared from geophysical data. This discussion is limited to the subsurface structural map with stratigraphic additions and modifications, its preparation, and its uses in the search for petroleum.

The subsurface structural map presents in plan view a three-dimensional pic-ture of the structure of a recognizable datum occurring mainly below the surface of the earth. The structure of this datum is represented by contours passing through points of equal elevation. These points of equal elevation are related to their vertical distance either above or below an arbitrary datum. The elevation on the datum bed is the algebraic difference between the surface elevation of the mapping datum, above or below the reference datum. In most parts of the coun-try where subsurface mapping is carried on in petroleum exploration, the sub-surface structural map can also be called a subsea map, since the reference datum is commonly sea-level, and the mapping datum is generally further below sea-level than the surface elevation of the well is above it.

BASE MAP

To prepare a useful subsurface map the geologist must have an accurate base map with his control points accurately spotted on it. These control points are obtained from logs of the wells that have penetrated to the mapping datum. The importance of a correctly spotted map can not be overemphasized. Huge amounts of capital have been spent buying leases and drilling wells on prospects based on incorrectly spotted wells, and undoubtedly much more money will be expended in the future because of similar errors. It is important for the best possible inter-pretation that the wells not only have the correct location in relation to the tract boundaries upon which they are located, but also they should be correctly related to wells on different tracts. In other words, the tracts themselves should be cor-

4

rectly dimensioned, shaped, and located. Misspotted or misshapen tracts can cause substantial errors in mapping.

Accuracy of the base map and its well locations is certainly the most important quality desired. However, a base map reasonably free from the clutter of land data is also desirable. The control points (wells) should be easily seen and not lost in a maze of lots and small tracts. The scale of the map used should be large enough so that the necessary detail of a complicated structure can be shown. It should be small enough so that sufficient area is covered to show the relation between several areas of present production and of possible future production. It follows, therefore, that the largest-scale maps will be used when the area mapped is complicated, highly developed, and/or mapped in extreme detail. It is desirable, when possible, to use the same base map for both leasing and mapping operations. Where a highly complicated ownership situation makes this impossible, a base with most of the ownership details removed is useful. Land ownership bases and regional map bases prepared by the same map company are desirable since ordinarily the relation of tracts between the bases is similar. The most accurate base maps are prepared from aerial maps and aerial surveys. Although maps prepared in this manner generally show tract relations correctly, the well locations are not nearly so reliable. This is because only a part of the wells drilled before the aerial survey can be located on the photographs. Locations of wells not pin-pointed on the photographs and wells drilled after the photographing must be spotted from other information. These locations that are spotted later are only as good as that information and as good as the draftsman who spots them.

WELL INFORMATION

Most of the information used by the geologist in preparing a subsurface map is obtained from wells drilled in the area to be mapped. This information is taken from logs of the wells, core descriptions, sample descriptions, completion information, *et cetera*. The more of this information that the geologist has at his disposal in readily accessible form, the more accurate will be his subsurface map.

WELL LOGS

Information taken from logs of the wells drilled in the area to be mapped provides most of the basic data required by the subsurface geologist. A well log may be defined as a record of the rocks and rock formations penetrated by the well or a record of certain physical properties of these rocks. The most common types of logs used are electric logs, sample logs, radioactivity logs, mud logs, core logs, temperature logs, drilling-time logs, drillers logs, caliper logs, *et cetera*. The most useful log for the exploration geologist is the electric log. Practically all wells now drilled in the search for petroleum are being electrically logged. No one appreciates this fact more than the geologist who has attempted to use drillers logs and sample logs for correlations where the stratigraphic section consists mostly of shale and sand. The electric log is now the basic correlation tool of the subsurface geologist. Most log libraries of companies engaged in the search for petroleum are made up mainly or wholly of electric logs. Indeed, many geologists in the Gulf Coast have never prepared a sample log.

Electric-log correlation.—Many volumes have been written about the electric log and its interpretation. It is beyond the scope of this paper and the ability of its writer to improve upon them. The discussion is limited to their uses in correlation. The elevation of the selected subsurface datum is determined by a sometimes not so simple comparison with another well log in which the datum has already been selected. This correlation is based on the characteristics of the datum and its position in a sequence of correlative events. The correlation may be relatively simple, as in the Denver-Julesburg basin, where wells tens of miles apart can be easily correlated; or very difficult, as in some parts of South Louisiana or South Texas, where wells only hundreds of feet apart can be correlated only with great difficulty. In these difficult areas micropaleontology is a great help—in some areas a necessity—in making the correlation. Difficulty of correlation is caused by two factors; first, rapid lateral stratigraphic variation and second, complicated structure. In areas where stratigraphic variations are small, and the structure is simple, the correlations and the mapping progress rapidly. An example of this is the "D" and "J" Cretaceous sand mapping in the Denver-Julesburg basin. In areas where stratigraphic variations are small, but the structure somewhat complicated (as in the Deep Wilcox trend of South Texas), or where the structure is simple, but the stratigraphic variations are great (as in some parts of the Yegua-Jackson trend of South Texas), the correlations and the mapping proceed somewhat slowly. When both structure and stratigraphy are complicated, progress is slow indeed. This condition is exemplified by the Miocene of South Louisiana and the Frio of the Rio Grande Valley of South Texas. The same geologist can map a county in the Denver-Julesburg basin faster than he can map some of the reasonably complicated fields in South Texas or South Louisiana.

The geologist who is familiar with correlation by electric log in areas of predominantly sand and shale deposition knows that the shales are more useful for minute correlations than are the sands. The electrical curves, particularly the so-called normal resistivity curve, exhibit what the geophysicist calls "character" in the shales. Most modern electric logs have an additional curve not present on the older logs, called the "amplified normal." This curve is just what it says it is, a horizontal amplification of the normal curve. In this manner, the minute variations in the normal resistivity curve are amplified, and correlation by "character" is greatly simplified.

Logging errors.—The elevation of a datum selected from any type of well log is subject to an error if the hole logged was not vertical. A hole that deviates greatly from the vertical will indicate abnormal thickness of formations and resultant greater distance from the surface to the datum point. A correction for this deviation must be made in order to determine the true elevation of the datum.

An error somewhat peculiar to the electric log is occasionally caused by improper splicing of different runs or parts of the log. This results in what might be called a "blueprint fault." A splice which omits a part of the log results in an apparent normal fault. A splice which repeats a part of the log results in an apparent reverse fault. These "blueprint faults" usually appear where an electric-log run is spliced onto the preceding run, or at a 100-foot or 50-foot depth-marker where the splice is often made.

Occasionally an apparent abnormal thickening or thinning of formations appearing on the electric log is due to differential stretching or shrinking of the material on which the log is printed. If the geologist encounters a log showing unusual thickening or thinning, he should check its vertical scale either by comparison with another log or by comparison with a ruler. In the ordinary log, this variation from true vertical scale is generally so gradual that it will not vitally affect the preparation of a subsurface map. In the preparation of a detailed structural cross section, however, it is necessary that this deviation be recognized and compensated at regular depth intervals. If this is not done the sum of the gradual scale errors compounded over the length of the log will result in a considerable error at the bottom of the log. Modern techniques of printing the log on a single strip of paper combined with modern printing materials have reduced blueprint errors to a minimum in recent years.

INDIVIDUAL WELL DATA

The geologist needs only a spotted map and some correlative well logs in order to prepare a subsurface structural map. However, the quality and reliability of this map can be considerably improved if the geologist has additional information at his disposal. This will generally consist of well information on the individual wells, geophysical maps of the area or parts of the area, and detailed field maps of fields in, or adjoining the area.

The importance of reliable well data on the individual wells in easily useable form can not be overemphasized. A reported show of oil or gas is generally direct evidence of a trap. These shows are as useful in the interpretation of a subsurface map as are the datum elevations themselves. Although reported shows are the most important information available from the individual well records, almost any information is useful. The record and description of cores and side-wall cores and the intervals and depths cored are especially useful, since they will eliminate from consideration zones that appear prospective on the electric log. The exact depths cored and a description of the results should be available to the geologist since he certainly can not condemn a zone as non-productive that has been cored without recovery. If a likely appearing zone has been side-wall cored without recovery, the condition of the bullets, whether broken, pulled off, or empty, will give the geologist valuable clues about characteristics of the formations cored. The records of drill-stem tests, potential tests, accumulated production, core-analysis data, and reserve estimates are particularly useful in evaluating a field, a prospect, an area, or a trend. In the absence of better data, casing-seat depths and mud weights can give clues to hazardous drilling conditions. Work-over dates will indicate the richness and life of individual reservoirs. Even the spud date and completion date can indicate the drilling difficulties encountered.

The large scouting staffs maintained by the major companies and the success of the many commercial oil-field reporting services indicate the importance of well information to these companies.

SELECTION OF MAPPING DATUM

The process of selecting a mapping datum may require much preliminary work, such as construction and correlation of regional cross sections and the determination and cataloging of the main producing zones and their characteristics, or it may consist simply of asking an experienced geologist what horizon he uses as a mapping datum. In either case several requirements should be met by the mapping datum selected. The map made on the datum should reflect the structure of the prospective zone or zones. It should occur over a fairly widespread area so that wells and fields in the area can be related. It should be shallow enough so that sufficient wells encounter the datum to supply enough control to construct an adequate map. And finally, the datum must be recognizable and the geologist must be able to correlate it.

Structure reflection.—The requirement that the map should reflect the structure of the prospective zone or zones is most important. For example, a map using the base of the *"Marginulina* lime" (Fig. 1) as a datum bears no recognizable resemblance to a map on an underlying producing zone (Fig. 2) at the Rayne field, Acadia Parish, Louisiana. The shallow map shows only regional dip. The difference is obvious between this shallow map and the underlying deep map with its large fault and the complete reversal and steepening of the rate of dip. One must conclude that a regional map on the base of the *"Marginulina* lime," which is the most commonly used mapping horizon in this area, would not be useful in the search for this deep accumulation. Its only use would be negative in nature, showing that in this trend a shallow anomaly is not a necessary indication for the presence of a deep structure.

Widespread occurrence.—The second requirement of widespread occurrence is not so important. Ideally, the entire mapped area can be related best by mapping on a single horizon. This is generally impossible except in stratigraphically simple areas of relatively flat rates of dip. An example of this is the Denver-Julesburg basin where several horizons can be correlated practically over the entire basin area. In the Gulf Coast, the more steeply dipping beds, abrupt lateral stratigraphic changes along strike, and the characteristic of progressively younger beds becoming the chief producing beds as the mapping proceeds from the inland areas toward the coast, generally prevents the geologist from mapping on a single horizon very far in either a dip or a strike direction. The geologist must remember that the mapping datum must reflect the structure of the main producing beds

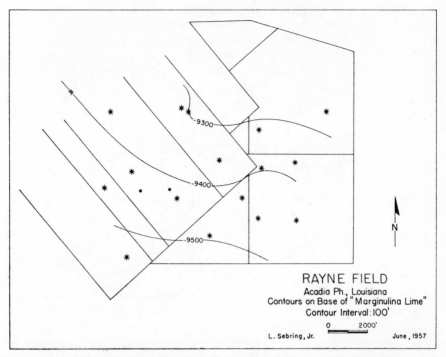

RAYNE FIELD
Acadia Ph., Louisiana
Contours on Base of "Marginulina Lime"
Contour Interval: 100'

0   2000'

L. Sebring, Jr.   June, 1957

Fig. 1.—Example of commonly used, relatively shallow, regional mapping horizon that does not give any indication of underlying deep accumulation. Depths in feet. Scale in feet.

and he should not hesitate to jump datum points when this condition is no longer met. Adjacent areas mapped on different datum points should be related by an overlap area where both datum points are selected and posted to the map.

Sufficient well control.—The third condition to be met by the mapping datum is that enough wells should encounter the datum to supply sufficient control for the construction of an adequate map. It can be argued that at the time of the discovery of the Rayne field there was insufficient deep well control to map an anomaly in this area. In this case, the geologist should recognize that the subsurface structural mapping method can not be applied to the search for that type of structure at that time, and other methods must be used. The geologist should review his shallow mapping over such an area carefully in order to see if there are any clues, whether structural or stratigraphic, to indicate the presence of the deep accumulation.

Some textbooks describe rather elaborate methods of preparing a deep map from shallow control. These methods use estimated or calculated thicknesses of beds between the desired deep map and the shallow datum. These thicknesses are presumably based on control from a few deep wells that penetrate both datums so that rates of convergence can be computed or an isopachous map constructed. Unfortunately, this geologist has never been fortunate enough to work

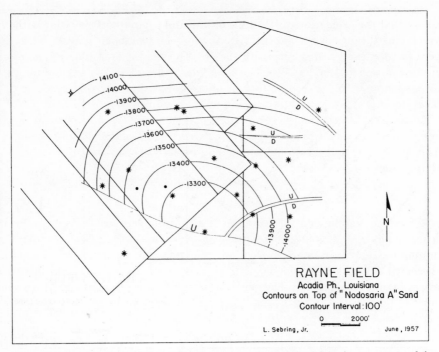

FIG. 2.—Map on deep producing zone bearing no recognizable resemblance to map on overlying regional mapping zone. Depths in feet. Scale in feet.

in an area where the rate of convergence can be determined with sufficient accuracy to compute a deeper subsurface map. Use of this method appears to impose a doubtful and possibly highly inaccurate thickness factor on a shallow subsurface map of substantially greater validity. The writer prefers to use the more reliable shallow map until sufficient deep control is obtained for the construction of a deeper map. In the case of the isopachous map, it is felt that if the data are sufficient to construct such a map, the data are sufficient to construct the deep structural map directly. This does not mean that the isopachous map is not a very useful tool in exploration, but that its use in the construction of a deep structural map would be questionable.

When a well does not reach the mapping horizon, its datum may be estimated, based on a shallow correlation. The geologist should remember that actually he is only projecting a shallow correlation and that unless the section is of a relatively uniform thickness, his projection may be subject to a large error. Estimating the datum by correlation with several deep wells will indicate the uniformity of the section and the validity of the estimated depth.

Recognition and correlation.—Finally, the geologist must be able to recognize and correlate the datum. He *must* be able to correlate it. The more easily recognizable the datum, the faster the mapping will progress. Correlations may be so easy that the geologist can select his datum and compute its elevation without even laying the logs beside another; or they may be so difficult that days may be spent in correlating a single log with the other logs in the area. There is no room for error in this matter; a miscorrelation is worse than none at all. All of the subsurface mapping and interpretation to follow is based on the correctness of the elevation of the datum and its location in respect to other control points.

Of course, the ideal mapping horizon would be a major producing zone in the area studied. It would be of widespread occurrence, a major drilling objective, easily recognized and correlated, and would reflect the structure of the other objective zones.

MAPPING TECHNIQUE

FIELD MAPPING

When the mapping datum has been selected, the geologist is ready to begin his mapping. He must remember his basic rules of contouring. These are that the contour line represents a line drawn along the intersection of a horizontal plane of a specified elevation with the datum bed, and that each contour line must pass between points of greater and less elevation than that of the contour. Also a contour can not cross over itself or another contour or branch, except it may cross over in the case of reverse faults or overturned anticlines.

The geologist should commence mapping in the areas of greatest control and work toward areas of less control. In other words, he should first map the fields or obtain detailed maps of these fields. These field maps should reveal the structure and producing characteristics of oil and gas fields already discovered in the area. They should show whether the trap which resulted in the known petroleum accumulation is of structural or stratigraphic origin or is a combination of the two. The development geologist must map all of the producing sands in the field, determine their thickness, their probable productive extent, and as much about their producing characteristics as he possibly can. He must do all this in order to help determine how to produce the field most economically to obtain the maximum oil and gas. The exploration geologist must map the field in sufficient detail to determine the trapping element or elements, and to determine if these elements can be extended beyond the limits of known production. The logical projection of these trapping elements can result in the discovery of additional reserves of oil or gas, either in an extension of the producing area, or in the discovery of a separate and distinct new field area where the entrapment conditions are similar.

If the structure and stratigraphy of the known producing area are complicated, the exploration geologist's study of the area may be long and tedious. Generally, the exploration geologist does not have at his disposal information as detailed or as reliable as does the development geologist. This is because the development geologist is usually charged only with studies of fields in which his company has production or contemplates acquisition of such production. Consequently his records of wells in these fields are likely to be excellent. The most reliable interpretation of any field is based on voluminous production information about fluid contacts, bottom-hole pressures, fluid characteristics, *et cetera*, as well as on the correlated well logs. The exploration geologist must study all of the producing fields regardless of ownership. His information on fields in which his company has no production will be gleaned from his electric logs, his probably sketchy core descriptions, a probably incomplete record of drill-stem tests made prior to completion, and the initial potential test. If the area is complicated, it is sometimes necessary for the exploration geologist to prepare maps on several horizons in order to determine the most nearly correct solution on any datum. An interpretation that satisfies all of the structural and productive requirements on numerous horizons is most probably a unique solution for the area studied. An example is a structural map on top of the "R" sand (Fig. 3), the chief producing zone in East White Lake field, Vermilion Parish, Louisiana. Although the production information available to the writer was meager, this interpretation is felt to be reasonably correct, as maps on seven horizons were prepared in arriving at the solution. Usually the exploration geologist is not required to prepare so many maps of a small area to fulfill his purposes. If his company has production in the area and a development department, he can generally obtain maps from them based on much more information than he has readily available. The geologist is often able to obtain more or less detailed field maps from sources outside his

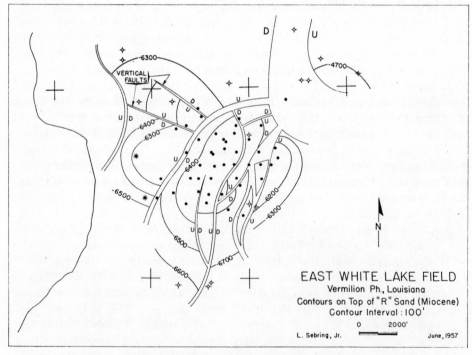

Fig. 3.—Example of complicated structure, solved by mapping on several zones. Depths in feet. Scale in feet.

company. However, he should use these with great care. He must carefully evaluate the quality of these acquired maps. In complicated areas, acquiring maps from three different sources may result in three different interpretations so different as to appear to be similar only in well names, locations, and title block. Under such conditions, the solution is for the geologist to prepare his own interpretation. If the geologist is not pressed for time, this is generally best, for in this manner he can attain the best understanding of the stratigraphy and structure of the area he is mapping. Areas of accumulation are anomalous and clues to their presence, either structural or stratigraphic, are best obtained by intense study of these already productive areas in order to determine what these clues may be. Presence of similar clues in areas not yet productive are signposts to new discoveries of similar type. Before the geologist makes his decision whether to map the fields himself or to obtain these field studies from other sources, he should remember that by acquiring these field maps from other sources he is probably eliminating them as a source of prospective production for his company. Although many geologists will be kind enough to supply their fellows with a field map, they will rarely supply him with such a map unless they have previously culled it for prospective areas of production. In most, if not all, of the older producing districts, the largest part of additional reserves found each year comes from extensions and new producing segments in and adjacent to the known producing areas, rather than from the discovery of entirely new and separate traps. Any exploration technique that ignores the additional possibilities in the fields themselves has questionable validity. This does not mean the exploration geologist should spurn completely the field maps prepared by others, because if he decides not to use them, they are still a valuable source of information to him, supplying him with alternate interpretations or the confirmation of his own. In extremely complicated fields they can show the geologist where he should commence his mapping by indicating which well is most likely to have a normal section. For example, in the East White Lake field, several of the rather shallow wells, 7,000–8,000 feet deep, are distinctly faulted in seven and eight places. A well log on one of these wells is certainly not the one to use to commence mapping.

EXPLORATION MAPPING

After the geologist has made a study of the producing fields in the area, he is ready to extend his study to adjacent areas of less control.

Contour interval.—The field studies will aid in selecting the contour interval to be used in regional mapping. If studies reveal that a closure of 25 feet will effectively trap hydrocarbons in commercial accumulations, obviously a regional map with a 200-foot contour interval would pass over many of these smaller anomalies with no hint of their presence. On the other hand, if the field studies reveal that several hundred feet of closure are required for production, a contour interval of 25 feet would just as obviously impose upon the geologist needless detailing in the preparation and maintenance of the map. The use of a contouring interval of less than 20 feet in regional mapping would not generally be recommended since a drill hole with only minor deviations from the vertical might show elevation variations that approach this figure. Contour intervals of less than 20 feet can reflect anomalies that are the results of errors in measurement and in exact correlation rather than actual subsurface structural variations.

Similar traps.—In exploratory mapping the geologist will first look for traps similar to those already producing in the area, and for the clues that might reveal their presence. The geologist should be cautioned, however, that although these familiar trap types should get his first attention, he should not overlook the possibility of a new type of trap that has not been known to produce in the area studied. This is particularly true in relatively new areas of production. A fairly recent example can be found in the Denver-Julesburg basin. The first oil found in this basin was in a closed anticlinal structure at the Gurley field in Nebraska. Later it was recognized that most of the accumulation in the basin and all of the best fields were actually caused by updip pinch-outs of permeability, either as a permeability re-entrant updip across essentially homoclinal dip, or as a permeability pinch-out across a nose. At first, probably because oil and gas were first found there, salt domes were thought to be the only structural type capable of producing petroleum in the Gulf Coast. In Southwest Texas, accumulation in fault traps was first discovered trapped against the upthrown side of up-to-the-coast faults. Early fault exploration sought this type of faulting to the exclusion of all others. Later it was discovered that oil could accumulate in anticlinal closures on the downthrown side to down-to-the-coast faults. Still later, and in relatively recent years, accumulation against the upthrown side of down-to-the-coast faults has been recognized and sought. The search for stratigraphic accumulation in South Louisiana is in its infancy and probably progress in this direction will be slow, although the geologist should not overlook the possibility in his search for oil. Companies will be understandably reluctant to drill 10,000-foot to 15,000-foot tests in search of an updip permeability pinch-out unless there is much improvement in exploration methods for this type of trap.

Rate of dip.—The rate of dip that the geologist will show will be determined in large part by the rate of dip in the areas of greatest control. The geologist should remember, however, that the rate of dip will be flatter over the highs and over the lows and will be steeper on the flanks of the structures. His contours will be parallel or nearly so, one with another, and any deviation from this or reversal or flattening of the dip indicates an anomalous structural condition that may be the clue to a structural trap containing hydrocarbons.

Oil and gas shows.—In areas of minimum control, one of the best clues to interpretation is an oil or gas show in a non-productive well. This is a direct indication of a trap and its importance can not be overemphasized. The trap that the show indicates is not necessarily of commercial value, but the geologist should study his evidence carefully before discarding the reported show as of no importance. In areas of little control, the geologist will base his interpretation of the structure in the vicinity of the show on predetermined producing structures in the trend. An example is shown in Figures 4 and 5. Geological conditions shown in plan view in the lettered diagrams of Figure 4 are shown in cross-section view, similarly lettered, in Figure 5. In Figure 4a are four non-faulted wells, A, B, C, and D. In the absence of any information on the normal rate of dip in the area, the geologist would contour this area as a simple structural nose with its axis through wells B and D. A cross-section view of this same condition is shown in Figure 5a. Suppose, however, that there is reported a show of "live" oil in well D in the top of the sand which is being used as a mapping datum. As a further condition, suppose that the mapping horizon is the top of a well developed permeable sand in all of the wells. Another rule must now be followed in contouring the area. Well D must now be separated in some manner from the other wells since it is structurally lower and contains a show, which they do not. Multiple interpretative possibilities now become apparent. The geologist will probably choose to keep the dip axis of this structure in about the same place since he now has two indications that the dip axis is as he has previously pictured it, that is, the structural contour passing through wells A, B, and C and the show in well D. A simple interpretation that could be applicable in almost any area of low dip and reasonably competent beds is found in Figures 4b and 5b. Here a high area is centered north of well D and separated from the other wells by a structural saddle. The high could just as well be south of well D, but a saddle must separate it from the

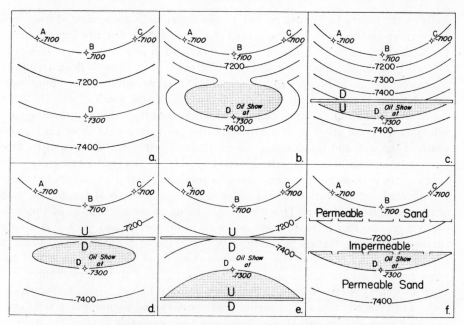

FIG. 4.—Examples of use of oil or gas shows in interpretation—map view. Prospective area, dotted pattern. Depths in feet.

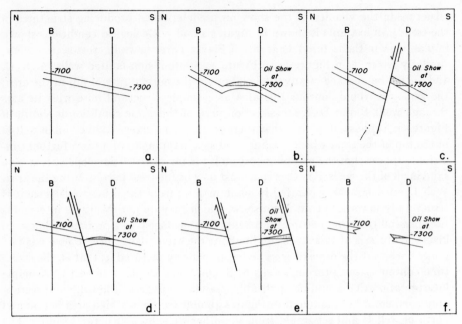

FIG. 5.—Examples of use of oil or gas shows in interpretation—cross-section view. Prospective area, dotted pattern. Depths in feet.

wells on the north. In Figures 4c and 5c, the upper wells are separated from well D by an up-to-the-south fault so placed that it would not cut any logged part of the wells. Such an interpretation might be favored in any trend where this type of trap is common, like the Edwards, "Pettus," or shallow Wilcox trends of Southwest Texas. In Figures 4d and 5d, well D is separated from the upper wells by a down-to-the-south fault with an indicated high area north of well D. If the log of well D did not go much below the datum, and if it were lengthening considerably over the other wells north of D, the geologist would show his high as south of the well and the down-to-the-south fault just north of the well D. This interpretation would be a favorite almost anywhere in the Gulf Coast where this type of trap is probably the most common producing feature. A somewhat more complicated interpretation would show down-to-the-south strike faults both north and south of the well D, with the well being on the north flank of the indicated stucture. This interpretation is shown in Figures 4e and 5e, and is a very common producing feature in the deep Wilcox trend of Southwest Texas and in some areas of the Frio and Miocene trends throughout the Gulf Coast. Still another possible interpretation is shown in Figures 4f and 5f. In these figures, the show in well D is caused by an updip permeability barrier which traverses the nose. Then since it was stated earlier that the mapping datum was a sand containing a show only in well D, but developed and permeable in all of the wells, it is necessary to draw a downdip permeability barrier south of the three structurally higher wells. This type of trap, caused by repeating bands of permeability approximately paralleling the structural strike, is found in some places in the Yegua-Jackson and Frio trends of Southwest Texas. It is the most common type of trap in the Denver-Julesburg basin, but there the figure would be turned 90° to the right since the regional dip in the main producing area of the basin is toward the west.

There are of course myriad other ways in which these data could be interpreted; however, these seem to be the simplest and most common. If there is a known producing field along strike with these wells, the geologist would probably extend the trapping element of the known field across this area in order to explain the show at D, unless there was good reason not to do so.

Absence of shows.—The absence of reported shows in the wildcat wells mapped is not so important as their presence. A prospective area should not be condemned because wells on the flanks of the prospective area have no reported shows. The flank wells may have actually had oil or gas shows that were not recorded. This may be because the well was not cored sufficiently, or because the well-sitting geologist was inexperienced, or because the type of drilling fluid obscured the show, or most commonly, the show was not recorded on the well data that the geologist has available. This is another example of the extreme importance of good well data. One geologist may map an area and indicate a highly questionable anomaly. Another may map the same area in much the same manner, but with much greater faith in the anomaly because he has well information that a well or wells on the flank of the indicated anomaly had shows.

In extremely permeable productive zones, like the Miocene of Louisiana, wells only a few hundred feet from production may penetrate the producing zone below the oil-water or gas-water contact and have no show. This phenomena is explained by the extreme permeability of the sands. In other areas, such as in parts of eastern Nueces County, some sands contain oil shows at their top, regardless of their proximity to a producing structure. This may be due to small irregularities in the top of the sand that traps minute amounts of oil during migration.

Computing datum across fault.—In areas where faults are present and effective at the mapping level, additional control may be obtained by computing the datum elevation on the opposite side of the fault. This is commonly done in the case of the normal fault by subtracting algebraically the amount of vertical displacement (throw) of the fault from the upthrown datum elevation in order to compute the elevation of the datum on the downthrown side of the fault. If the well is faulted by a normal fault below the datum, an upthrown elevation can be computed by adding algebraically the throw to the downthrown datum. Commonly, the only direct evidence of faulting found in subsurface maps made with electric logs, is an omission of section in a well cut by normal fault and a repetition of section in a well cut by a reverse fault. The amount of throw of the normal fault is assumed equivalent to the thickness of the section missing.

A datum computed in this manner is only as reliable as the throw computation. And since this computation is subject to many errors, the computed datum generally should be used cautiously in the preparation of the map. This is due to several causes. If faulting went on contemporaneously with deposition, the value of the throw of the fault will generally vary with each normal well with which the faulted well is compared. Usually the well is not faulted exactly at the datum, and it is a rare fault indeed that has the same amount of displacement at all levels. Rigorous use of computed throws in figuring datum elevations across faults can result in structural interpretations that are misleading, no matter how carefully the geologist made the throw computation. An example is shown in Figure 6, where well A is faulted at −6,900 feet. Suppose that by comparison with well B the fault throw is computed to be 200 feet. The downthrown point for well A would be − 7,330 feet and the geologist would indicate a possible high area east of A on the downthrown side of the fault (Fig. 6a). By comparison with well C suppose that the fault throw is determined to be 300 feet. In that case, the downthrown point would be −7,430 feet and essentially regional dip would be indi-

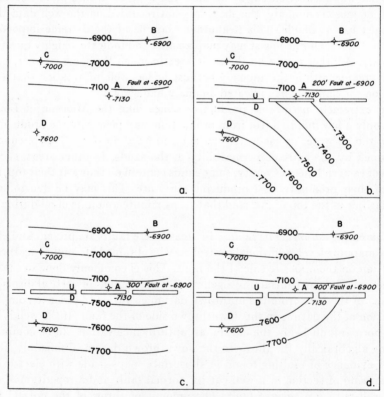

FIG. 6.—Example of effect of computed value of fault throw on interpretation. Depths in feet.

cated (Fig. 6c). Finally, suppose by comparison with well D the throw is found to be 400 feet. In this case, a high would be indicated west of well A on the downthrown side of the fault (Fig. 6d). In this quite common case where the value of the computed throw is so dependent on the normal well used in the computation, other clues such as shows, evidence of regional nosing, and geophysical anomalies would certainly be more reliable than a datum computed across a fault. If the throw is computed with the aid of a normal well along the stratigraphic strike with the faulted well, the value of the throw is much more likely to be accurate than when the normal wells are considerably updip or downdip from the faulted well. Another problem that arises with this type of datum computation is caused by the well being cut by more than one fault. Are the faults dipping the same directions such that their throws should be added in order to compute the datum across them? Or is one or more of the faults a fault dipping against the others? In this case, some of the faulting may act as compensating faulting so that its throw should be subtracted from the aggregate. This problem seems insoluble without adequate well control, and if the geologist has that much control, he usually does not need the additional information gained from the computed fault datum.

In a great many areas in the Gulf Coast, a large amount of reverse dip or turnover into the downthrown side of regional down-to-the-coast faults is a common characteristic. Almost everywhere, in this case, a computed downthrown point will be much higher than the true value of the datum. This is because a well drilled adjacent to the downthrown side of the fault will penetrate the steeply dipping beds at an angle, and each formation will show an indicated thickness greater than if the well were drilled normal to the bed surfaces. When deposition and faulting are contemporaneous, the upthrown section of the faulted well will be more similar to normal wells on the upthrown side of the fault. Also, the downthrown section of a faulted well will be more similar to normal wells on the downthrown side of the fault. Therefore, the geologist who uses upthrown normal wells in computing the upthrown datum, and downthrown normal wells in computing the downthrown datum, will get the best results with this technique.

Careful study of faulting in the areas of maximum control in the mapping area will indicate the reliability of the computed or restored datum. If the values are about the same regardless of the well used in the computation; if the fault throw is fairly constant in wells that are cut both shallow and deep; if the section is very similar in both upthrown and downthrown normal wells; and if the computed values in the areas of dense control fit into the well controlled structural picture; then, they may be considered fairly reliable in the wildcat areas. If these conditions are not met, the geologist is probably better off using the computed throw value with considerable latitude in his interpretation. He should never, of course, have his upthrown side lower than the downthrown side, and he should always have a finite difference in elevation across the fault. The fact that his fault indicates a throw of 200 feet at one point and 300 feet at a point a mile farther along its strike, should not be the cause of much concern. If he studies his fields care-

19

fully, he will probably find examples showing just this. He should remember that a fault as well as a fold is a simple result of a force or forces; that when he combines the two features into a faulted and folded, contoured, structural map his end result must explain the combined geological result of two structural features. In other words, if the geologist must traverse an area between two wells with 1,000 feet elevation difference, he will do it with structurally contoured dip, or faults, or both, as he may deem most reasonable. His only definite restriction is that whatever his interpretation, the final result must show by a combination of faulting and contours the correct difference in elevation. Field studies of most faulted fields will show that faults are commonly sinuous, have variable amounts of throw along their strike and along their dip, and show variations in degree of dip as well. When faults die out, they do so into a fold that may be minute or large according to the competency of the beds. The forces that form a fault must first fold the beds until they break. If the beds are incompetent, only a very small force is necessary to cause them to fault; if they are very competent, a very large force may be necessary.

In the special case when the datum is faulted out, it is necessary for the geologist to compute an upthrown and a downthrown datum. Fortunately, in this case, his computation should be fairly reliable. This is because the fault occurred exactly at the datum, and the throw computation is necessarily made at exactly the desired point.

Fault location.—In regional mapping, the position of the fault with respect to the well that is faulted should be shown as accurately as possible. This position is a function of the position of the fault in the well, of the direction of strike, and of the degree of dip of the fault plane. Generally, in areas away from the fields, it is difficult to find the necessary three faulted wells in a suitable geographical location so that the geologist can determine dip and strike by the three-point method. Fault studies in the fields will best determine reasonably accurate dip and strike of the faults. If there are more than three faulted wells in any area studied, the most accurate determination of dip and strike can be made by the construction of a map contoured on the fault plane. Multiple faulting in the wells may make it impossible to determine the dip and strike of the faults due to difficulty in relating the individual faults in the various wells.

Ordinarily, the scale of the map used in regional mapping is so small that it is not necessary for the geologist to show fault gap (zone where the datum surface is missing due to normal faulting). In the case of very large faults or large-scale maps, or both, it should be shown. The width of the gap shown will not only be a function of the throw of the fault but its dip as well. The larger the throw and the lower the angle of dip of the fault, the larger the fault gap.

When tracing tensional faults from areas of good control into areas of less control, the geologist will generally show them striking approximately parallel with the strike of the beds. They will commonly be downthrown in the direction of dip of the beds. This is a simple application of the theory that the faulting is a result of forces applied to the beds. The more anomalous dip faulting will be found in areas of great structural hiatus, such as around salt domes or in areas where there is a sharp change in the regional strike. The common graben structural form in salt-dome areas is a well recognized phenomenon in the Gulf Coast. The up-to-the-coast strike fault, common in some areas, can be regarded as essentially a compensating fault in many places developing in lieu of the sharp reverse dip into the more common down-to-the-coast fault.

ADDITIONAL INFORMATION

The subsurface geologist has now used most of the direct information available in interpreting the structure in areas of considerable control. He should now turn to any additional information, no matter how obscure, in his interpretation of the areas of less control. This information may range from detailed seismic maps, surface maps, core-drill maps, and shallow subsurface maps to rumors of another company's seismic high. This information will be of varying degrees of reliability, but it should all be carefully weighed before it is used or discarded in preparing the final interpretation.

A. I. Levorsen, in his book, *Geology of Petroleum*, has stated: "The petroleum geologist is in many ways like a detective—he is forever following up and evaluating clues that might lead to the discovery of a pool of oil or gas." It follows that the geologist who has the most clues or who uses the most clues will be the most successful in his search for petroleum.

Here again, a comparison of the available additional data in an area of considerable well control is probably the most useful way in which to determine their value. If the surface structural map, core-drill map, or shallow subsurface map indicates an anomaly over a producing area, then similar anomalies in wildcat territory may also indicate areas of future production. Gravity and magnetometer maps may be used in the same way. This type of evaluation of the additional information at the geologist's disposal will also help determine what type of additional exploration methods will be desirable in the future evaluation of some of his more nebulous prospects.

REFLECTION SEISMOGRAPH MAPS

Usually the most reliable deep additional information that the geologist can get without the drilling of wells is obtained from the reflection seismograph. This tool gives him a final map which attempts to show the subsurface structural conditions of the area explored at the desired depth in exactly the same manner as does the subsurface geological map. The subsurface geologist's attitude toward the seismograph map may range from sublime confidence in the map to complete ridicule of it. He may assiduously trace all of the seismic contours onto his map, or he may ignore them altogether. A compromise between the two extremes is usually the best solution. Although the seismograph is generally the best deep tool available to the subsurface geologist short of a wildcat well, it does have its limitations. These limitations and sources of error should be recognized by the geologist. It is generally stated that the seismograph's limit of error is approximately 50 feet. In areas of good records, constant weathering, and good velocity control, it may be considerably less than this. In other more difficult shooting areas, it may be considerably more than this. Probably the seismograph is most accurate in determining the direction of dip. It is also very good in determining the amount of dip. Estimations of actual depth to any horizon will vary in reliability with the available velocity control and the number of well ties. Excluding velocity and weathering problems, the seismologist generally has the most trouble with fault interpretations. Direct evidence on the records may be lacking and in this case he will usually compensate any mis-ties by drawing the suspected fault through areas of poor record quality. His value for the fault throw is often in error. Frequently he may reverse the upthrown and downthrown blocks. This is often a result of attempting to draw a preconceived structural picture, possibly foisted on him by an overzealous subsurface geologist, who assures him that a suspected fault traversing the area can be upthrown only in a specified direction.

When the subsurface geologist sends a seismograph crew into a prospect area, he will often base his opinion on their efficiency by a comparison of their map with his preconstructed subsurface map. If they picture a high, all is well; if not, he may frequently condemn the shooting as unreliable. Before condemning the shooting, the geologist should determine if the seismograph's subsurface picture is reasonable. Though the geologist's interpretation of the structure before shooting, based on all available information, is in his opinion both reasonable and possible, he should still recognize that there are other possible interpretations, probably more pessimistic than his, that will fulfill all of the geological requirements. If the seismograph interpretation is one of these, he should accept it. He most probably might as well. His company has engaged the crew at considerable expense in the belief that the crew can furnish them with a more detailed and reliable survey of the prospective possibilities of the area than can the subsurface geologist. If, on the other hand, the final seismic map of the area is geologically unreasonable, it is the geologist's duty to bring this to the attention of his management. Management should insist that the final seismic interpretation should be reasonable geologically, since in correcting any of these geological inconsistencies, evidence for new prospective possibilities may be uncovered.

Strangely enough, a seismograph map of ancient vintage may be of as much or more actual use to the exploration geologist as a seismic map recently completed. This is because with the older map his management will usually allow him considerably more latitude in its interpretation. Subsequent drilling will generally reveal inconsistencies and errors in the original seismic survey. If his company and its competitors have been active exploration-wise, there will be very few undrilled closed structural anomalies on the old map. These undrilled anomalies will attract the geologist's attention first. If similar anomalies are shown on the map and are already productive, then of course the untested anomalies become prospects of the first order. The geologist will have more confidence in the anomalies shown if some of those proved productive by drilling were proved after the completion of the shooting.

After the more obvious anomalous areas have been examined, the geologist will examine the old map carefully for the more numerous less obvious clues. Is there an area of anomalous dip that can be recontoured more logically and optimistically? Have recent developments indicated that accumulation is occurring without structural closure, but rather by updip pinch-out across a regional nose? In this case, areas of indicated structural nosing, immediately become prime prospecting grounds. The geologist may indicate faulting in areas of abnormally steep dip. He may indicate structural closure in areas of abnormally flat dip. Only his experience and his imagination will limit him in the myriad possibilities of this type of study.

COMBINATION AND STRATIGRAPHIC TRAPS

The essential requisite of any geological map used in exploration is that it must indicate the anomalies that trap the hydrocarbons. A shallow subsurface map on a datum considerably higher than the producing level is useful in exploration for petroleum at the lower level, only if an anomalous condition is revealed at the shallow level. In the case of structural accumulation, this may take the form of actual closure or a definite and recognizable flattening or nosing of the beds at the shallow level that can be definitely related to the deep structure. In the case of petroleum accumulations trapped by a combination of structure and stratigraphy or by stratigraphy alone, the subsurface structural map will not by itself satisfy the essential requirement of indication of the petroleum accumulation.

PERMEABILITY BOUNDARY LINES

The essential requirement of indication of the petroleum accumulation will generally be satisfied if the basic subsurface structure map is modified by the addition of boundary lines, commonly called pinch-out lines, showing the updip extent of permeability in the various potential producing zones. Where the bands of permeability repeat themselves along a dip profile in strike bands, as in the Denver-Julesburg basin, or where production occurs in lenticular sands regardless of the structure, the actual final map will be simplified by showing all of the limits of permeability, both up and down dip as well as lateral limits. The permeability of the individual zones is determined by core analysis, drill-stem tests, log analysis, and core description.

It can be argued correctly that if the producing reservoir is occurring in strictly stratigraphic traps, where the structure is not a trapping factor, that the addition of structure contours would serve only needlessly to clutter the map. Although this is true by definition, in the case of strictly stratigraphic traps, generally in the same area there are also petroleum accumulations where the structural element is definitely a trapping factor. The ideal exploration tool will be useful in exploration for all types of trap in the area studied.

In areas where there are numerous updip pinch-outs of different potential producing zones at various levels, these may be superimposed on a basic structural map contoured at any depth. However, the structural map must reflect the structure adequately for all of the levels at which the pinch-outs occur. The final map may become complicated if a great many of these pinch-outs occur, and the geologist must carefully differentiate the separate zone pinch-outs by coloring or by some other means.

An example of the use of permeability boundaries in exploration is shown in Figure 7. A four-township area is shown in central Kimball County, Nebraska.

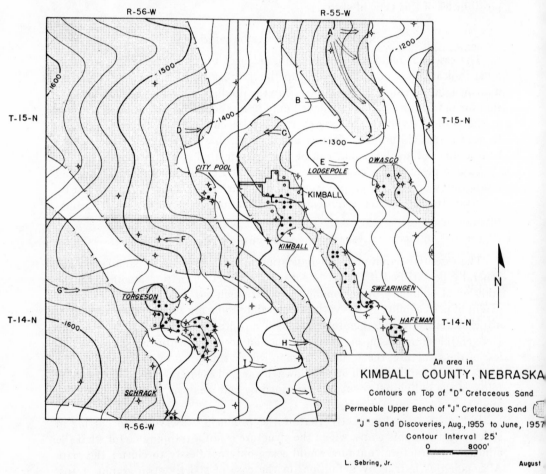

FIG. 7.—Example of effective use in exploratory mapping of permeability boundaries superimposed on subsurface structure map. Depths in feet. Scale in feet.

around the town of Kimball. The map was contoured on a 25-foot interval at the top of the "D" Cretaceous sand and was completed August 11, 1955. There are no closed structural contours on the map and, in fact, the writer mapped only one 25-foot closure in the western three-fourths of the county. At the time of the mapping, he believed that the "D" sand acted as a single permeability unit and that only the uppermost bench of the "J" sand was productive in the area. Therefore, only permeability limits for the upper bench of the "J" sand and for the entire "D" sand were superimposed on the structural map. Since the "J" sand accounts for the bulk of the production in the area, and all of the discoveries since the completion of the map, only permeability limits for what was considered its uppermost bench are shown on Figure 7.

Discoveries made in the mapped area from the date of completion of this map through May, 1957, were taken from the weekly issues of *The Oil and Gas Journal*. Their position is shown on the figure by arrows and letters, A through J.

The multiple discoveries at A, now named Lukassen, Griffith, and Aue fields, were mapped as prospective. The discoveries at C, Morton field, D, Heideman field, and H, Marian field, were also mapped as prospective. At J, although the discovery well was outside the prospective area, this well is the farthest downdip producer in the field. Subsequent development has extended updip across the prospect area and has joined the two discoveries at H, Marian field, and J, Hein field, with the already productive Sloss field at the east. Sloss field is located under the title block in the southeast corner of T. 14 N., R. 55 W., and is not shown. This is by far the most important development in the area mapped, since there are now more than 50 new producing wells completed since the mapping was done and active development is continuing. This entire producing area is now called Sloss field. The discovery at G, Kleinholtz field, could be described as a near-miss, since the area updip from the discovery was mapped as prospective. At I, Evertson field, the discovery might be termed an economic success and a geologic failure. The area was mapped as a "D" sand prospect and it actually produced from the "J" sand. Here the "J" sand possibilities were incorrectly evaluated, because all of the wells surrounding the prospect area showed the "J" sand to be tight by core and drill-stem test. The fact that all of these wells also had shows in the "J" sand should have been given more consideration.

At B, Strain field, and at E and F, which have not been named and may be considered extensions, the map did not show the areas as prospective. At B and E, the writer should have been more optimistic in extending the well defined northwest-southeast Kenton-Houtby-Owasco field trend. (Only the farthest northwest field, Owasco, occurs in the mapped area.) At F, he was almost certainly a victim of incorrect well information. An offset to the discovery well showed the producing sand to be tight by core description and containing no show. The sand appeared permeable on the electric log, and in fact was so interpreted. The failure to report a show condemned the area.

By June 1, 1957, only those new discoveries at D, Heideman field, with eight producing wells, and at H and J, the previously mentioned Sloss area, appear to be commercially significant. At A, there are seven producing wells in scattered, but associated areas, and eventually this area may attain some commercial significance. All of the other new fields had a maximum of two producing wells.

This type of evaluation of a subsurface map has its limitations, because it gives an exaggerated picture of the quality of the mapping. Some dry holes were drilled in these areas that were mapped as prospective. However, if the map had been carefully maintained as the well logs on all of the new wells in the area were obtained, a more accurate delineation of the prospect areas would be possible.

MAP MAINTENANCE

In these days of intense competition, it is essential that the maps used in exploration be maintained currently. As soon as any new information about an area previously mapped becomes available, it should be evaluated immediately and the map changed accordingly. This maintenance should be on the basis of intensity equal to the original study of the area. In other words, if intense field studies were the basis for the original mapping, then these field maps should be revised when additional wells are drilled. On the other hand, if the fields were not worked originally, there would be little point in working the recently completed field wells.

The geologist must use his information as it becomes available to him. If he allows the logs to be filed before he works them, then the chances are that some of these logs will be overlooked and never worked. His map will then suffer in comparison with that of his competition.

For the geologist himself, map maintenance is the best way he can evaluate his own work. If too many of his prospects are being tested unsuccessfully, then he is undoubtedly mapping too optimistically. If, on the other hand, and this is a much more serious error, discovery after discovery is being made in areas where he has mapped no prospects, then his approach is at best too pessimistic, and may be completely in error. The geologist who works a newly received log and finds that it fits his map with only minor modifications of the interpretation, should feel a glow of real pleasure. After all, in a small way, he has justified his place in the industry and he will keep this hard-won position only so long as his successful predictions outnumber his failures by a considerable margin.

Erasure of contours and complete recontouring over large areas because of additional well control will have a sobering effect on any geologist. Indeed it should. There is no better indication that the geologist is mapping incorrectly and is failing at his appointed task.

Basically the best maps will require the least changes and the least time for maintenance. The perfect map would require no changes since all new well datum elevations would fit the map perfectly.

EXPLORATION ATTITUDE

The perfect map is, of course, impossible of attainment. As long as there is drilling, geologists will have to change their maps to compensate for the errors revealed. If the geologist has to err, he should err on the side of optimism. After all, if a geologist condemns every wildcat well to be drilled, he will on the average be correct seven times out of eight. Fortunately, although this pessimist is correct most of the time, he will never add a barrel of oil to the reserves of this country. The geologist whose prospect is drilled and found dry certainly will not feel happy about it. He should feel infinitely worse, however, if a prospect that he has condemned is proved productive.

It has been said that new oil reserves will be found only so long as there are people who believe that additional oil reserves remain to be found. The geologist must be optimistic and positive in his approach. This attitude should apply throughout his exploration technique and even to his basic concepts of the origin, migration, and accumulation of petroleum. Each limitation that the geologist places on these concepts limits the areas where he may search for petroleum. If he believes that oil can originate only in marine beds, he immediately eliminates from his consideration large areas of predominantly non-marine beds. If he believes in long-distance migration of petroleum, he will generally eliminate from exploration consideration areas adjacent to any sizeable accumulation. If he believes that only certain traps can contain hydrocarbons, he will certainly overlook any other type of trap.

The proper attitude for the exploration geologist might be described as reasonable optimism. If he is mapping in an established area of only small oil accumulations, he can reasonably expect to find only small accumulations, as long as he is searching for the same pay zones and using the same basic techniques as his competitors. Perhaps he is searching for oil in a heretofore barren area and by some type of regional mapping discovers a number of structural anomalies. If a large percentage of these anomalies have been drilled unsuccessfully, he can not reasonably hope to be any more successful unless he uses a different technique or a different basic approach.

In the older producing districts it would seem reasonable, therefore, to concentrate the search for large reserves in those districts where they have been, and are being found. Most companies, of course, do this. In virgin areas, which have been explored without success, a new and different basic approach and techniques not previously applied to the area would seem to be more likely of success than a repetition of methods previously applied unsuccessfully.

PREVIOUS MAPPING

Heretofore, this paper has been concerned with subsurface mapping procedures where the geologist started with a blank map and built his map from the ground up. Most major companies, and a great many of the more progressive independents, who have maintained offices in certain areas, ordinarily have basic structural maps of the district. The geologist who has been transferred into the district will generally be required to maintain these maps. This new geologist will have a tendency to interpret these maps in the light of his own experience in other areas. He will often prematurely condemn the maps as useless and his predecessor as an incompetent. The succeeding geologist should not be harsh in his judgment too soon. He should maintain these maps for a reasonable length of time and see how they stand up before condemning them. He is probably in a better position to evaluate the mapping abilities of his predecessor than any other person, but he should reserve judgment until he becomes aware of the peculiarities of his new district. He should judge his predecessor's map in much the same way that he does his own. It should fill all the the basic requirements of any geological map used in exploration. These have already been discussed.

"SELLING" THE PROSPECT

The best geological map ever made is useless unless it is actively used in exploration. The geologist must actively "sell" his map and the prospects it reveals to his management. To a very large degree, the exploration success of any geologist and his company is in direct ratio to the activity of the company. If the geologist is unable to persuade his company to drill or acquire acreage, he will certainly find no oil or gas. The exploration geologist who does not "find production," for whatever reason, has failed at his appointed task.

No honest effort should be spared by the geologist to "sell" his company on his prospects. Probably there are more variations in the manner of presenting and successfully "selling" a prospect among the competing companies than in any other matter. The geologist may be able to acquire acreage or commit his company to a well by picking up the telephone; or he may be required to prepare a leatherette-bound prospect folder complete with brightly colored maps and cross sections and resplendent with fluorescent tape. Whatever the effort required, a successful result justifies it.

Ideally the geologist should be able to sell his prospect on its merits alone, honestly appraised and set forth in a straightforward manner. This will enable his management to appraise it as well, and come to a decision about its merit by a fair comparison with prospects from the other districts.

A written recommendation describing and identifying the prospect, its location, its geology, type and reliability of data, land situation, productive possibilities, expected cost, and recommended action is desirable. Not only will this written recommendation serve to consolidate the thoughts of the geologist, but it will be an adequate record of the original reasons for the action recommended. This recommendation should be accompanied by maps and some other illustrations that will show the geologist's interpretation of the area recommended and its relation to the surrounding area.

It is obvious that the management that requires the least extraneous material, over and above the basic geological requirements of the prospect recommendations, will obtain the most exploration effort from the geologist. A basic company policy that frees the geologist from as many routine non-geological duties as possible, will allow him to devote more of his time to exploration.

Periodic replacement of the original copy, by a revised up-to-date version of the regional subsurface map furnished to the management, should be a very effective long-term selling method. If a reasonable number of the geologist's prospects, which he has been unable to sell his management, produce and the periodic revisions show only minor changes from the preceding version, then the management's resistance to his new recommendations should be worn down by simple attrition.

CONCLUSION

The subsurface structural map, modified by the addition of permeability boundaries, when it is prepared and maintained by a competent geologist with a complete log file and adequate well records, will be an increasingly effective and relatively inexpensive tool for use in exploration for petroleum. Barring the discovery of a direct method of finding oil, the various subsurface methods based on well control will gradually tend to supplant all other methods as more and more wells are drilled in the never-ending search for petroleum.

REFERENCES

KRUMBEIN, W. C., AND L. L. SLOSS, 1951, *Stratigraphy and Sedimentation.* 497 pp. W. H. Freeman and Company, San Francisco.
LAHEE, FREDERIC H., 1931, *Field Geology.* 789 pp. McGraw-Hill Book Company, Inc., New York.
LEROY, L. W., 1950, *Subsurface Geologic Methods.* 1156 pp. Colorado School of Mines, Golden.
LEVORSEN, A. I., 1956, *Geology of Petroleum.* 703 pp. W. H. Freeman and Company, San Francisco.
Oil and Gas Journal. All issues: August, 1955, through May, 1957. The Petroleum Publishing Company, Tulsa.

Reprinted from:
BULLETIN OF THE AMERICAN ASSOCIATION OF PETROLEUM GEOLOGISTS
VOL. 50, NO. 2 (FEBRUARY, 1966), PP. 363-374, 11 FIGS., 1 TABLE

THEORETICAL CONSIDERATIONS OF SEALING AND NON-SEALING FAULTS[1]

DERRELL A. SMITH[2]
Metairie, Louisiana

ABSTRACT

Differentiating between sealing and non-sealing faults and their effects on the subsurface is a major problem in petroleum exploration, development, and production. The fault-seal problem has been investigated from a theoretical viewpoint in order to provide a basis for a better understanding of sealing and non-sealing faults. Some general theories of hydrocarbon entrapment are reviewed and related directly to hypothetical cases of faults as barriers to hydrocarbon migration and faults as paths for hydrocarbon migration. The phenomenon of fault entrapment reduces to a relation between (1) capillary pressure and (2) the displacement pressure of the reservoir rock and the boundary rock material along the fault. Capillary pressure is the differential pressure between the hydrocarbons and the water at any level in the reservoir; displacement pressure is the pressure required to force hydrocarbons into the largest interconnected pores of a preferentially water-wet rock. Thus the sealing or non-sealing aspect of a fault can be characterized by pressure differentials and by rock capillary properties.

Theoretical studies show that the fault seal in preferentially water-wet rock is related to the displacement pressure of the media in contact at the fault. Media of similar displacement pressure will result in a non-sealing fault to hydrocarbon migration. Media of different displacement pressure will result in a sealing fault, provided the capillary pressure is less than the boundary displacement pressure. The trapping capacity of a boundary, in terms of the thickness of hydrocarbon column, is related to the magnitude of the difference in displacement pressures of the reservoir and boundary rock. If the thickness of the hydrocarbon column exceeds the boundary trapping capacity, the excess hydrocarbons will be displaced into the boundary material. Dependent on the conditions, lateral migration across faults or vertical migration along faults will occur when the boundary trapping capacity is exceeded. Application of the theoretical concepts to subsurface studies should prove useful in understanding and in evaluating subsurface fault seals.

INTRODUCTION

Faults are recognized as an important control in the distribution of hydrocarbons in many hydrocarbon provinces. Some faults are known to be sealing to the migration of hydrocarbons under present subsurface conditions; other faults are indicated to be non-sealing to hydrocarbons and commonly are postulated to have provided the path for migration of hydrocarbons in the geologic past. Differentiating between sealing and non-sealing faults and their effects in the subsurface is a major problem for the petroleum geologist and engineer in petroleum exploration, development, and production.

This paper treats the fault-seal problem from a theoretical viewpoint in order to provide the basis for a better understanding of sealing and non-sealing faults. Some general theories of hydrocarbon entrapment are reviewed and related directly to hypothetical cases of faults as barriers to hydrocarbon migration and faults as paths for hydrocarbon migration. Application of the theo-

retical concepts to subsurface studies is considered.

GENERAL THEORY AND EQUATIONS

Hubbert (1953), in deriving the theoretical concepts of a petroleum trap, has shown that the boundary of a reservoir rock is a barrier to hydrocarbon migration because of its capillary properties. Specifically, the boundary represents a difference in the displacement pressure of the reservoir rock and the boundary rock. Displacement pressure is defined as the pressure required to force hydrocarbons into the largest interconnected pores of a hydrophilic (preferentially water-wet) rock. Below this displacement pressure, no hydrocarbons can enter the water-wet rock. The boundary rock necessarily has a higher displacement pressure than the reservoir rock.

Some calculations by Hubbert (1953) give an order of magnitude of the displacement pressures for sediments of various grain sizes in the oil-water system, as shown in Table I. Clay, for example, is a good boundary rock because a large differential pressure is required to force oil into its very minute pores. Clean, uncemented sand is a poor boundary rock (and thus a good reservoir rock) because a very small differential pressure is

[1] Presented before the 15th Annual Meeting of the Gulf Coast Association of Geological Societies, Houston, Texas, October 28, 1965. Manuscript received, February 3, 1965.

[2] Shell Oil Company. (ERP Publication 393).

TABLE I. CAPILLARY DISPLACEMENT PRESSURE FOR
SEDIMENTS OF VARIOUS GRAIN SIZES IN THE
OIL-WATER SYSTEM
(Modified after Hubbert, 1953)

Sediment	Grain Diameters d (Millimeters)	Capillary Displacement Pressure p_d (Atmospheres)
Clay	Less than 1/256*	Greater than 1
Silt	1/256 to 1/16	1 to 1/16
Sand	1/16 to 2	1/16 to 1/500
Granules	2 to 4	1/500 to 1/1000

* The value 1/256 mm. for clay particles is maximum; much finer clays are known. In clay with particle size of 10^{-4}, for example, p_d would be about 40 atmospheres.

required to force oil into the pores. Regardless of the rock type, hydrocarbons must be subjected to a pressure equal to or greater than the displacement pressure of the rock before they will enter the pores of the rock. To state this in different terms, hydrocarbons will be trapped unless they are subjected to a differential pressure equal to or greater than the displacement pressure of the rock. This principle of entrapment of hydrocarbons has been pointed out by Hobson (1954, p. 7–10), Levorsen (1954, p. 436–438), Hill *et al.* (1961), and Roach (1965, p. 133).

In the common subsurface situation in which water is the wetting phase and hydrocarbons the non-wetting phase, Hill *et al.* (1961) have shown that hydrocarbons will be trapped if the capillary pressure is less than the displacement pressure of the reservoir boundary material. Capillary pressure is defined as the differential pressure between the hydrocarbons and the water at any level in the reservoir. Pressure in the hydrocarbons will exceed the pressure in the water at any given level because of the difference in density between the hydrocarbons and the water (Hubbert and Rubey, 1959). The differential pressure or capillary pressure increases at increasing elevations above the free water level of the accumulation. Hill *et al.* (1961) have shown that a boundary rock with a finite displacement pressure has a finite trapping capacity in terms of the thickness of hydrocarbon column. A rock that serves as a boundary to a certain thickness of hydrocarbon column may not necessarily be a boundary to a greater thickness of hydrocarbon column.

Capillary pressure may be expressed by several different equations as shown by Leverett

(1941), Thornton and Marshall (1947), Pirson (1950), Levorsen (1954), and Cole (1961). It is commonly expressed by the equation

$$P_c = (\rho_w - \rho_h)gz \qquad (1)$$

where P_c is capillary pressure in dynes per square centimeter, ρ_w and ρ_h are, respectively, the densities of the water and the hydrocarbons in the reservoir in grams per cubic centimeter, g is the acceleration of gravity in centimeters per second squared, and z is the height in centimeters above the free water level or zero capillary pressure plane of the hydrocarbon accumulation (Leverett, 1941). Expressed in units of measurement commonly used in oil-field practice, the equation for capillary pressure also may be written as

$$P_c = (\rho_w - \rho_h)0.433z \qquad (2)$$

where 0.433 is a derived constant of suitable dimensions to allow an expression of capillary pressure in pounds per square inch, density in grams per cubic centimeter, and height in feet.

An example of capillary pressure in a hypothetical oil reservoir is given in Figure 1. In this example, the maximum capillary pressure is 20 pounds per square inch; *i.e.*, the pressure in the oil phase at the top of the reservoir exceeds the pressure in the water phase by 20 pounds per square inch. Based on the concepts of Hill *et al.* (1961), the boundary rock overlying this hypothetical oil reservoir must have a displacement pressure of at least 20 pounds per square inch in order to trap the thickness of oil column shown. In more general terms, the hydrocarbons will be trapped if the capillary pressure at all points is less than the displacement pressure of the reservoir boundary material. This concept of entrapment of hydrocarbons is directly applicable to the problem of sealing and non-sealing faults.

For convenient reference, the discussions of sealing and non-sealing faults will deal with hypothetical examples of faulting in an alternating sequence of sandstone and shale such as that found in the Tertiary section of the Gulf Coastal Plain. However, the significant parameter is the capillary property of the rock and not the rock type. The theoretical concepts are equally applicable to other rock sequences such as carbonate rocks which exhibit a wide variation in capillary properties, as shown by Archie (1952), Murray (1960), and others. Stout (1964), for example,

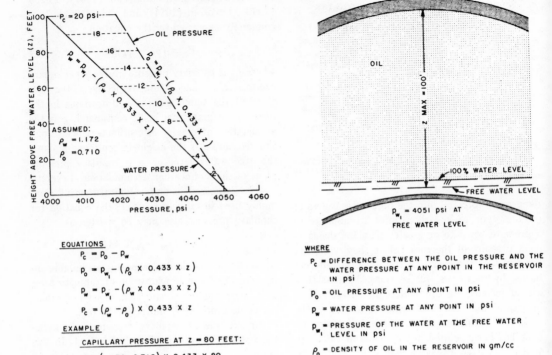

EQUATIONS

$$P_c = p_o - p_w$$

$$p_o = p_{w_1} - (\rho_o \times 0.433 \times z)$$

$$p_w = p_{w_1} - (\rho_w \times 0.433 \times z)$$

$$P_c = (\rho_w - \rho_o) \times 0.433 \times z$$

EXAMPLE

CAPILLARY PRESSURE AT $z = 80$ FEET:

$$P_c = (1.172 - 0.710) \times 0.433 \times 80$$

$P_c = 16$ psi AT 80 FEET ABOVE THE FREE WATER LEVEL

WHERE

P_c = DIFFERENCE BETWEEN THE OIL PRESSURE AND THE WATER PRESSURE AT ANY POINT IN THE RESERVOIR IN psi

p_o = OIL PRESSURE AT ANY POINT IN psi

p_w = WATER PRESSURE AT ANY POINT IN psi

p_{w_1} = PRESSURE OF THE WATER AT THE FREE WATER LEVEL IN psi

ρ_o = DENSITY OF OIL IN THE RESERVOIR IN gm/cc

ρ_w = DENSITY OF WATER IN THE RESERVOIR IN gm/cc

z = HEIGHT ABOVE THE FREE WATER LEVEL IN FEET

Fig. 1 - Capillary pressure in a hypothetical oil reservoir under hydrostatic conditions.

has applied these concepts in his work on pore geometry as related to carbonate stratigraphic traps.

In theory, there are two general ways in which a boundary to the lateral migration of hydrocarbons might result from faulting: (1) by juxtaposed sedimentary lithologic types of different capillary properties and (2) by emplaced fault-zone material formed by mechanical or chemical processes related directly or indirectly to faulting. Sandstone in contact with shale is an obvious example of the first; perhaps less obvious is the second situation, which is illustrated in Figure 2a. Here, two sandstone bodies of different capillary properties are juxtaposed by faulting. These two bodies are termed here Sand 1 and Sand 2. If the displacement pressure of Sand 1 (p_{dR}) is less than the displacement pressure of Sand 2 (p_{dB}), hydrocarbons are trapped, provided that the capillary pressure (P_c) at all points in Sand 1 along

the interface of the two beds is less than the displacement pressure (p_{dB}) of Sand 2. Similarly, hydrocarbons can be trapped against fault-zone material of high displacement pressure, as illustrated in Figure 2b. In both situations the fault must be considered to be a sealing fault because hydrocarbons are trapped at the fault.

From previous discussions, the trapping-capacity limit of the boundary material is determined by the elevation at which the capillary pressure equals the displacement pressure of the boundary, i.e., where

$$P_c = p_{dB}. \qquad (3)$$

The maximum height above the free water level at which hydrocarbons can be trapped by the boundary material is found by substituting equation (2) into equation (3) and solving for z:

$$z\,\text{max} = \frac{p_{dB}}{(\rho_w - \rho_h) \times 0.433}. \qquad (4)$$

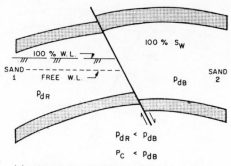

(a) SEALING FAULT FORMED AS A RESULT OF
DIFFERENT CAPILLARY PROPERTIES OF
JUXTAPOSED SEDIMENTARY LITHOLOGIES
(SANDS).

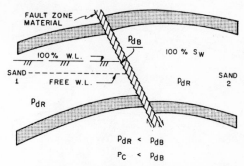

(b) SEALING FAULT FORMED AS A RESULT OF
DIFFERENT CAPILLARY PROPERTIES OF
RESERVOIR ROCK AND FAULT ZONE
MATERIAL.

Fig. 2 - Schematic sections illustrating some hypothetical situations of fault entrapment of hydrocarbons under hydrostatic conditions.

The maximum thickness of hydrocarbon column is somewhat less than z max and can be derived from equation (4). The schematic sections in Figure 3 illustrate some hypothetical examples of fault traps filled to capacity. In these situations, z max in the reservoir can be expressed in terms of the height of the continuous phase of hydrocarbon column (h_o max) and the height from the free water level to the 100 per cent water level (h_w) by

$$z \max = h_o \max + h_w . \qquad (5)$$

Substituting equation (5) into equation (4) gives

$$h_o \max + h_w = \frac{p_{dB}}{(\rho_w - \rho_h) \times 0.433} . \qquad (6)$$

The height from the free water level to the 100-per cent water level (h_w) is determined by the

point at which the capillary pressure (P_c) equals the displacement pressure of the reservoir rock (p_{dR}). For the hydrostatic case, this is expressed by

$$P_c = p_{dR} = (\rho_w - \rho_h) \times 0.433 h_w , \qquad (7)$$

and the equation for h_w is

$$h_w = \frac{p_{dR}}{(\rho_w - \rho_h) \times 0.433} . \qquad (8)$$

Substituting equation (8) into equation (6) and solving for h_o max yields

$$h_o \max = \frac{p_{dB} - p_{dR}}{(\rho_w - \rho_h) \times 0.433} . \qquad (9)$$

The maximum thickness of hydrocarbons that can be trapped by the boundary is related to the

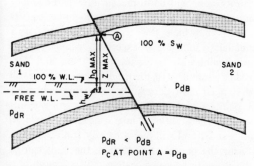

(a) DIFFERENT CAPILLARY PROPERTIES OF JUXTAPOSED
SEDIMENTARY LITHOLOGIES (SAND).

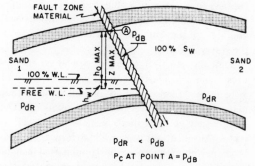

(b) DIFFERENT CAPILLARY PROPERTIES OF RESERVOIR
ROCK AND FAULT ZONE MATERIAL.

Fig. 3 - Schematic sections illustrating some hypothetical situations of fault traps filled to capacity under hydrostatic conditions.

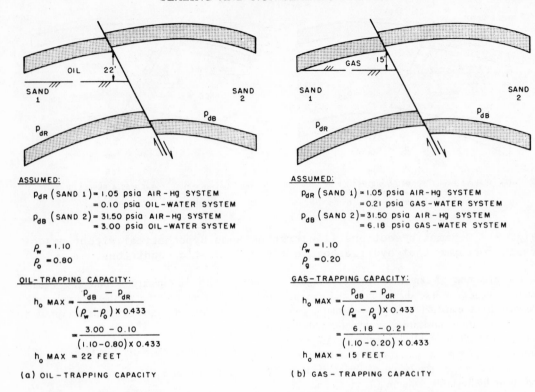

ASSUMED:

p_{dR} (SAND 1) = 1.05 psia AIR-Hg SYSTEM
            = 0.10 psia OIL-WATER SYSTEM

p_{dB} (SAND 2) = 31.50 psia AIR-Hg SYSTEM
            = 3.00 psia OIL-WATER SYSTEM

ρ_w = 1.10
ρ_o = 0.80

OIL-TRAPPING CAPACITY:

$$h_o \text{ MAX} = \frac{p_{dB} - p_{dR}}{(\rho_w - \rho_o) \times 0.433}$$

$$= \frac{3.00 - 0.10}{(1.10 - 0.80) \times 0.433}$$

h_o MAX = 22 FEET

(a) OIL-TRAPPING CAPACITY

ASSUMED:

p_{dR} (SAND 1) = 1.05 psia AIR-Hg SYSTEM
            = 0.21 psia GAS-WATER SYSTEM

p_{dB} (SAND 2) = 31.50 psia AIR-Hg SYSTEM
            = 6.18 psia GAS-WATER SYSTEM

ρ_w = 1.10
ρ_g = 0.20

GAS-TRAPPING CAPACITY:

$$h_o \text{ MAX} = \frac{p_{dB} - p_{dR}}{(\rho_w - \rho_g) \times 0.433}$$

$$= \frac{6.18 - 0.21}{(1.10 - 0.20) \times 0.433}$$

h_o MAX = 15 FEET

(b) GAS-TRAPPING CAPACITY

Fig. 4 - Hypothetical examples illustrating the hydrocarbon-trapping capacity of a boundary sand under hydrostatic conditions.

difference in the displacement pressures of the boundary rock and the reservoir rock. Hydrocarbons can not be trapped unless there is a difference in displacement pressures at the interface of the media. Further, if fault-zone material is present with a displacement pressure less than that of the reservoir sandstones, or if the fault consists of a series of connected, open fractures of low displacement pressure, hydrocarbons will migrate up the fault rather than accumulate in the reservoir beds.

Some hypothetical examples are given in Figures 4 and 5 to illustrate the trapping capacity of a boundary sandstone and boundary fault-zone material, based on the principles discussed. The distribution of the boundary material as well as its displacement pressure will govern the maximum height of hydrocarbon column that can be trapped at a fault. If the boundary material is distributed uniformly along the fault, the displacement pressure of the boundary material will be the controlling factor. In some situations, the distribution of the boundary material may be the

controlling factor. For instance, if the fault-zone material in Figure 5 were distributed across only the upper 20 feet of Sands 1 and 2, it is obvious that only 20 feet of hydrocarbon column could be trapped against the fault in Sand 1.

If the thickness of the hydrocarbon column exceeds the boundary trapping capacity, the excess hydrocarbons will be displaced into the boundary material. Figure 6 illustrates two hypothetical situations in which the trapping capacity of a boundary sandstone is exceeded; capillary pressure equilibrium is assumed in these and succeeding examples. In Figure 6a, hydrocarbons will be trapped until the capillary pressure in Sand 1 at point A equals the displacement pressure of Sand 2. The additional migrating oil is displaced into Sand 2 and will migrate away from the fault. During the period of migration, the fault can be considered to be a non-sealing fault, because oil is migrating across the fault. Ultimately, after migration ceases, the fault will be sealing to some height of hydrocarbon column. The final height of the hydrocarbon column trapped in Sand 1

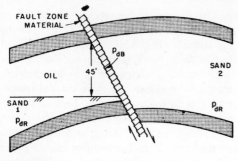

ASSUMED:

p_{dR} (SANDS) = 1.05 psia AIR-Hg SYSTEM
　　　　　　 = 0.10 psia OIL-WATER SYSTEM
p_{dB} (FAULT ZONE) = 62.48 psia AIR-Hg SYSTEM
　　　　　　 = 5.95 psia OIL-WATER SYSTEM

ρ_w = 1.10
ρ_o = 0.80

OIL-TRAPPING CAPACITY:

$$h_o \text{ MAX} = \frac{p_{dB} - p_{dR}}{(\rho_w - \rho_o) \times 0.433}$$

$$= \frac{5.95 - 0.10}{(1.10 - 0.80) \times 0.433}$$

$$h_o \text{ MAX} = 45 \text{ FEET}$$

(a) OIL-TRAPPING CAPACITY

ASSUMED:

p_{dR} (SANDS) = 1.05 psia AIR-Hg SYSTEM
　　　　　　 = 0.21 psia GAS-WATER SYSTEM
p_{dB} (FAULT ZONE) = 62.48 psia AIR-Hg SYSTEM
　　　　　　 = 12.25 psia GAS-WATER SYSTEM

ρ_w = 1.10
ρ_g = 0.20

GAS-TRAPPING CAPACITY:

$$h_o \text{ MAX} = \frac{p_{dB} - p_{dR}}{(\rho_w - \rho_g) \times 0.433}$$

$$= \frac{12.25 - 0.21}{(1.10 - 0.20) \times 0.433}$$

$$h_o \text{ MAX} = 31 \text{ FEET}$$

(b) GAS-TRAPPING CAPACITY

Fig. 5 - Hypothetical examples illustrating the hydrocarbon-trapping capacity of boundary fault zone material under hydrostatic conditions.

will be equal to the trapping capacity as determined by use of equation (9), regardless of the volume of hydrocarbons that might have migrated into Sand 1. With a suitable trapping environment in the downthrown block, a non-sealing fault to lateral migration could result, as illustrated in Figure 6b. The difference in water levels between the two blocks is related to the

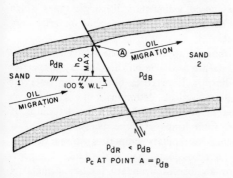

(a) FAULT TRAP FILLED TO CAPACITY; ADDITIONAL MIGRATING OIL IS DISPLACED INTO SAND 2. THICKNESS OF THE TRAPPED HYDROCARBON COLUMN IN SAND 1 REMAINS CONSTANT.

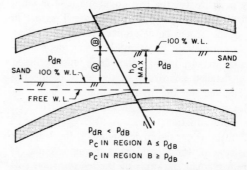

(b) NONSEALING FAULT TO LATERAL MIGRATION WITH DIFFERENT WATER LEVELS IN JUXTAPOSED RESERVOIRS. DIFFERENCE IN WATER LEVELS IS RELATED TO DIFFERENCE IN CAPILLARY PROPERTIES OF JUXTAPOSED SANDS.

Fig. 6 - Schematic sections illustrating some hypothetical situations in which the trapping capacity of a boundary sand is exceeded under hydrostatic conditions.

difference in displacement pressures of Sands 1 and 2. The fault is non-sealing to lateral migration because there is a continuous phase of hydrocarbons between the fault blocks, and hydrocarbons will produce as a single accumulation as long as the continuous hydrocarbons phase exists. A hypothetical migration, accumulation, and production history of such a faulted reservoir is illustrated in Figure 7.

Figure 8 illustrates some hypothetical situations in which the trapping capacity of the fault-zone material is exceeded. In Figure 8a, the additional migrating oil is displaced into the fault-zone material. The oil would then tend to be expelled into Sand 2, because the capillary forces on the oil, as discussed by Hubbert (1953, p. 1977), would be less in Sand 2 than in the fault-zone material. In Figure 8b, the oil displaced into the fault-zone material would migrate vertically through the fault-zone material because the shale is a boundary to lateral migration of the oil displaced into the fault-zone material.

The equations have been developed for determining the boundary trapping capacity under hydrostatic conditions. The principles of entrapment by differences in displacement pressure are valid, however, for hydrocarbons in both hydrostatic and hydrodynamic environments. As shown in Figure 9, hydrocarbons will be trapped in a hydrodynamic situation if the capillary pressure at all points does not exceed the boundary displacement pressure. The determination of capillary pressure for the hydrodynamic case is more difficult than for the hydrostatic case. Pressure in the oil phase can be determined by the equations for a hydrostatic situation; however, determination of pressure in the water phase must be based on a detailed knowledge of the vertical pressure distribution in the dynamic water, which can differ significantly from the vertical distribution in a static situation.

For the traps shown in Figure 9, the trapping capacity of the boundary sandstone, Sand 2, (the maximum thickness of the hydrocarbon column directly below point A) would be the same for the hydrostatic case and for the hydrodynamic case in which water flow across the interface is horizontal. However, horizontal water flow would in very few instances, if ever, be associated with a hydrocarbon reservoir in the subsurface. Water usually will move nearly parallel with the bed-

ding through permeable rocks, as has been pointed out by Hubbert (1953, p. 1970). The dip of the stratum will produce a vertical component of flow in the dynamic water phase. The presence of a hydrocarbon accumulation in a permeable stratum also contributes to some vertical component of flow in the water phase, because the accumulation will act as a restriction to water flow, and water will tend to divert around it. The maximum thickness of the hydrocarbon column below point A (Fig. 9) trapped by the boundary will be greater than in the hydrostatic case if there is a downward component of water flow at the interface, and less than in the hydrostatic case if there is an upward component of water flow at the interface. These examples illustrate that a fault can be non-sealing to water movement without necessarily being non-sealing to hydrocarbon movement.

APPLICATION

Application of the theories of entrapment could resolve many aspects of the fault-seal problem if data were available on the capillary properties of media in contact at the fault. There is, however, a paucity of data on the character of fault zones, fault-zone materials, and associated features in sediments near faults in the subsurface. This presents a considerable problem because a determination of the materials in contact at the fault is not usually possible in the subsurface. The determination that can be made is one of juxtaposed sedimentary lithologic types which would be in contact if no fault-zone material were present.

An indirect approach to the problem is possible. Data are available on the capillary properties of sediments from cores obtained for reservoir and petrophysical determinations. Evaluations can be made from these data to determine if known fault traps are the likely result of a difference in capillary properties of sedimentary lithologic types juxtaposed at the fault. The presence or absence of boundary fault-zone material along the fault might be inferred from these investigations. In addition, the thickness of hydrocarbons trapped at a fault provides a means for estimating the minimum displacement pressure of the fault boundary material, which might be useful for evaluating faults.

The capillary properties of rocks usually are

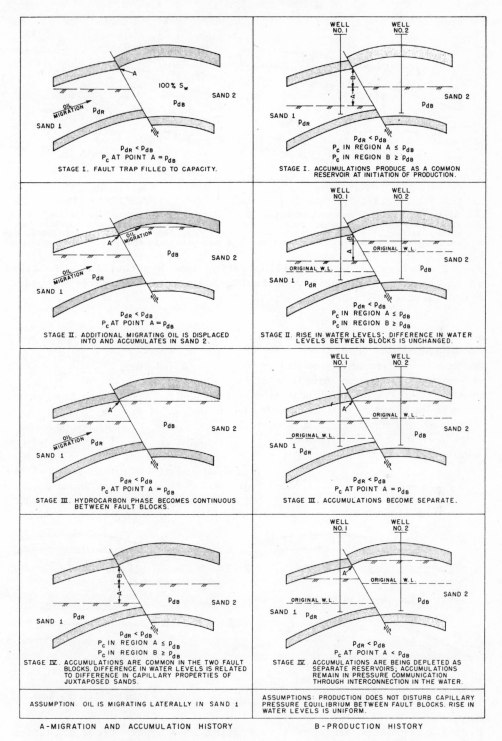

Fig. 7 - Schematic sections illustrating a hypothetical migration, accumulation, and production history of juxtaposed reservoir sands of different capillary properties.

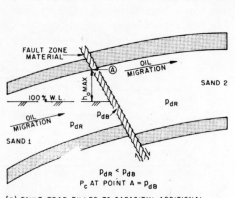

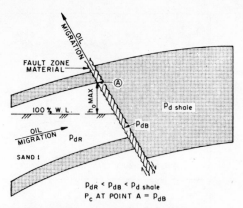

(a) FAULT TRAP FILLED TO CAPACITY; ADDITIONAL MIGRATING OIL IS DISPLACED THROUGH THE FAULT ZONE MATERIAL INTO SAND 2. THICKNESS OF THE TRAPPED HYDROCARBON COLUMN IN SAND 1 REMAINS CONSTANT.

(b) FAULT TRAP FILLED TO CAPACITY; ADDITIONAL MIGRATING OIL IS DISPLACED INTO THE FAULT ZONE MATERIAL. THICKNESS OF THE TRAPPED HYDROCARBON COLUMN IN SAND 1 REMAINS CONSTANT.

Fig. 8 - Schematic sections illustrating some hypothetical situations in which the trapping capacity of fault zone material is exceeded under hydrostatic conditions.

measured in the laboratory by injecting mercury into dry rock samples (Purcell, 1949). The displacement pressure can be determined from the air-mercury capillary pressure curve (Fig. 10). In subsurface evaluations, data from the air-mercury system of the laboratory must be converted to the gas-water and oil-water systems encountered in the reservoir rock. Approximate conversion factors are given for the gas-water system as

$$P_c \text{ (gas-water)} = \frac{P_c \text{ (air-mercury)}}{5.1}, \quad (10)$$

and for the oil-water system as

$$P_c \text{ (oil-water)} = \frac{P_c \text{ (air-mercury)}}{10.5}. \quad (11)$$

The conversion factors are based on interfacial tensions of 35 dynes per centimeter for the oil-water system, 72 dynes per centimeter for the gas-water system, and 480 dynes per centimeter for the air-mercury system. The interfacial tension is affected by the composition of the fluid as well as by other factors, including pressure and temperature; therefore, the conversion factors

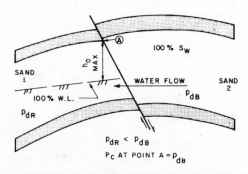

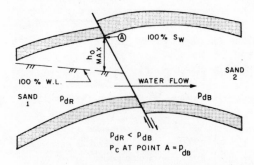

(a) WATER FLOW FROM THE BOUNDARY SAND INTO THE RESERVOIR SAND.

(b) WATER FLOW FROM THE RESERVOIR SAND INTO THE BOUNDARY SAND.

Fig. 9 - Schematic sections illustrating some hypothetical situations of fault entrapment of hydrocarbons under hydrodynamic conditions with sands juxtaposed across the fault.

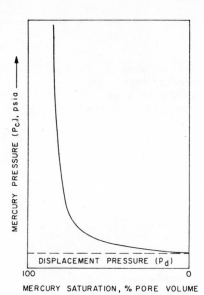

MERCURY PRESSURE (Pc), psia

DISPLACEMENT PRESSURE (Pd)

100   0

MERCURY SATURATION, % PORE VOLUME

Fig. 10 - Idealized air-mercury capillary pressure curve.

given above may be only an approximation in specific cases.

As shown in Table I, sand generally has low displacement pressures and a small hydrocarbon-trapping capacity, whereas clay has high displacement pressures and a large hydrocarbon-trapping capacity. Not all sandstone bodies have low displacement pressures; shaly sandstone, poorly sorted sandstone, and sandstone with calcite or silica cement can have high displacement pressures and a large hydrocarbon-trapping capacity. Figure 11 shows the displacement pressures of some cores from Gulf Coast Tertiary sandstones. Many of the sandstone cores have displacement pressures of sufficient magnitude to trap significant thicknesses of hydrocarbons. A comparison of these observed sandstone displacement pressures with the data used in the hypothetical examples in Figures 4 and 5 will give a general idea of the significance of the data.

Other data necessary to determine the trapping capacity commonly are available from measurements made on a routine basis in field operations. Reservoir-pressure data are available from drill-stem tests and from pressure measurements made during the producing life of a reservoir. PVT analyses provide data on the density of the hydrocarbons under reservoir conditions. In addition, the subsurface density of oil, gas, and water can be estimated in some cases from surface measurements commonly available on well effluent.

SUMMARY AND CONCLUSIONS

1. In hydrophilic rock, the displacement pressure of the media in contact at the fault determines whether a fault is sealing or non-sealing to the migration of hydrocarbons. If the media have similar displacement pressures, the fault will be non-sealing to hydrocarbon migration. If the media have different displacement pressures, the fault will be sealing, provided that the capillary pressure is less than the boundary displacement pressure.

2. The trapping capacity (the thickness of the trapped hydrocarbon column) of a fault boundary is related to the difference in displacement pressures of the reservoir rock and the boundary rock. If the thickness of the hydrocarbon column exceeds the boundary trapping capacity, the excess hydrocarbons will be displaced into the boundary rock. Dependent on the conditions, lateral migration across faults or vertical migration along faults will occur when the boundary trapping capacity is exceeded.

3. In theory, hydrocarbons can be trapped at a fault where sedimentary lithologic types of different displacement pressure are in contact or where fault-zone material with a displacement pressure greater than that of the reservoir beds is present along the faults. The distribution and displacement pressure of the boundary material along the fault are factors controlling the thickness of hydrocarbon column that can be trapped at a fault.

4. A fault can be non-sealing to vertical migration if fault-zone material has a lower displacement pressure than that of the reservoir beds or if the fault consists of a series of connected, open fractures of low displacement pressure.

5. If the capillary properties of media in contact at the fault (including sedimentary lithologic types and fault-zone material) are known, it can be determined whether a fault is sealing or non-sealing, and the trapping capacity of the fault can be calculated. If the presence or absence of boundary fault-zone material is unknown, data usually are available to determine if known fault traps are the likely result of a difference in capillary properties of sedimentary lithologic types

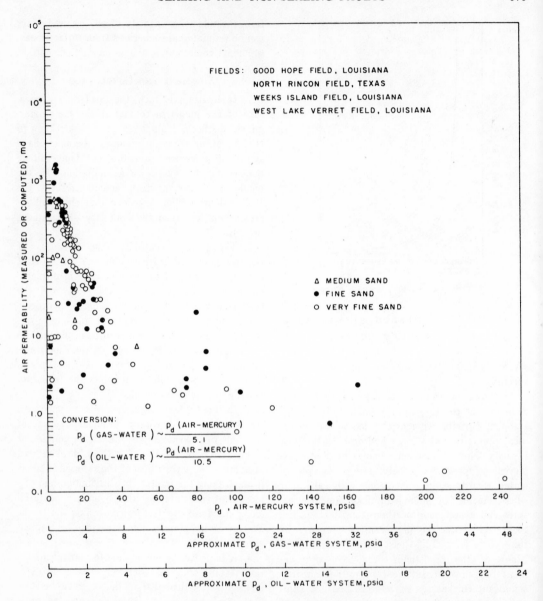

Fig. 11 - Displacement pressure (p_d) from air-mercury capillary pressure curves of some cores from Gulf Coast Tertiary sands.

juxtaposed at the fault. The presence or absence of boundary fault-zone material along the fault can be inferred in some cases from these investigations.

REFERENCES

Archie, G. E., 1952, Classification of carbonate reservoir rocks and petrophysical considerations: Am. Assoc. Petroleum Geologists Bull., v. 36, no. 2, p. 278–298.

Cole, Frank W., 1961, Reservoir engineering manual: Houston, Gulf Publishing Co., p. 5–9.

Hill, Gilman A., Colburn, William A., and Knight, Jack W., 1961, Reducing oil-finding costs by use of hydrodynamic evaluations, *in* Economics of petroleum exploration, development, and property evaluation: International Oil and Gas Educational Center, Southwest Legal Foundation: Englewood Cliffs, N. J., Prentice-Hall, Inc., p. 38–69.

Hobson, G. D., 1954, Some fundamentals of petroleum geology: London, Oxford University Press.

Hubbert, M. King, 1953, Entrapment of petroleum

under hydrodynamic conditions: Am. Assoc. Petroleum Geologists Bull., v. 37, no. 8, p. 1954–2026.

—— and Rubey, William W., 1959, Role of fluid pressure in mechanics of overthrust faulting: Geol. Soc. America Bull., v. 70, no. 2, p. 149–150.

Leverett, M. C., 1941, Capillary behavior in porous solids: Trans. A.I.M.E., v. 142, p. 152–169.

Levorsen, A. I., 1954, Geology of petroleum: San Francisco, W. H. Freeman and Co., p. 433–439.

Murray, R. C., 1960, Origin of porosity in carbonate rocks: Jour. Sed. Petrology, v. 30, no. 1, p. 59–84.

Pirson, Sylvan J., 1950, Elements of oil reservoir engineering: New York, McGraw-Hill Book Co., Inc., p. 245–272.

Purcell, W. R., 1949, Capillary pressures—their measurement using mercury and the calculation of permeability therefrom: Trans. A.I.M.E., v. 186, p. 39–48.

Doach, J. W., 1965, How to apply fluid mechanics to petroleum exploration: World Oil, April, p. 131–134.

Stout, John L., 1964, Pore geometry as related to carbonate stratigraphic traps: Am. Assoc. Petroleum Geologists Bull., v. 48, no. 3, p. 329–337.

Thornton, O. F., and Marshall, D. L., 1947, Estimating interstitial water by the capillary pressure method: Trans. A.I.M.E., v. 170, p. 69–80.

Reprinted from:
BULLETIN OF THE AMERICAN ASSOCIATION OF PETROLEUM GEOLOGISTS
VOL. 50, NO. 10 (OCTOBER, 1966), PP. 2058-2067, 13 FIGS., 3 TABLES

THE OBSCURE AND SUBTLE TRAP[1]

A. I. LEVORSEN[2]
Tulsa, Oklahoma

ABSTRACT

The tremendous expanding demand for petroleum and its products that continues to develop means that we must take a hard look at our future sources of petroleum supply. In spite of the fact that most exploration has been and is directed toward the search for petroleum in local structural traps, many of the largest oil and gas pools in the Western Hemisphere are trapped by non-structural phenomena. Structural traps are so obvious that they are the first to be tested, but we are now facing a situation in which the supply of structural traps in the United States seems to be limited; untested anticlines are becoming more difficult to find. Does this indicate an impending shortage of petroleum? The answer would seem to be *no*—but this means that the search will have to be for more obscure and subtle trapping situations. The search will continue for the purely structural trap, out there will be added stratigraphic variations and fluid-flow phenomena, all operating either together or independently.

We have "stumbled" into many great non-structural oil and gas pools while looking for purely structural traps, but the time seems to have arrived when we must start looking directly for combination traps of all kinds involving different proportions of structure, stratigraphic change, and fluid-flow phenomena. Such traps may contain very large petroleum pools, as past experience has shown.

There are in the Rocky Mountain region many such untested potential combinations of large structures, stratigraphic changes, and favorable fluid-flow conditions to justify the belief that this region has a continued great future as a petroleum-producing region of importance to our national needs. The fact that the Rocky Mountain Section of the Association is dedicating a full meeting to the obscure and subtle trap is a sure indication of a change in our thinking. Once we start actively to look for traps that combine structure, stratigraphic change, and fluid phenomena instead of to look only for local structure, there is no reason why discoveries in the United States should not continue to meet the demand. The Rocky Mountain region has as bright a future for petroleum discovery as any other.

INTRODUCTION

A study of the abstracts for this meeting suggests that there are many oil pools in non-structural traps in the Rocky Mountain region. For decades the search has been concentrated in looking for structural traps—local anticlines, domes, and faulted folds. As structural traps become more difficult to find, some believe that this implies that the end of our supply of petroleum is coming. The program to follow, however, shows an awareness of the fact that there are other traps than the purely structural—in other words, the exploration road is beginning to turn. Local and regional structures still are important, but their effect is lessened and in some cases overshadowed by stratigraphic and fluid anomalies. We are entering the era of combination traps— where structure, stratigraphy, and fluid phenomena *combine* to make the trap.

[1] Read before the Rocky Mountain Section of the Association at Billings, Montana, September 27, 1965, by Orlo E. Childs. This keynote address was prepared by Dr. Levorsen. The illustrations were prepared from Dr. Levorsen's slides by Ezio Fanelli and Arnold Ewing. Manuscript received, September 27, 1965; accepted, March 30, 1966.

[2] Consulting petroleum geologist. Deceased, July 16, 1965.

COMBINATION TRAPS

GENERAL

There are several aspects of combination traps that should be mentioned. First, they generally are obscure, and a real detective ability is required to recognize the commonly subtle geologic changes that are significant. It is a lot of fun searching for them. However, we must be careful not to get into the situation of the lipstick manufacturer. He had the bright idea of making a lipstick that would glow in the dark—but it didn't sell. He found out that half the fun was in the search!

A second aspect of combination traps is their size. In the Western Hemisphere they hold some of the largest oil and gas pools yet discovered. A few examples will show the size of the stakes involved.

		Bbls. Recoverable	
1. Bolivar Coastal field, Lake Maracaibo, Venezuela	17 billion	(Fig. 1)
2. Poza Rica field, Mexico	2 billion	(Fig. 2)
3. East Texas field, U.S.A.	6 billion	(Fig. 3)
4. Pembina field, Alberta, Canada	2 billion	(Fig. 4)

In pools such as these, there is little or no local

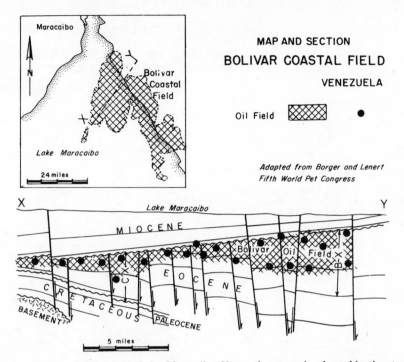

FIG. 1.—Bolivar Coastal field, Lake Maracaibo, Venezuela: example of combination trap.

structural anomaly; the position of the pool is determined chiefly by some stratigraphic and fluid anomaly—by a combination of several elements, no one of which is sufficient to trap a pool but which together are adequate.

For the third aspect of the combination trap we might consider an area that has been developed for a much longer period of time and more intensively than the Rocky Mountain region. This is eastern Kansas and eastern Oklahoma, where Eugene Weirich (1953) has discussed the oil pools trapped in the shelf deposits of the lower Pennsylvanian System. There are many hundreds of oil pools in this region, as shown on the map (Fig. 5), which are located within the Cherokee and Atoka units of the Pennsylvanian rock column.

This is the region in the Mid-Continent where the search for anticlines had its greatest boost in the 1910s and 1920s and where, today, in county after county, there is hardly a non-producing square mile without one or more dry holes. The Pennsylvanian sandstone bodies of this region are almost completely developed. Yet Weirich tells me that three quarters of all the pools shown on this map are in stratigraphic-type traps—sand-stone patches, lenses, bars, channel sandstone, reefs, facies changes, and truncated reservoir rocks.

Closer to those of you from the United States Rockies is western Canada. I have been told that the number of pools in western Canadian stratigraphic traps is about the same as in the Cherokee and Atoka of Kansas and Oklahoma—three-fourths to four-fifths—but that the amount of oil produced is even greater; 90–95% is stratigraphic.

In developing the principles of oil accumulation, we should consider the possibility that many of the pools located on anticlines and other local structures are simply accumulations localized by younger folding of larger, pre-existing oil pools or were localized by pre-existing lenses and reefs that already had accumulated a pool of oil. If, for example, local folds had been superimposed on the East Texas pool or the Pembina pool, they would have been full of oil. However, anticlines in the vicinity of both are barren because they do not overlie a reservoir rock or a stratigraphic anomaly able to trap oil.

Early availability of a stratigraphic trap to accumulate a pool probably is a most significant

A. I. LEVORSEN

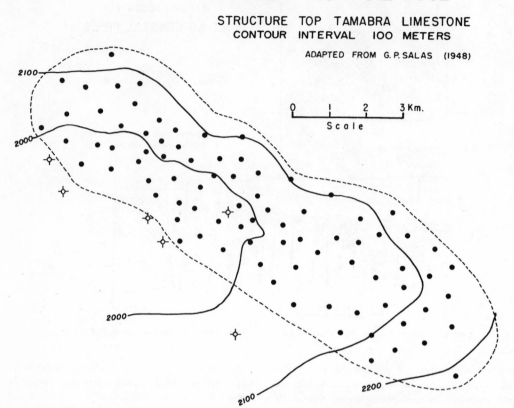

POZA RICA OIL POOL

STRUCTURE TOP TAMABRA LIMESTONE
CONTOUR INTERVAL 100 METERS

ADAPTED FROM G. P. SALAS (1948)

Fig. 2.—Poza Rica field, Mexico: example of combination trap.

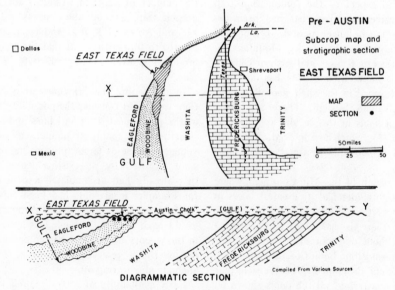

Pre - AUSTIN

Subcrop map and
stratigraphic section

EAST TEXAS FIELD

Fig. 3.—East Texas field: example of combination trap.

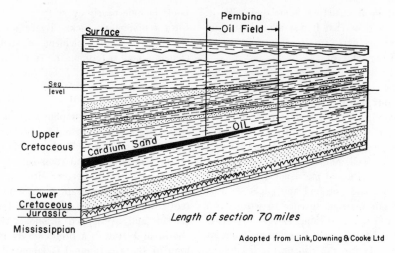

FIG. 4.—Pembina field, Alberta, Canada: example of combination trap.

and important feature. In short the trap was there and ready to receive the petroleum when first covered and sealed and when the fluids were squeezed out of the surrounding shales. Furthermore, the stratigraphic trap generally was available for a pool of petroleum to accumulate millions of years before local structures were formed.

My first experience with stratigraphic traps

came early. In 1919 I was sent to Butler County, Kansas, to look for structures like those trapping the Augusta and Eldorado pools. The Fox-Bush field was being developed in a sandstone at the base of the Pennsylvanian, and I started to work near it. About 4 mi. away, I discovered an anticline several miles long, with 20 ft. of closure and about 50 ft. of structural relief. Were we excited! This was especially true because the Fox-Bush

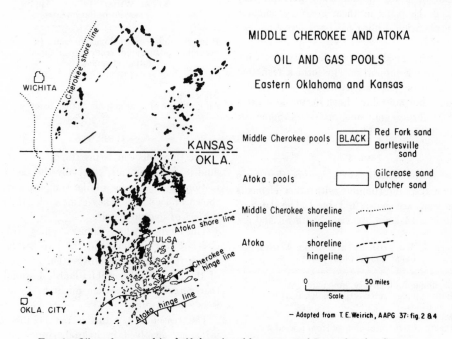

FIG. 5.—Oil pools trapped in shelf deposits of lower part of Pennsylvanian System.

field had no local structure associated with it. Management told us that it was all right to buy some royalty (my part came to around $50 for a 1/128th, as I remember it).

Drilling of the No. 1 Steinhoff well was begun on the closure, and I went to Minnesota to be married. On our return, my wife and I stopped in a hotel in Kansas City, the headquarters town for the company. A wire from one of the boys in the field came saying, "Steinhoff No. 1 looks like a 100-barrel-a-day well. Shut down the tankage." I can still see my wife and me dancing around the hotel room! I could not find any paper so used the back of our marriage license to write down the figures: 100 times 30 times 12 times 10 years times $2.75 a barrel—well, it came to a lot of money and the figures are still on the license!

We traveled to the well as fast as possible only to find that it had made all its oil during the first hour or so. When tanks were erected and the well drilled deeper, it was found that shale of the Pennsylvanian rested directly on the Mississippian limestone, and the small spurt of oil came from the unconformity zone. We had the structure all right, but Fox-Bush had the sandstone— and the oil!

Thus we see that the search for combination traps can be a lot of fun; they may be very large, and the pools in them are very common where exploration drilling has been active and aggressive.

As I see it, the three chief elements that combine in varying proportions to form a trap that will hold an oil pool are structure, stratigraphic variations, and fluid flow. Each occurs in a great variety of forms, gradients, and sizes. Some are shown in Tables I and II.

STRUCTURE

Some of the varieties of structure that form petroleum traps or combine with other elements to form traps are shown in Table I. The structural conditions that are associated with oil and gas

Table I. Varieties of Structure Which Form Petroleum Traps

Anticline, dome, fold, terrace, monocline, *etc.*
Fault, fault complex, intersecting faults, fractures
All combinations
All scales from regional to local
All gradients of structural change from low to high

pools include nearly every type and size that deform sedimentary rocks. Every oil-and-gas pool trap contains some element of structure. Such elements may determine the trap or may be only incidental to the trap.

STRATIGRAPHIC VARIATIONS

Some of the stratigraphic variations of reservoir rocks that trap or aid in the trapping of petroleum are shown in Table II. Many variations in the stratigraphy and composition of the sedimentary rocks may become either a dominant or a partial cause of a trap. An example of the potential of one of these variations—the truncated, eroded, and overlapped secondary-type trap—is shown in Figure 6, a pre-Pennsylvanian subcrop map of the area around Oklahoma City. Half a dozen oil pools are situated along the updip truncated edges of the pre-Pennsylvanian formations in the area, and none has a local structure. This subcrop map may be expanded to cover the entire area of the pre-Pennsylvanian of the United States, and Figure 7 gives an idea of the potential of the many thousands of miles of truncated pre-Pennsylvanian sediments which are available to explore.

Table II. Stratigraphic Variations of Reservoir Rocks Which Trap or Aid in Trapping Petroleum

Primary: uniform lithologic character to facies change
Secondary: truncated, eroded, overlapped
　　　　　Solution channels
　　　　　Cementation
　　　　　Fracture systems
All scales from regional to local
All gradients of stratigraphic change from low to high

FLUID FLOW

The third element is the trapping effect of the fluid in the sediments, especially the rate and direction of flow. Some of the trapping effects are shown in Table III. They come under the general subject of hydrodynamics.

The term hydrodynamics comes from the Greek *hydor* (water) and the Greek *dynamikos* (powerful). From my observation of the impact of the term on the explorationist, I would judge that a realistic translation might be "your guess is as good as mine."

The reason I say this is not because of the principles involved in hydrodynamics, for they

TABLE III. FLUID FACTORS WHICH TRAP OR
AID IN TRAPPING PETROLEUM

Wide composition variation, chemical and physical
Gradients, change direction and rate with:
    Folding, faulting, mountain building
    Regional tilting, erosion, overlap
    Volcanism, heating, cooling
    Solution and recementation
All scales from regional to local, and all rates and
    directions. Always changing with normal geologi-
    cal processes

are sound, but rather because of their application to petroleum exploration. Much of the reason for the confusion and misunderstanding is that the industry is not yet willing to obtain the necessary fluid-pressure information. Just try to get a company to take pressure measurements on the water-bearing sandstone bodies found in a dry hole! Yet these measurements are essential if there is to be a quantitative approach to hydrodynamics. At present, with the exception of a few areas, the approach is qualitative and subjective.

A crude illustration of the importance of hy-

drodynamics as a trapping agency is shown in Figure 8. In *A*, water is flowing down the tube and buoyant corks or balloons are rising against the flow. It might be said to be in dynamic equilibrium. In *B*, however, a slight constriction is placed in the tube, and in the constricted zone the water flows faster. It flows enough faster to cause the rising balloons to congregate below the constriction. We might liken the tube to the permeable sedimentary layer and the balloons to buoyant oil and gas. The oil and gas congregate below the constriction until enough buoyancy develops to force them through the faster-flowing water within the constricted zone. The faster flow of the water is a trapping mechanism.

The trapping effect of fluid flow is determined by several variables—such as the density and the relative amounts of water, oil, and gas; amount of the constriction; and pressure gradient, which controls the rate of flow of the water. Thus even a small constriction or change in slope may be sufficient to increase the rate of flow down the dip or to decrease the up-the-dip buoyancy effect

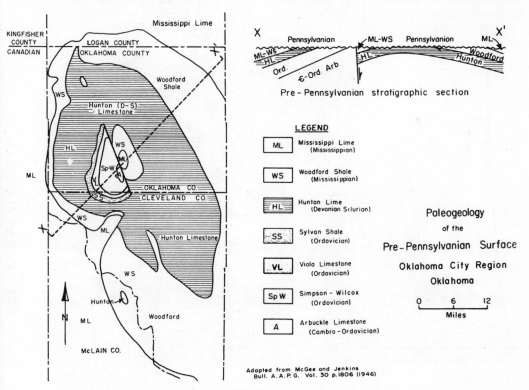

FIG. 6.—Pre-Pennsylvanian subcrop map, Oklahoma City region: example of truncated, eroded, and overlapped secondary trap.

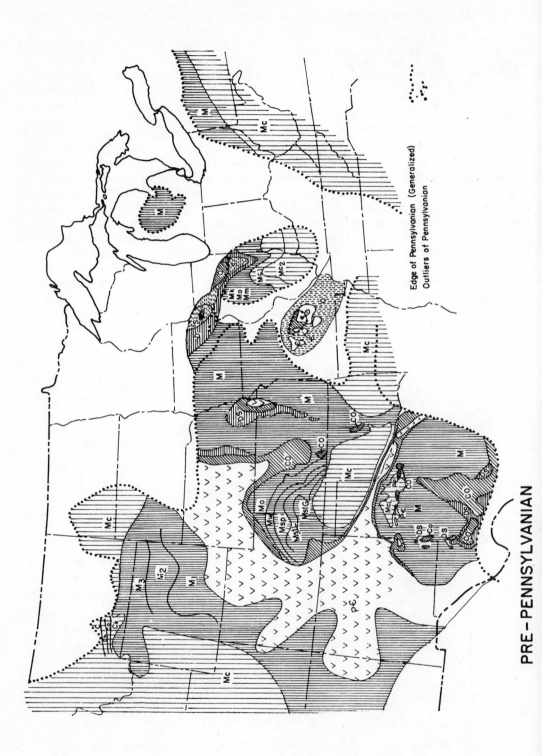

PRE-PENNSYLVANIAN

Edge of Pennsylvanian (Generalized)
Outliers of Pennsylvanian

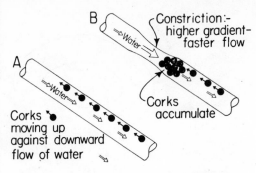

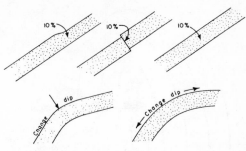

Fig. 8.—Hydrodynamics as a trapping agency.

Fig. 9.—Structural and stratigraphic changes that increase fluid flow downdip or decrease up-the-dip buoyancy. These changes affect fluid equilibrium and, as a result, trap oil.

until the equilibrium is upset and the oil is trapped. Some examples are shown in Figure 9.

ROCKY MOUNTAIN REGION

There are several pools in the Rocky Mountains that occur in a structural environment such as that shown in Figure 10. They may be trapped by some stratigraphic variation or by a down-the-dip fluid gradient. There is more to the situation than is shown in Figure 10, for when we extend the area mapped it is seen to be down the flank of a local anticline (Fig. 11). One question which might be asked is this: are pools such as these in structural traps or are they in hydrodynamic traps? The curving contours suggest that the slight arching on the flank of the anticline may be enough to provide favorable gradient conditions.

In each of the trapping elements, we are deal-

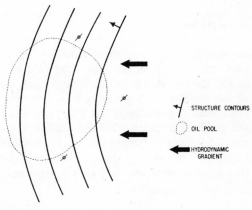

Fig. 10.—Structural environment of several Rocky Mountain pools.

Fig. 7.—Pre-Pennsylvanian map of United States.

M	Lower and middle Mississippian
Mk, Mm	Kinderhook, Meramec
M_1, M_2, M_3	Lower, middle, upper Madison
Mo, Msp, Mw	Osage, Spergen, Warsaw
MstG, MstL	Ste. Genevieve, St. Louis
Mc	Upper Mississippian (Chester)
Mc_1, Mc_2	Lower, upper Chester
Mct, Mch, McO, McK	Chester—Tyler, Heath, Otter, Kibbey
D	Devonian
S	Silurian
O	Ordovician
CO	Cambro-Ordovician
C	Cambrian

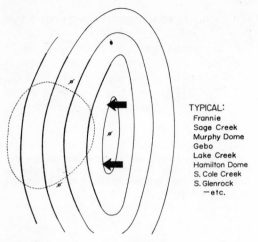

TYPICAL:
Frannie
Sage Creek
Murphy Dome
Gebo
Lake Creek
Hamilton Dome
S. Cole Creek
S. Glenrock
    —etc.

FIG. 11.—View of structural environment of several Rocky Mountain pools, covering larger area than shown in Figure 10.

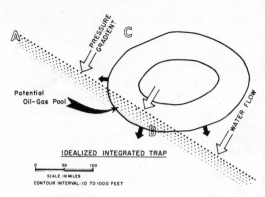

IDEALIZED INTEGRATED TRAP

SCALE IN MILES
CONTOUR INTERVAL: 10 TO 1000 FEET

FIG. 12.—Idealized example of combination or integrated oil traps.

ing with gradients of different kinds—structural gradients, stratigraphic gradients, and fluid gradients. They may combine in all proportions on all scales, and in all directions, and some of these combinations become traps that hold pools of oil and gas. Such traps are called combination or integrated traps; an idealized example is shown in Figure 12. The relative trapping effect may range from proportions of one-third each for structure,

stratigraphy, and fluids to 10–10–80 or 40–40–20, or any other combination.

Figure 13 is an attempt to classify some of the pools in the Rocky Mountain area, which have been described in geologic literature, by using a triangle with the corners labeled "structural," "stratigraphic," and "fluid pressure." The dots at the corners and along the sides of the triangle show the pools, and the blank area shows the remaining potential to be explored for combination traps.

The crests of the many anticlines of the Rocky Mountain region have been explored, and some

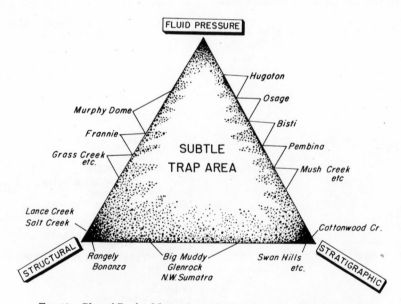

FIG. 13.—Plot of Rocky Mountain pools according to trapping controls.

anticlines have been found to be productive. These pools would correspond to the pools at the "structural" corner of the triangle. There is sufficient arching, minor faulting, and change of rate of slope on the flanks of the anticlines with dry holes at the crest, however, to combine with favorable stratigraphic and hydrodynamic elements to trap many undiscovered combination pools. Every dry anticline should be re-examined on all flanks to the deepest parts of the bounding synclines.

Similarly, there are thousands of miles of stratigraphic variations that occur within the stratigraphic section of the Rocky Mountain region that might combine with favorable structural and fluid phenomena to make traps—some of which might well be of large size.

Finally, with the many wide differences in elevation of each of the aquifers—or potential petroleum-reservoir rocks—there must be many sloping potentiometric surfaces. These indicate the existence of a fluid flow, which, combined with even minor changes of stratigraphic permea-bility and structural attitude, can provide a trap that will hold an oil or gas pool.

It is going to take some really fine detective work to find these undiscovered pools, and it is going to be fun. There may be many of them, and some may be very large. All I can say is, let's get our detective badges out, shine them up, and get going!

SELECTED BIBLIOGRAPHY

Borger, H. D., and E. F. Lenert, 1959, The geology and development of the Bolivar Coastal field at Maracaibo, Venezuela: Proc. 5th World Petroleum Cong., New York, Sec. 1, Paper 26, p. 481–498.
Levorsen, A. I., 1960, Paleogeologic maps: San Francisco and London, W. H. Freeman and Co., 174 p.
McGee, D. A., and H. D. Jenkins, 1946, West Edmond oil field, central Oklahoma: Am. Assoc. Petroleum Geologists Bull., v. 30, p. 1797–1829.
Salas, G. P., 1948, Geology and development of Poza Rica oil field, Veracruz, Mexico: Oil and Gas Jour., v. 47, no. 25 (Oct. 21), p. 129.
—— 1949, Geology and development of Poza Rica oil field, Veracruz, Mexico: Am. Assoc. Petroleum Geologists Bull., v. 33, p. 1385–1409.
Weirich, T. E., 1953, Shelf principle of oil origin, migration, and accumulation: Am. Assoc. Petroleum Geologists Bull., v. 37, p. 2027–2045.

Reprinted from:
BULLETIN OF THE AMERICAN ASSOCIATION OF PETROLEUM GEOLOGISTS
VOL. 50, NO. 10 (OCTOBER, 1966), PP. 2277-2311, 25 FIGS.

PALEOGEOMORPHOLOGY AND ITS APPLICATION TO EXPLORATION FOR OIL AND GAS (WITH EXAMPLES FROM WESTERN CANADA)[1]

RUDOLF MARTIN[2]
Calgary, Alberta

ABSTRACT

Under the term *paleogeomorphology* are grouped all geomorphological phenomena which are recognizable in the subsurface. Buried relief features with a marked three-dimensional geometry are of importance to the petroleum geologist whenever they lead to the trapping of hydrocarbons. *Paleogeomorphic traps* form a third and distinct group which ranks in importance with stratigraphic and structural traps as a major mechanism for localizing hydrocarbon occurrences. They are not simply another type of stratigraphic trap, but form by themselves a category of considerable economic significance. Paleogeomorphic traps can not be analyzed, nor can their occurrence be predicted, by stratigraphic or structural methods of study, but must be treated as a geomorphological problem. Of the many types of buried relief features, some are of greater interest to the petroleum geologist than others. In this paper, the morphology of buried erosional landscapes is discussed in greater detail. Other types of buried relief features are fossil reefs, barrier beaches, and submarine canyons.

Hydrocarbons may be trapped either directly or indirectly as a result of paleogeomorphological processes. In either case, the traps may occur below or above (against) the morphological surface. Numerous examples of the effects of erosion are known from the Paleozoic and Mesozoic buried landscapes of western Canada and the Mid-Continent area of the United States. Hydrocarbon traps occur below the highs on such erosional surfaces as well as in sandstone bodies deposited in the lows on these surfaces. The rules which govern the formation of ancient landscape forms are worked out in detail, with particular emphasis on the application of quantitative geomorphology to the pattern of the ancient drainage system and on such features as summit levels and the influence of geological factors (erosion-resistant levels, influence of faults and fractures, *etc.*). Weathering and underground solution also play a role in providing both reservoirs and buried traps for oil and gas.

The method of analysis must take into account (1) the structural attitude of the strata below and above the erosional surface, (2) the lithology of the formation overlying the unconformity (if used for isopachous mapping), (3, 4) synsedimentary structural movements and the compaction factor in the formation overlying the unconformity, (5) problems of correlation, (6) reservoir development, (7) presence of a "seat seal," and (8) other factors.

Paleogeomorphology provides an interesting evaluation of "modern" geomorphological thinking. The fossil relief forms "frozen" by the transgression of younger rocks neither disprove the validity of the classical peneplain concept, nor do they support fully the newer ideas regarding slope retreat. The actual conditions found are described by a re-definition of the old term *paleoplain*. The summit level is an intrinsic part of this feature, and not a dissected earlier peneplain. A new geomorphological concept introduced here concerns the alternation of obsequent and resequent interfluve spurs.

Buried landscapes should provide a high percentage of the future oil and gas fields yet to be discovered in North America. The most important stratigraphic levels at which buried landscapes occur are those that formed after major periods of orogenesis. Their geographical locus corresponds to the broad belts of subsequent transgressions.

PALEOGEOMORPHOLOGY AND GEOMORPHOLOGY

The term *paleogeomorphology*, as far as the writer has been able to ascertain, first was used in its proper sense, as a sub-science of geomorphology, by Thornbury (1954, p. 30–33). In an

[1] Read before the Rocky Mountain Section of the Association at Billings, Montana, September 27, 1965. An earlier paper given at the 49th Annual Meeting of the A.A.P.G., Toronto, Ontario, on May 21, 1964, under the title "Techniques of exploration for buried landscapes," is included in the present publication. Manuscript received, January 10, 1966; accepted, May 2, 1966.

[2] Consulting geologist, Rudolf Martin & Associates Ltd. The writer is indebted to J. T. Hack, A. D. Baillie, D. P. McGookey, C. W. Spencer, and M. Berisoff for reading the manuscript and making helpful suggestions.

earlier reference to *paleogeomorphic maps*, Kay (1945) used the word only as a special type of *paleogeographic maps*. Thornbury (1954, p. 582–587) devoted a special section to the "Application of Geomorphology to Oil Exploration," and a considerable part of this subchapter dealt with what the writer subsequently has termed *paleogeomorphic traps*. About the same time that the writer first used the term *paleogeomorphology* in connection with subsurface studies in oil fields (Martin, 1960), Harris (1960) used it in connection with field observations. Since then, the term has been used in two additional publications by the writer (R. Martin and Jamin, 1963; Martin, 1964b) and in another by McKee (1963).

Enlarging on Thornbury's concept, the writer

groups under the term *paleogeomorphology* the study of all geomorphic phenomena which are recognizable in the subsurface and in outcrops of previously buried formations. Geomorphology is the science of the earth's relief features and of the processes which created them. Hence, paleogeomorphology is the science of buried relief features. This includes submarine features, such as submarine canyons and those parts of reefs or volcanic islands that are below sea-level and can not very well be considered separate from their subaerial parts.

The petroleum geologist's emphasis must be on the three-dimensional shape or *form* that has created the hydrocarbon trap. Yet, a proper interpretation of this form, especially from scattered subsurface data, is not possible without an understanding of the *process* which has created the form. Considering how difficult it is in the study of modern landscapes to indicate with absolute certainty the process or processes which created a certain relief feature, one may well doubt the practical value of paleogeomorphological studies, especially as a means of finding accumulations of oil and (or) gas. Nevertheless, the studies made by the writer during the last 7 years, principally in western Canada, have shown that certain deductions can be made from the study of buried landscape forms which lend themselves to geomorphological interpretation; these deductions, then, can serve as a basis for future exploration.

Because of the tremendous number of factors that shape each feature, many of which are of purely local character, only average parameters can be established, and no mathematically exact rules can be formulated. However, once these averages are known, they become determining factors in the reconstruction of a buried landscape from a limited number of control points. Such a reconstruction may then serve as a basis for a lease-acquisition program, geophysical surveying, and eventually for drilling. By defining the objective and limiting the area within which a feature should occur, a paleogeomorphological study can result in considerable savings of effort, time, and money spent searching for these types of hydrocarbon traps.

The geomorphological processes that are involved in paleogeomorphology were defined previously by the writer (Martin, 1960). In that paper, the writer made a broad subdivision into *constructive* and *destructive* geomorphic processes. Apart from the effect of endogene processes, the earth's relief is formed by material constantly being removed in some places and being added elsewhere. On this basis, Lobeck (1939, p. 7, 9), building on an earlier definition by Davis (1884), distinguishes *constructional* and *destructional* forms, and Thornbury (1954, p. 34–35), *aggradation* and *degradation*. The writer's subdivision is similar to, but not exactly the same as, Lobeck's; his depositional forms are included with the first instead of the second group. Thornbury does not include under aggradation such constructive processes as the work of organisms, vulcanism, and the impact of meteorites. The recent trend in geomorphology textbooks is toward emphasis on landforms (mostly destructive processes) alone. To the petroleum geologist, however, the constructive processes (especially organic-reef growth) are of equally great importance.

The most important topics for the petroleum explorationist are the morphology of organic reefs, buried erosional landforms, submarine canyons, and weathering and underground solution. Other features, such as (river) point bars and (marine) barrier beaches, usually are discussed as sedimentary processes, although their truly three-dimensional aspects render them particularly subject to paleogeomorphological analysis. Thornbury (1954, p. 31–32, p. 584–586) devotes several pages to "shoestring sands" and observes that probably no phase of petroleum exploration can be used to better advantage than a knowledge of the detailed characteristics of specific topographic features.

PALEOGEOMORPHIC TRAPPING OF HYDROCARBONS

Thornbury (1954, p. 553) stated[3]: "No claim is made that even the most thorough knowledge of geomorphology equips one to become a . . . petroleum geologist . . ., but it is the author's belief that too often these 'practical geologists' fail to make maximum possible use of basic geomorphic concepts"; and elsewhere (1954, p. 33): "When geomorphologists . . . fully realize . . . this use which can be made of geomorphic principles and knowledge, the subject will become a true working tool in the practical application of geology."

[3] Quotations from Thornbury (1954) are reproduced with permission of John Wiley and Sons, Inc.

LeGrand (1960) emphasized another side of the problem when he stated that ". . . the intricate work of natural agencies operating to form landscapes is poorly understood. This not only has a retarding influence on geomorphic studies, but is damaging to important economic considerations in several fields of geology." Kay (1945, p. 427), discussing "paleogeomorphic maps," did not really do justice to Thornbury's later definition of paleogeomorphology, because he omitted all mention of the use of subsurface data in constructing such maps; his main concern was the separation of seas from lands. Levorsen (1960, p. 22) mentions "paleotopography" in passing, but does not deal with the subject in detail. Thus, although paleogeomorphology plays an important role in the accumulation of hydrocarbons, it is difficult to find specific details on this subject in petroleum geology textbooks, even though many descriptions of oil and gas fields have been published in which paleogeomorphological phenomena are the major factors in entrapment. The literature on reefs, buried hills, and other paleogeomorphic traps is voluminous; what has been lacking thus far is the treatment of the subject as a problem in geomorphology.

The problems posed by paleogeomorphology, especially in areas where buried relief features abound, are as important to the petroleum geologist as those of sedimentation and structure; however, they do not appear thus far to have received equal attention. Most authors, including Levorsen (1954), treat paleogeomorphologic hydrocarbon traps as another type of stratigraphic trap—which they are not. An important difference between stratigraphic and paleogeomorphologic traps is the pronounced three-dimensional aspect of the latter. The writer prefers to limit the use of the term *stratigraphic trap* to those traps which are caused by a lateral change in reservoir properties within a given stratum. Erosion surfaces, reef buildups, and other geomorphic phenomena are bounded by air or water at the time of their formation. Subsequent deposition of younger strata adjacent to but different from those that constitute such a morphological surface does not create a stratigraphic trap but a paleogeomorphic trap. The following new classification of hydrocarbon traps lists the trapping mechanisms in approximately the order in which they originated.

1. *Stratigraphic*—Trap formed by lateral change in reservoir-rock properties.
2. *Paleogeomorphic*—Trap formed by shape of land (or underwater) surface.
3. *Structural*—Trap formed by structural deformation of reservoir rock.
4. *Mixed*—Combination of any two or three of the above. This group includes traps formed by any combination of the preceding together with hydrodynamic effects.

Paleogeomorphic trapping of hydrocarbons caused by the presence of buried relief features may be classified as either direct or indirect in character.

A. *Direct* trapping of hydrocarbons may occur either below the morphological surface or above (against) it. Several examples are given in Figure 1.

I. Accumulations *below* the morphological surface may be caused by the following.

(a) Erosional (destructive) processes, *i.e.*, the formation of traps by alternating buried hills and valleys or by the erosion of a submarine canyon.

(b) Constructive processes, such as organic reef building, grading into bioclastic carbonate bars, and the formation of sand barriers (shoestring sands), dunes, and river terraces.

II. Accumulations *above* or *against* a morphological surface may be caused by the following.

(a) Erosion of a river bed or marine channel, followed by local deposition of sands (point bars).

(b) Deposition of reservoir rocks in an existing valley or against the slope of a hill, a reef, or a volcano (buttress sands).

A combination of I and II (paleogeomorphic surfaces potentially bounding a hydrocarbon accumulation both above and below) may be exemplified by river terraces deposited against the sides of a valley. In this situation, the lower erosional surface is composed of the bottom and sides of the valley in which the reservoir sands were deposited; the upper surface is the valley subsequently cut into these deposits and later filled with clay.

B. *Indirect* trapping of hydrocarbons because of paleogeomorphological process occurs in several forms. As in direct trapping, a distinction is made between accumulations below the morphological surface and those above it (Fig. 2).

I. *Below* the morphological surface, altera-

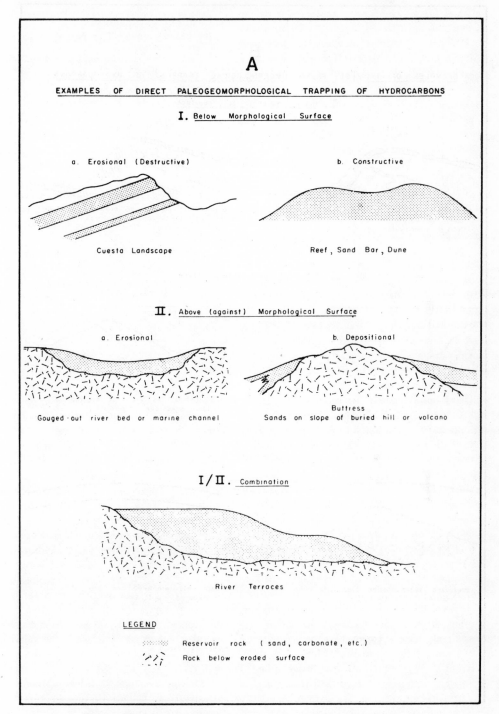

FIG. 1.—Examples of direct paleogeomorphic trapping of hydrocarbons.

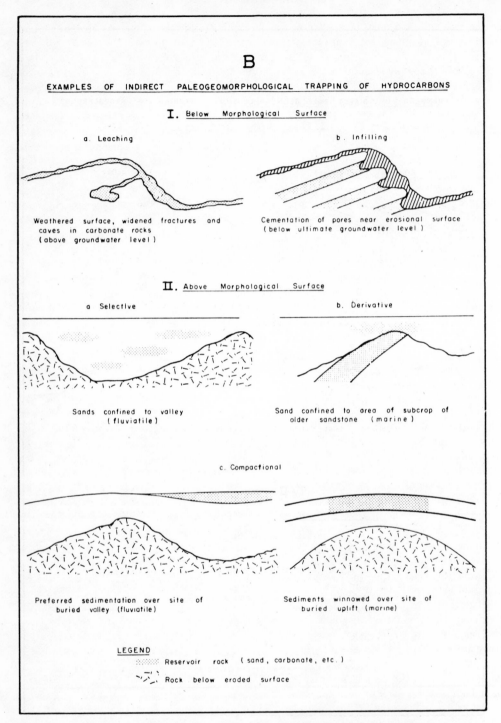

B

EXAMPLES OF INDIRECT PALEOGEOMORPHOLOGICAL TRAPPING OF HYDROCARBONS

I. Below Morphological Surface

a. Leaching

Weathered surface, widened fractures and caves in carbonate rocks (above groundwater level)

b. Infilling

Cementation of pores near erosional surface (below ultimate groundwater level)

II. Above Morphological Surface

a Selective

Sands confined to valley (fluviatile)

b. Derivative

Sand confined to area of subcrop of older sandstone (marine)

c. Compactional

Preferred sedimentation over site of buried valley (fluviatile)

Sediments winnowed over site of buried uplift (marine)

LEGEND

Reservoir rock (sand, carbonate, etc.)

Rock below eroded surface

Fig. 2.—Examples of indirect paleogeomorphic trapping of hydrocarbons.

tions in the reservoir rock may take place as a result of the action of ground water on carbonate rocks and anhydrite.

(a) While the morphological feature concerned is above ground-water level, there will be leaching, weathering at the surface, solution widening of fractures and formation of cavities below the surface, solution of anhydrite, and formation of collapse breccias.

(b) Below ground-water level (usually at a later stage than (a), because it would result from renewed subsidence), circulating waters carrying dissolved carbonates, sulfates, etc. from higher levels will deposit them, causing cementation, infilling of porous rock with secondary calcite, anhydrite, or silica, and formation of a secondary caprock.

II. *Above* the morphological surface, the accumulation of hydrocarbons may be influenced by several factors related to this surface.

(a) Selective deposition of reservoir sands above the valleys of a subsiding old land surface.

(b) Derivation of reservoir sands from subcropping older sandstone bodies during a marine transgression.

(c) Differential compaction of sediments which overlie paleogeomorphological elements. Such compaction will cause the sea floor or a later land surface to be slightly higher above old elevations, and slightly lower above old depressions. Below sea-level, this may lead to winnowing of sediments above the old highs; on land, it will cause rivers to flow over the old lows.

As in studies of sedimentation and structural trends, the problem for the petroleum geologist in making a paleogeomorphological study is to recognize trends and shapes and to find ways of extrapolating these from known points of control. The rules governing trends and forms of earth-relief features are in most respects completely different from those governing either sedimentation or structure; however, in some respects they are interrelated. As Horberg (1952, p. 188) has said, an historical approach to geomorphology ". . . imposes the responsibility of developing a body of principles by which geomorphic history can be interpreted. These principles as such are seldom considered and are yet to be defined and organized."

The following pages deal with several paleogeomorphological phenomena observed in the subsurface, primarily in the oil and gas fields of western Canada. The examples given illustrate subsurface features of purely morphological character which have received too little attention because their true nature was not properly recognized. Comparison with present-day geomorphic phenomena not only leads to better understanding of such features, but also facilitates their analysis and thereby opens the way to extrapolation and prediction. This in turn may result in commercial application of the principles involved. In an earlier paper (Martin, 1960), the writer considered some of the problems of reef morphology. Because this subject will be discussed in a future paper, the present paper is limited to a discussion of buried landscapes.

BURIED EROSIONAL LANDFORMS

Buried hills and valleys are important to the petroleum geologist not only because of the hydrocarbons accumulated in the hills, *i.e.*, below the unconformity (type A-I-a), but also because they help determine the occurrence and shape of sand bodies that have accumulated in the valleys (types A-II-a and B-II). Buried erosional landforms are widely known from the Paleozoic and Mesozoic of the Mid-Continent area of the United States and of the provinces of Alberta, Saskatchewan, and Manitoba in western Canada. Surface outcrops also have drawn attention to the existence of unconformities with up to 1,000 feet of paleotopographic relief. Examples include relief observed on the Precambrian surfaces of Wisconsin and northern Michigan, in the Adirondacks, in the Ozarks, and in the Black Hills (Meyerhoff, 1940), and on the Lower Ordovician surface of southwest Virginia and northern Tennessee (Harris, 1960).

TIME OF FORMATION OF BURIED LANDSCAPES

Buried landscapes, as might be expected, are most prevalent at those times in stratigraphic history that followed major periods of mountain building. As a result of each of these orogenic movements, vast areas that previously were part of sedimentary basins were uplifted to positions well above the prevailing base level of erosion (Fig. 3). A further worldwide lowering of this base level must have occurred whenever a period of extensive glaciation followed a major orogeny. Whereas many lesser orogenies also have led to the formation of buried landscapes, it is believed

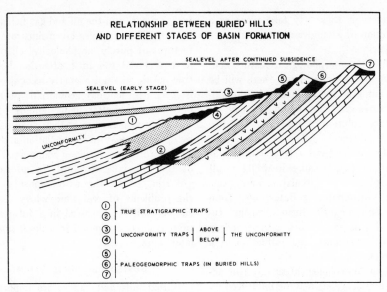

Fig. 3.—Relationship between buried hills and different stages of basin formation.

that the formations in which a search for paleo-geomorphic hydrocarbon traps of this type may be most fruitful correspond to the major epochs of landscape formation which coincided in part with widespread glaciation. The major epochs of orogeny—glaciation—landscape formation are tabulated.

Major Orogeny	Glaciation	Landscape Formation
Alpine	Pleistocene	Tertiary-Recent
Hercynian	Permo-Penn.	Post-Miss.
Caledonian	Ord.-Silurian	Ord.-Silurian
Huronian	Precambrian	Post-Precambrian

Sloss (1963) added to these two additional unconformities of continental scope (post-Early Ordovician and post-Early Jurassic). Wheeler (1963) subsequently added a third (post-Devonian). Some of these lesser episodes of landscape formation are more important to the exploration for oil and gas than others. However, a complete list would include many additional episodes not mentioned by these two authors. Until such a complete survey is made, the short list given above will help to define the episodes of landscape formation of major interest.

LOCUS OF BURIED LANDSCAPES

Whereas the preceding considerations help to limit the occurrence of buried landscapes in time, another factor limits their occurrence in space. In the experience of the writer, the areas in which such landscapes occur generally are bounded on the basinward side by a structural hinge line, beyond which the inclination of the strata involved tends to increase markedly into the basin. The hinge line also may be associated with one or more boundary faults, and commonly (but not in every place) coincides with the shoreline of at least one post-unconformity depositional sequence. The locus of buried landscapes formed during any one interval of time therefore tends to be limited on the basinward side by the hinge line corresponding to that time, and on the landward side by the shoreline of the maximum subsequent transgression (Fig. 3). Because the slope of the old pre-transgression landscape, before the time of renewed basin subsidence and transgression, must have been very slight in most cases (i.e., a few feet per mile), the width of the denuded area buried by younger sediments may amount to many hundreds of miles.

Any analysis of the components of a buried landscape must proceed along purely geomorphological lines of reasoning. Serious efforts to prospect for oil and gas traps in buried hills or valleys must be based on such a paleogeomorphological analysis and on the extrapolation of the results according to geomorphological principles.

For a complete analysis, it is necessary to determine the climate that prevailed at the time the

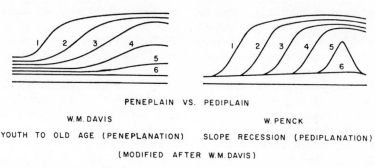

PENEPLAIN VS. PEDIPLAIN

W.M.DAVIS   W.PENCK

YOUTH TO OLD AGE (PENEPLANATION)   SLOPE RECESSION (PEDIPLANATION)

(MODIFIED AFTER W.M.DAVIS)

FIG. 4.—Different stages in slope development according to W. M. Davis and W. Penck.

landscape was formed, the nature of the rocks that formed the landscape, the structure of these rocks, *etc.* In addition, one must deal with the geomorphological parameters that determined the final shape of the landscape, such as drainage pattern, drainage density, slope angles, valley gradients, summit levels, and other factors.

NATURE OF PROCESS OF PLANATION

While making such a paleogeomorphological analysis of certain buried landscapes in western Canada, the writer has found that "modern" geomorphology can gain considerably from the experience gathered from such an analysis of "fossil" or "frozen" landscapes. One of the most important principles involved is that of the process of planation. Here, the classical American geomorphological concept of the *cycle of erosion* or *geographical cycle,* as formulated by Davis (1889), first lost considerable ground to the supporters of the European concept of *scarp retreat* formulated by Penck (1924) (see Fig. 4). Later both theories were tacitly abandoned by most American geomorphologists in favor of quantitative geomorphological concepts and the study of processes. However, neither of these has contributed to a better knowledge of the end product of planation. It is this end product which is represented by the buried landscapes of the past, and it turns out to look neither quite like that which Davis visualized, nor quite like Penck's idea.

According to Davis, a landscape passing from youth through maturity to old age finally would be denuded to a *peneplain* approaching base level. At most, some scattered monadnocks, or disconnected hills, would be left on this surface. Davis did not claim that a peneplain had to be flat; in his description (1899a), he referred to the sub-

Cambrian contact (miscorrelated in the quotation) seen in the Grand Canyon of the Colorado River with the following words from Dutton (p. 369 in 1954 Dover reprint): "In the Kaibab division . . . we may observe . . . a few bosses of Silurian strata rising higher than the . . . sandstone which forms the base of the Carboniferous. These are Paleozoic hills, which were buried. . . . But they are of insignificant mass, rarely exceeding *two or three hundred feet* in height" (writer's italics). Hill (1901, p. 363), who first introduced the term *paleoplain* (or buried peneplain), had the following to say about the buried pre-Cretaceous landscape of the Wichita Falls, Texas, area: "That irregularities of configuration once existed in this pre-Cretaceous land is also shown by some of their degraded remnants that still persist, like the Ouachita Mountains, which were not completely buried beneath the Cretaceous sediments, and the Burnet uplift, which was finally buried before the close of the Lower Cretaceous."

The writer's studies have shown that such a paleoplain still contains all the elements of the preceding mature landscape; there is a recognizable drainage pattern, with valleys at a fair gradient, separating a coherent system of ridges that are by no means isolated "monadnocks." Paleogeomorphological data thus might appear to lend support to the theory of Penck (1924), who believed that a slope, once formed, retains its original angle and therewith its identity, while responding to the forces of erosion by retreating gradually parallel with its previous position and losing some of its height. At the foot of such a slope, a footslope or *pediment* develops, which maintains the minimum slope consistent with the local erosional facies. Eventually, when the steep scarps on two sides of an uplifted area have re-treated to the point where the ridge between them becomes obliterated and the pediments from adjoining valleys meet, a *pediplain* results which is not really very different as an ultimate landform from Davis' peneplain, but which has reached this end form by very different means. However, the buried landscapes of western Canada definitely have not been reduced to a "pediplain." There is a pronounced difference in slope between the dip slopes and the scarps or "face slopes." Yet, the steep scarps of Penck are gone; the inclination of the buried Mississippian surface in southeastern Saskatchewan ranged from only 15 to 233 ft./mi.

Thus, paleogeomorphology appears to prove that both Davis and Penck were partly right; the slopes, instead of retaining the same angle at all times, do become reduced (to 3° or less), as predicted by Davis; but the framework of the landscape as a whole also retains its general shape, instead of becoming a featureless peneplain. Hill's term *paleoplain* appears well suited to the definition of such a buried landscape. The present writer's re-definition of Hill's term would further reflect the fact that this paleoplain has retained sufficient relief to be an important trapping factor in the accumulation of hydrocarbons, ores, water, *etc.* The closest among modern landscapes to this new definition of paleoplains are the *saucer-shaped valleys* of Tanganyika described by Louis (1964, p. 47), which exhibit slopes of 2–3 per cent (106–158 ft./mi.), locally increasing to 5–10 per cent (264–528 ft./mi.). These are peneplains in the Davisian sense; on the other hand, King (1951, p. 58) cites gradients of only 6 in.–1 ft./mi. for some of the pediplains of South Africa.

CUESTA LANDSCAPES

The paleoplains studied by the writer have retained all or most of the aspects of a cuesta landscape. Here again, the nomenclature of "modern" geomorphology can bear some adjustment. Davis (1899b) and subsequent authors regarded a cuesta landscape as typical of a coastal plain. In reality, a cuesta landscape is bound to develop on the homoclinal flank of any subsiding basin as the tilted strata on its periphery are eroded into parallel, outward-facing, "belted" escarpments. A good example of a cuesta landscape that is not

directly connected with any coastal plain is the Paris basin (Fig. 5); another is the Jurassic "Alb" of Schwaben in southern Germany, which even Davis (1899b) describes as a typical cuesta landscape.

The formation of cuestas is a logical corollary to new basin subsidence after a period of epeirogenic uplift (Fig. 3). However, in areas where structural deformation has taken place on any noticeable scale, a more complex landscape will result. An example of such a complex feature from the Soviet Union, taken from Khutorov (1958), is shown in Figure 6. Though the anticline itself is not productive, a paleotopographical hydrocarbon trap of considerable extent occurs on the topographically highest flank. In addition, an accumulation of "seepage" oil occurs in the younger sandstone beds which cover this buried ridge. The differential compaction above the ridge which has affected these sandstone bodies also should be noted.

A typical cuesta landscape was developed on the northeast flank of the Williston basin in southeastern Saskatchewan, southwestern Manitoba, and northern North Dakota on rocks of Mississippian age (lower Frobisher beds, Alida beds, and Tilston beds), which are overlain by Triassic (?) "Red Beds" of the Lower Watrous Formation in the area studied by the writer (Martin, 1964a, b). The oil has accumulated beneath the unconformity and the accompanying "caprock" wherever the subcrop trend of a reservoir bed crosses a paleotopographical ridge. Accumulations in the upper Frobisher beds and particularly in the overlying Midale beds appear to be less dependent on paleotopography, and are therefore not included in this study; neither are those in the Souris Valley beds which underlie the Tilston beds in southwestern Manitoba.

A map of the southeastern Saskatchewan part of this area, on which the shape of the pre-Triassic (?) landscape has been reconstructed by the use of an isopachous map of the "Red Beds," is shown in Figure 7. The cuestas in this landscape correspond to the subcrops of the beds most resistant to erosion. The regional pre-"Red Beds" dip of the Mississippian strata in southeastern Saskatchewan was about 20–30 ft./mi., which is about 10 times the gradient of the major consequent valley in the area. A schematic cross section at right angles to the general strike of the

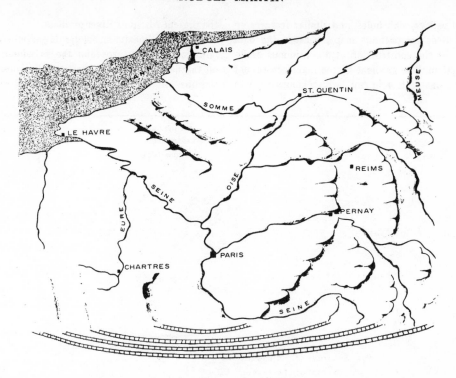

Fig. 5.—Cuesta belts surrounding Paris basin.

Mississippian strata, using as datum the top of the "Red Beds," shows that the cuestas which developed on resistant beds of the Mississippian were sloping at a slightly smaller angle than the beds themselves (Fig. 8, top).

Borrowing a classical diagram from Lobeck (1939), a few words may be said here regarding the nomenclature of valleys and ridges on a homoclinal basin flank (Fig. 9). Streams flowing down the original topographic slope toward the basin (*i.e.*, in this case, down the dip slope) are known as *consequent streams*. As these cut into the stratigraphic sequence, tributaries developed at right angles to them along the exposures of weaker rock, creating *subsequent streams*. Shorter tributaries, parallel with the consequent system but flowing into the subsequent streams, are known as *obsequent* if flowing in a direction opposite to the consequent streams (*i.e.*, down the face slope, or the escarpment of the resistant rock), and as *resequent* if flowing in the same direction as the consequent streams.

On both the map (Fig. 7) and the topographical profile shown in cross section (Fig. 8, top),

the following main features may be noted from north to south.

1. The wedge-out of the Lower Watrous "Red Beds" at their zero isopachous contour line. This does not appear to have any special significance in terms of Mississippian topography.

2. A longitudinal, west-northwest-trending subsequent valley, which is referred to as the Tilston lowland.

3. A parallel ridge, the Tilston cuesta, corresponding to the subcrop of the Tilston beds.

4. A second subsequent valley, here called the Alida lowland.

5. A second ridge, the Alida cuesta, corresponding to the subcrop of the Alida beds.

6. A gradually diminishing southward slope over the Frobisher beds subcrop.

7. A "nickpoint" at the northern end of the Midale beds subcrop, corresponding to a flexure in the "Red Beds" and resulting in an apparent steeper gradient south of this point.

VALLEY GRADIENT

The cross section (Fig. 8, top) shows that the slope between (1) and (2), and the valley bottoms of (2) and (4), are connected with (6) by a smooth gradient, indicating the existence of a graded stream system preceding "Red Beds" time. This would eliminate the possibility of including closed valleys, sink holes, and similar features on "Red Beds" isopachous maps, because the data allow for the construction of a continuous drainage system. The gradient of the main consequent stream was 2 ft. 4 in./mi., which compares with the present Missouri River gradient.

Both the maximum slope observed on the northeast-facing scarps and the maximum slope on the flanks of the northeast-trending obsequent streams are 160 ft./mi. on the "Red Beds" iso-

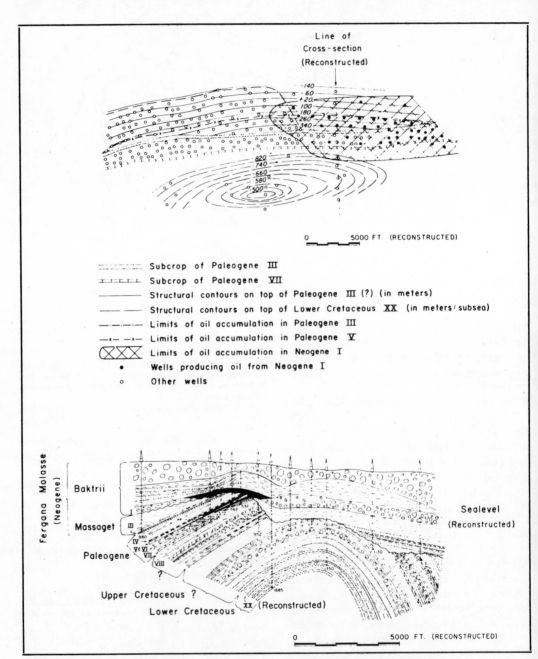

Fig. 6.—Structure map and cross section of South Alamyshik field, Uzbek S.S.R., Turkestan, U.S.S.R. (after Khutorov, 1958), showing escarpment on the flank of an anticline.

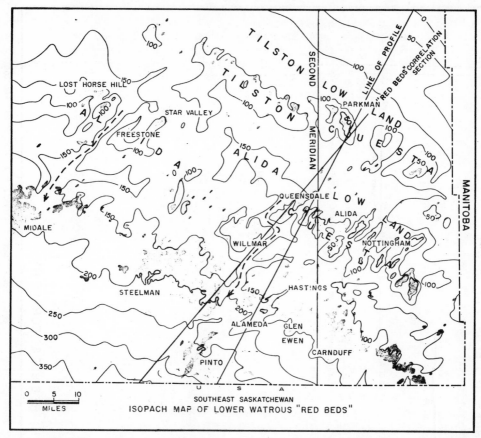

Fig. 7.—Southeast Saskatchewan. Isopachous map of Triassic (?) lower Watrous Formation "Red Beds."
Thin isopachous contour values correspond to high elevations of underlying buried Mississippian landscape.
Two oblique lines correspond to lines of section of Figure 8. Contours in feet.

pachous map (233 ft./mi. before compaction). This is more than seven times the dip of the Mississippian beds toward the southwest. Therefore, it appears most unlikely that any of the northeast-trending valleys would have been formed by resequent (*i.e.*, in this case, downdip-flowing) streams. The writer thus is confident that a large proportion of the valleys that separate the individual oil fields of the Tilston and Alida cuestas was formed by obsequent streams, which flowed northeast into the subsequent streams of the northwest-trending Tilston and Alida lowlands, and which cut back by headward erosion (scarp recession) into the corresponding cuestas.

COMPACTION FACTOR AND SYNSEDIMENTARY MOVEMENTS

In order to establish the amount of compaction, for which all data based on measured "Red Beds" thicknesses must be corrected, two areas in southeastern Saskatchewan were selected for study, one of 12 townships in the Freestone-Star Valley area, and one of 9 townships around the Alida and Nottingham fields. The method used was to construct a structural contour map across a large area, using the top of the "Red Beds" as datum, and establishing the average dip and strike in the area. Next, the departure of each well-control point from this average (residual structure) was measured and plotted against the local "Red Beds" thickness (Fig. 10). The average compaction value thus established is 32 per cent for both areas. Therefore, pre-compaction thickness of the "Red Beds" was, on the average, 147 per cent of the present thickness. Because of the relatively uniform lithologic character of this formation in most of the area studied, the effects of compaction may be disregarded wherever only

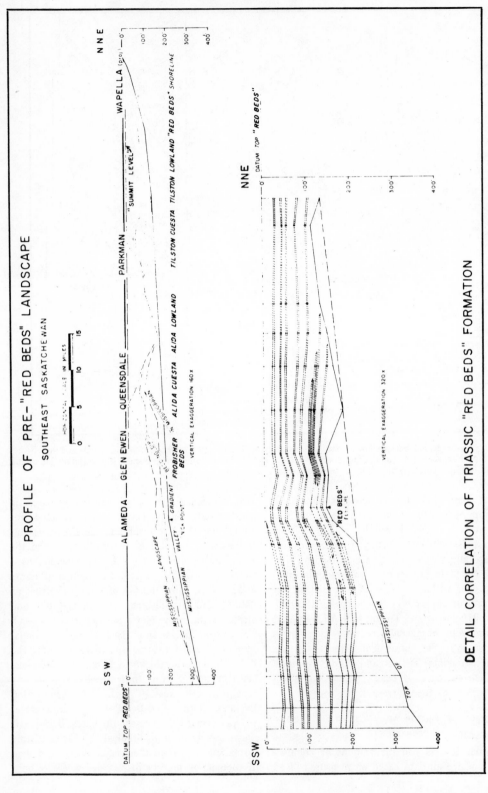

FIG. 8.—Southeast Saskatchewan. Profile of pre-"Red Beds" landscape (top) and detail correlation of Triassic (?) "Red Beds" (bottom). Location of profile and correlation section on Figure 7.

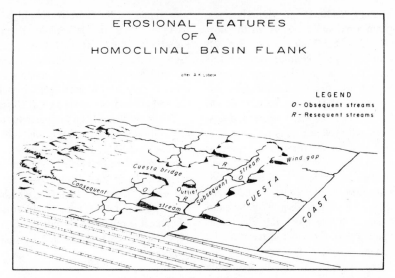

FIG. 9.—Erosional features of a homoclinal basin flank, after Lobeck (1939). (Published with permission of McGraw-Hill Book Company.)

relative thicknesses need to be considered (*e.g.,* on maps such as Fig. 7).

A further check, made to establish whether the "Red Beds" had been affected by tilting or other structural processes during their deposition, involved a detailed correlation along the approximate position of the deepest consequent valley shown on Figure 7. Thin siltstone markers within the "Red Beds" can be correlated through surprisingly long distances and exhibit very little change, indicating that these beds were deposited in extremely quiet, or stable, conditions. The cross section (Fig. 8, bottom) shows the bedding planes within the "Red Beds" to be practically parallel with the top of the formation (which is conformable with the overlying Jurassic), except

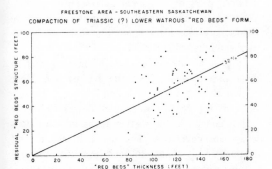

FIG. 10.—Freestone area, southeastern Saskatchewan. Compaction of Triassic (?) Lower Watrous "Red Beds."

for a flexure toward the south in the northern part of T. 3, R. 3 W. 2 M. The detailed correlation of all logs in this general area (only a few of which are shown on the cross section) shows that a downwarping occurred during late "Red Beds" time along a hinge line that coincides approximately with the present northern limit of the Midale beds. In order to visualize correctly the pre-"Red Beds" Mississippian topography, the area of the Midale beds subcrop (and even farther south) must be restored to a correspondingly higher original elevation. The apparent steepening of the slope south of the "nickpoint" (7) (Fig. 8, top), resulting from the intra-"Red Beds" downwarping, is unimportant in the discussion of the paleotopographical features north of the area affected by this phenomenon. A real nickpoint, or sudden steepening of the gradient, would indicate the encroachment of a younger, steeper drainage system on an older one; the position of such a nickpoint might have been fixed temporarily in space by the outcrop of a resistant layer in the corresponding valley. In the present case, however, the "nickpoint" is not a Mississippian geomorphological feature, but a post-Mississippian structural feature.

SUMMIT LEVEL

The highest points along the Tilston, Alida, and Frobisher cuestas seem to be at a similar elevation, *e.g.,* 6–10 ft. below the top of the "Red

Beds" at Parkman, 0–30 ft. in the Hastings-Alida-Nottingham area, and 40–41 ft. in the Lost Horse Hill-Freestone-Star Valley area (Fig. 7). Figure 11 is an attempted contour map of this "summit level." The classical Davisian geomorphological concept of such a summit level, or "base surface," would be to interpret it as an old peneplain that has been uplifted anew and dissected by a second cycle of erosion (Fig. 12, left). This also was the interpretation expressed by Siever (1951), who described the Mississippian erosion surface which is covered by Pennsylvanian rocks in Illinois. However, it should be noted, first, that the southeastern Saskatchewan summit level, south of the hinge line that coincides approximately with the subcrop trend of the Midale beds, is no longer a near-horizontal surface. Second, the summits of the buried Alida and Frobisher beds hills are lower in the vicinity of the main consequent valley that runs along the west side of the Queensdale field. Nevertheless, although the hills become lower in the direction of this valley in terms of absolute elevation, the amount of relief above the valley floor (80–120 ft.) is approximately the same. Even in the Steelman field, south of the hinge line, where the Mississippian is at a much lower level than farther north, the relief is of the same order. The writer concludes, therefore, that the buried Mississippian landscape of southeastern Saskatchewan supports the idea of Rich (1938, p. 1701) and others that ". . . such structural ridges must develop an even crest line entirely independent of any peneplanation. Briefly, this is because the height of the crest is determined by the meeting of the slopes that rise from the weak-rock areas on either side. Since . . . these slopes tend to have a constant angle, they must meet at a relatively uniform elevation above the grade profiles of the subsequent streams in the adjacent weak-rock area." More recently, Hack (1960, p. 91) observed that ". . . regularity of the landscape and the rather uniform height of the hills owe their origin to the regularity of the drainage pattern that has developed over long periods, by the erosion of rocks of uniform texture and structure." The summit level, according to these ideas, would not be caused during a previous erosion cycle but would be the result of contemporaneous leveling processes (Fig. 12, right). The paleogeomorphological data from southeastern Saskatchewan support this view. It should be stressed in particular that the paleoplain discussed here is essentially equivalent to Davis' peneplain. If a "summit level" were a dissected peneplain, then a paleoplain could not have a summit level. Because it does, the summit level must be an essential and contemporaneous part of the paleoplain.

The summit level is an important concept in the exploration for prospective buried hills. A lowering of this level by 80 ft., as observed at Queensdale, corresponds to a basinward displacement of the escarpment by at least 4 mi. (based on a formation dip of 20 ft./mi.). Therefore, whether the summit level is a remnant of an old erosion level, or whether it is related to the current drainage system of the area, is a problem that has many practical implications.

RESISTANCE OF DIFFERENT LAYERS TO EROSION

In addition to the main escarpments of the Alida and Tilston cuestas, there are numerous scarps that correspond to resistant levels of lesser importance. Figure 13 shows diagrammatically the three subsidiary escarpments which are typical of most of the buried hills along the Alida cuesta and which correspond to resistant layers within the Alida beds. The same features have been described by Vogt (1956, Fig. 4) from the Alida and East Alida (North Nottingham) fields, and also appear on a cross section of the Nottingham field published by Edie (1958, Fig. 7). However, escarpments of this subsidiary type do not form the long cuestas which constitute the divides between the principal subsequent valleys. They occur instead as short strike ridges on otherwise southwest-northeast-trending spurs lying between pairs of obsequent or resequent valleys. Where particularly well developed, such strike ridges may give these spurs the shape of "hammerhead hills" (Fig. 18).

In general, the Alida (MC_2) shale, as well as most porous rocks (e.g., the Kisbey and other (dolomitic) sandstone beds and the dolomitic sections of the Middle Alida and Lower Frobisher beds), constitute the weaker rocks in which valleys were developed. The dense limestone and evaporite beds appear to have been the more resistant, ridge-forming beds. There is some evidence that lateral facies changes caused variations in resistance of these rock units, with the

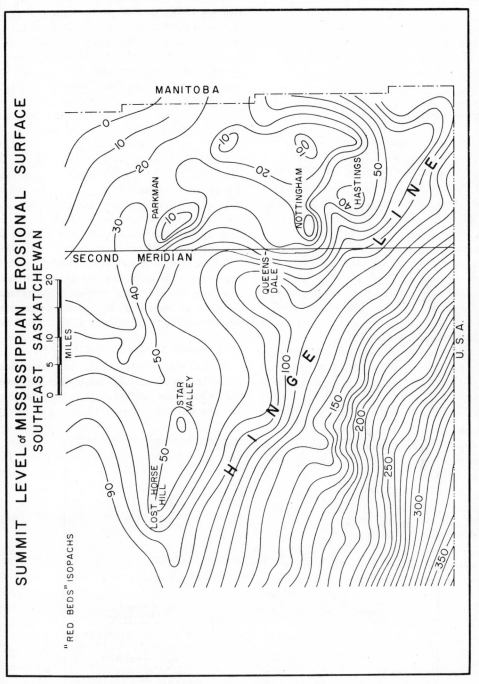

FIG. 11.—Summit level of Mississippian erosional surface, southeast Saskatchewan ("Red Beds" isopachous contours). Lowest iso-pachous contour values correspond to highest summits of buried Mississippian topography. Summit level is tangential to erosional surface illustrated by Figure 7. Contours in feet.

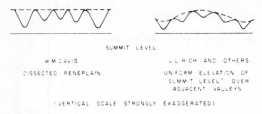

SUMMIT LEVEL

W M DAVIS   J L RICH AND OTHERS

DISSECTED PENEPLAIN   UNIFORM ELEVATION OF
                                             "SUMMIT LEVEL" OVER
                                             ADJACENT VALLEYS

(VERTICAL SCALE STRONGLY EXAGGERATED)

FIG. 12.—Nature of summit level, according to
W. M. Davis and newer theories. Paleogeomorphic
data support more recent view.

result that some cuestas and lowlands disappear
or reappear along strike. For example, the Alida
lowland, which is developed in the Alida shale
east of the Queensdale field, is in the basal Middle Alida beds farther west, north of Queensdale
(Fig. 18). The most important relief-forming section is in the lower part of the Frobisher beds,
just above the Kisbey sandstone or its equiva-
lents (Upper Alida beds). There is some evidence in the South Hastings pool that the Carle-
vale evaporite is more resistant than the underlying carbonate pay zone, but this zone in the Frobisher beds definitely forms a much lower ridge
than the Alida cuesta north of it.

One method of determining which sections are
erosion resistant is to construct a frequency
graph of the interval thickness between a particular marker (e.g., top of Alida beds) and the top
of the Mississippian found in all wells within a

relatively large area, as, for example, between the
Freestone and Queensdale fields (Fig. 14). This
area was chosen because, at the time of the survey, it contained no major oil fields but only
wildcat wells that were fairly uniformly scattered. Where small fields or clusters of wells occurred, only one representative well was used for
constructing the graph. The correlation of the resistant intervals with the lithologic character is
evident from this graph, but only the graph indicates the relative frequency of occurrence (c.q.,
relative area of distribution) of each resistant
bed. One of the main problems facing the geologist in making such a graph is the exact correlation of logs (commonly of different types) run in
wells of relatively shallow penetration into the
Mississippian. In addition to the graph, a series of
cross sections was constructed from all wells in
the same area. Although these cross sections illustrated the concept as well as the graph, it is obviously impossible to incorporate all data from a
large area in one cross section; this method,
therefore, becomes unwieldy and is less accurate
than that of the frequency graph.

QUANTITATIVE GEOMORPHOLOGY

The fact that only two of the many erosion-resistant intervals in the Mississippian of southeastern Saskatchewan formed major cuestas, whereas
the remainder form only short strike ridges,

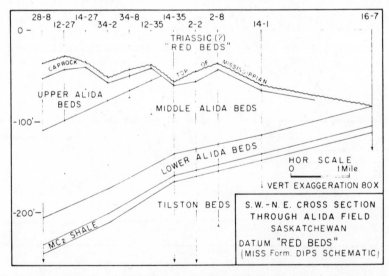

FIG. 13.—Southwest-northeast cross section through Alida field, Saskatchewan, illustrating occurrence of
subsidiary escarpments. Main cuesta at left. Location of cross section on Figure 18.

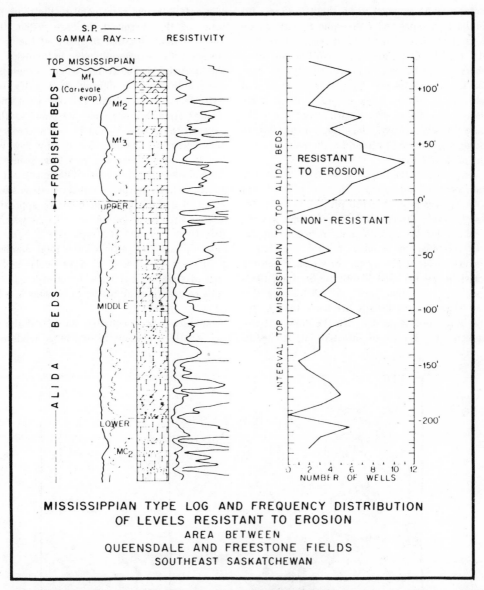

MISSISSIPPIAN TYPE LOG AND FREQUENCY DISTRIBUTION
OF LEVELS RESISTANT TO EROSION
AREA BETWEEN
QUEENSDALE AND FREESTONE FIELDS
SOUTHEAST SASKATCHEWAN

FIG. 14.—Mississippian type log and frequency distribution of beds resistant to erosion, area between Queensdale and Freestone fields, southeast Saskatchewan. Main Alida cuesta corresponds to the +35 ft. level; the —45, —70, and —105 ft. levels correspond to three subsidiary cuestas of Figure 13.

points up the importance of considering this buried landscape from a true geomorphological rather than a limited geological viewpoint. A landscape is formed by the interaction of climatic factors (such as rainfall) and the geological framework. Flowing water may adapt itself to geological factors, but first of all is controlled by hydrophysical laws. These laws and the nature of the under-

lying geology determine the ultimate shape of the drainage pattern (Fig. 15).

Horton (1945) has formulated several quantitative geomorphological laws which are extremely useful in landscape analyses directed toward the search for hydrocarbon traps. He defines *drainage density* as the sum of the lengths of all streams in a basin, divided by the area of that basin. This factor may be assumed to be about the same for similar rock types under the same climatic conditions. Thus, having determined the drainage density for the Queensdale-Alida-Nottingham area, where obsequent streams appear to have an average spacing of 3.3 mi., it may be assumed that other parts of the Alida cuesta (and probably also of the Tilton cuesta) will have about the same drainage density. Thus, the drainage-density factor may be used to interpret the paleotopography of these areas.

According to the *law of drainage composition*, the number and average length of tributary streams vary in geometrical progression with the stream order. Stream order, in Horton's system, is No. 1 for the smallest unbranched tributaries and increases in number toward the largest (main) stream. (This system of stream-order designations was modified slightly by Strahler, 1952a and 1960.) Horton (1945, p. 298) states that first-order streams usually are more than ⅓ mi. and less than 2–3 mi. long. In the buried Mississippian landscape of southeastern Saskatchewan,

first-order streams correspond to the short tributaries of the obsequent drainage system. These are the tributaries that originate at the foot of the short strike ridges observed in the fields along the Alida cuesta. The average length of these is 1½ mi. These are not difficult to identify where well spacing is ¼–½ mi. The obsequent and resequent streams flowing northeast and southwest between these fields would be second order; the subsequent streams into which they flow would be third order; and the two large consequent streams flowing into the Williston basin sea, fourth order. The total number of streams of one order in a drainage basin tends to approximate closely an inverse geometric series in which the first term is 1 and the ratio is the *bifurcation ratio*. This is the ratio of the total number of branchings of streams of a particular order to that of streams of the next lower order. It usually is constant for all orders of streams in a basin (Fig. 16, right).

Another Horton equation states that the average length of streams of any order tends to ap-

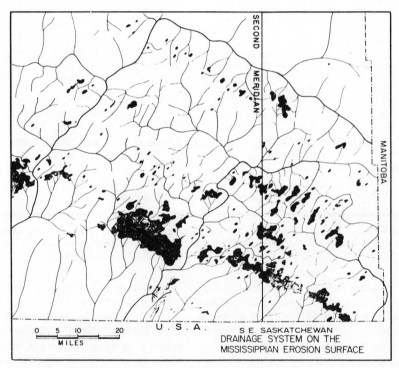

Fig. 15.—Southeast Saskatchewan. Drainage system on Mississippian erosion surface, derived from "Red Beds" isopachous maps.

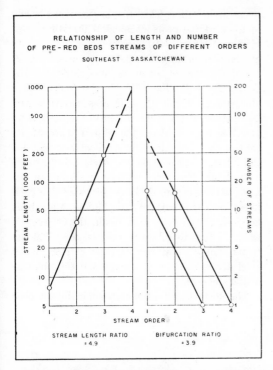

RELATIONSHIP OF LENGTH AND NUMBER
OF PRE-RED BEDS STREAMS OF DIFFERENT ORDERS
SOUTHEAST SASKATCHEWAN

STREAM LENGTH RATIO = 4.9

BIFURCATION RATIO = 3.9

FIG. 16.—Relationship of length and number of pre-"Red Beds" streams of different orders, southeast Saskatchewan. Stream orders 1 to 4 refer to streams of increasing length. Ratio of average length of streams of a given order to that for next lower order is known as *stream length ratio*. *Bifurcation ratio* is ratio of total number of branchings of all streams of a particular order to that for next higher order.

proximate closely a direct geometric series in which the first term is the average length of first-order streams and the ratio is the *stream length ratio,* or the ratio of the average length of streams of a particular order to that of streams of the next lower order (Fig. 16, left). If stream lengths and stream numbers are plotted against the stream orders on a semi-log scale, straight lines result (Fig. 16). These can be used to analyze less densely drilled parts of the same drainage area.

The *law of stream slopes* indicates an inverse geometric relation between average channel slope and stream order. (Strahler, 1952a, also introduced a relation between *basin areas* and stream orders.)

These quantitative concepts are very useful in interpreting an area paleotopographically; *i.e.,* they keep the number of tributaries, lengths of

streams, and channel slopes within specific limits according to geomorphological principles. However, deviations from these "laws" do occur, mostly because of geological factors. During the last 15 years, a tremendous volume of work has been done in the field of quantitative geomorphology, much of it under the direction of A. N. Strahler at Columbia University, and of L. B. Leopold at the U. S. Geological Survey. Their studies have confirmed fully Horton's concepts and have refined further the techniques of this new sub-science. However, many of their findings go far beyond the particular sphere of interest of the paleogeomorphologist. Because this branch of science is so young and its findings are being augmented continuously, only one textbook (Strahler, 1960, p. 376 *ff.*) devotes considerable space to quantitative geomorphology.

The logic of the quantitative approach to geomorphology is so overwhelming that it seems almost unbelievable that its laws were formulated only as recently as 1945. Given a region of simple geological texture, and located within the boundaries of one climatic province, it can be said that about an equal volume of rain will fall per unit of time on each unit of surface area in that region. Working its way down the slope of the original virgin landscape (*e.g.,* immediately after emergence from the sea), this rain water, starting out as sheet flow, gradually combines into several definite courses and carves valleys into the landscape. Where, locally, there are too few valleys in the beginning to handle the rainfall volume per unit area, it is reasonable to assume that additional valleys will form between the first valleys. It is equally reasonable to assume that, where too many valleys form in the beginning, two or more eventually will combine into one. Thus, a certain equilibrium, which can be analyzed statistically, tends to be created. If this equilibrium is upset by external forces, such as a change in climate, uplift or subsidence of the land, or other relative changes in base level, the forces involved will reshape the landscape until it is again in equilibrium. This concept of "dynamic equilibrium" is presented eloquently by Strahler (1950; 1952a, b) and by Hack (1960).

In the buried landscape of southeastern Saskatchewan, other geomorphological observations may be made which reflect this law of dynamic equilibrium. Although these have received practically no attention from "modern" geomorphologists, they are of considerable importance to the paleogeomorphologist in the exploration for oil and gas. One phenomenon, that of the "hammerhead hills," was mentioned earlier in this paper. Another, which may be termed the *law of alternating obsequent and resequent interfluve spurs*, was hinted at briefly by Horton (1945, p. 355–357), and is illustrated in Figure 17. In essence, the principle involved is that, where resequent or obsequent drainage is established on one side of a cuesta, the drainage on the opposite side tends to avoid draining the same part of this ridge, with the result that valleys on one side of the cuesta are opposite spurs on the other side and *vice versa*. Figure 18 illustrates this principle with an example from the area of the Alida and Nottingham fields.

A third principle involves the *oblique angle of entry* of subsequent tributaries into the consequent main valley. This, too, is mentioned briefly in some of the geomorphological literature (*e.g.,* Horton, 1945, p. 349), but assumes considerably greater importance in paleogeomorphology. A practical case in point is illustrated by the Queensdale field (Fig. 18). The subsequent valley along the north side of the Queensdale, Alida, and Nottingham fields is developed in the MC$_2$ shale and lower beds along the eastern part of its course, thus providing an excellent seat-seal for the oil accumulations in the Alida and Nottingham fields. However, because it strikes at an angle to the strike of the beds, this valley cuts back into the lower part of the Middle Alida beds north of Queensdale. As a result, the Queensdale reservoir has an oil-water contact corresponding to its spill-point, whereas farther east the water level occurs much lower because of the presence of the seat-seal.

The main types of non-dynamic interference

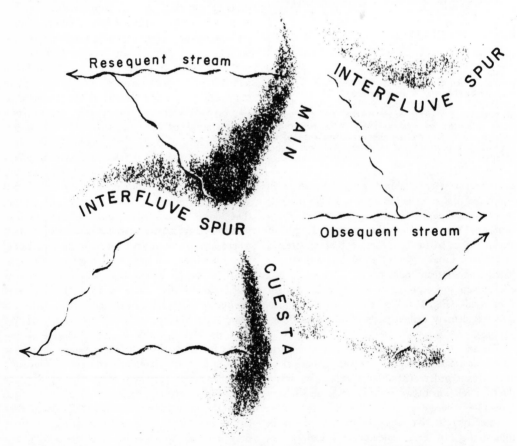

FIG. 17.—Principle of alternating obsequent and resequent interfluve spurs. Modified from Horton (1945).

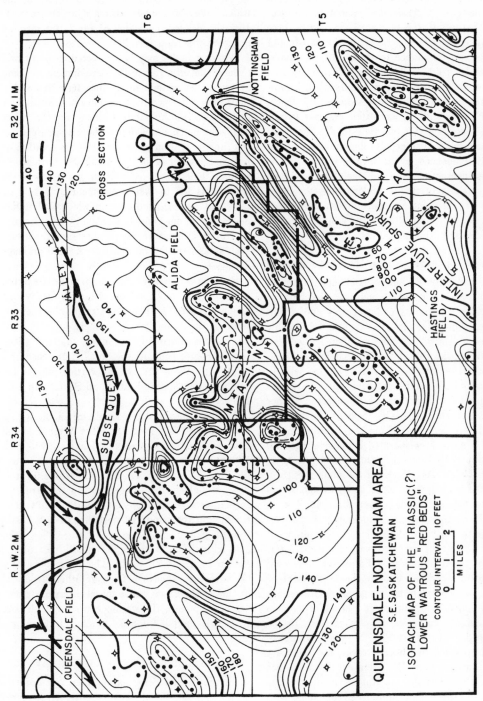

Fig. 18.—Detailed isopachous map of "Red Beds" formation, Alida-Nottingham area, southeast Saskatchewan, showing shape and location of Mississippian buried hills. Note "hammerhead" hills and alternation of interfluve spurs. Line of cross section in Alida field indicates location of Figure 13. Contours in feet.

QUEENSDALE-NOTTINGHAM AREA
S.E. SASKATCHEWAN

ISOPACH MAP OF THE TRIASSIC(?)
LOWER WATROUS "RED BEDS"

CONTOUR INTERVAL 10 FEET

0   1   2
MILES

with the forces of denudation with which geomorphology is concerned are related to the geological nature of the landscape on which these forces act. Interference takes the forms of: (1) differences in resistance to erosion of the underlying rocks (discussed earlier); (2) faults and fractures; a compound form results when a fault separates rocks of different resistance (cf. Hack, 1960, Fig. 4); and (3) karst-type drainage in carbonate rocks. The last two types of landscape are discussed below. The geomorphic parameter most strongly influenced by geological factors is the drainage density (Strahler, 1952a, Fig. 4). In an area of relatively simple geology, however, such as the Mississippian of southeastern Saskatchewan or southwestern Alberta, the drainage density should be fairly constant.

FAULTS AND FRACTURES

The Mississippian buried landscape of southwestern Alberta includes good examples both of the influence of faults and fractures and of underground drainage. In this area, a comparable situation exists in the Mississippian Turner Valley and Pekisko Formations to that in southeastern Saskatchewan (Bokman, 1963). The fields in the Sundre-Harmattan-Crossfield-Calgary area are separated by valleys that cut through the Turner Valley into the underlying Shunda Formation; the fields that produce from the Pekisko on the east and north are separated by valleys cut into the Lower Mississippian Banff Group. As a result, within reservoir rocks of the same age, there are large variations in fluid contacts and in oil columns between fields (Bokman, 1963, Fig. 8, top). Here, too, the paleotopography is not the only factor that has determined the trapping of hydrocarbons, but changes in reservoir characteristics of the Turner Valley Formation have contributed to a certain extent to this entrapment.

The Mississippian landscape of southwestern Alberta underwent one more period of erosion than that of southeastern Saskatchewan. In the western part of this area, the Mississippian paleoplain is covered by Jurassic rocks, but these are absent farther east. Pre-Cretaceous erosion has exhumed some of the pre-Jurassic landscape, has left some other valleys untouched and filled with Jurassic sediments, and has created additional valleys. Detailed correlation of the thin Jurassic cover, where present, is required to determine which valleys are pre-Jurassic and which are post-Jurassic in age, and which of the latter may be exhumed pre-Jurassic valleys.

Whereas faults may be suspected to occur and to have played a role in the development of some of the buried valleys of southeastern Saskatchewan, their existence there is more suspected than proved. In southwestern Alberta, on the other hand, the much lower dip of the Mississippian at the time of erosion has resulted in the creation of resistant Turner Valley and Pekisko Formations subcrop belts that are two or more townships wide. This led to the development of a sort of tableland (mesa) landscape in which valley formation could not take place solely by streams concentrating in areas of less resistant rocks, because such areas were too far apart for normal drainage density to develop. Consequently, other geological flaws, such as zones of structural weakness, exerted an important influence on the drainage system which formed (Fig. 19). The straight alignment of some of the major valleys strongly suggests fault or fracture controls. For example, available structural evidence supports the possible presence of a fault with about 100 ft. of vertical throw between the Harmattan-Elkton and Harmattan East fields.

Another subsurface example of a fault-controlled valley is known from the West Edmond field, Oklahoma (McGee and Jenkins, 1946, Fig. 10, etc.). A pre-Pennsylvanian stream cut through the Siluro-Devonian Hunton Limestone in this field, at right angles to the subcrop trend of this formation, and generally followed a fault zone trending in the same east-west direction. Whereas the upper and lower reaches of the valley coincide with the fault trace, the middle part of the stream swings ½ mi. north. This swing illustrates the fact that, though a fault or other structural line of weakness may give initial direction to valley development, the stream course may deviate from this direction for unknown subsidiary reasons.

UNDERGROUND DRAINAGE IN CARBONATE ROCKS

Intraformational breccias (caused by roof collapse) and sand-filled cavities, observed in cores or samples from several wells in the area between the Sundre field and Calgary (Fig. 19), indicate

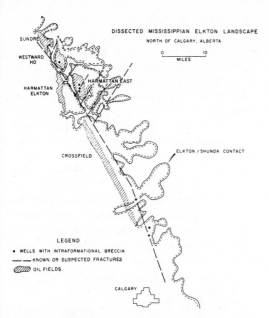

DISSECTED MISSISSIPPIAN ELKTON LANDSCAPE
NORTH OF CALGARY, ALBERTA

SUNDRE

WESTWARD HO

HARMATTAN ELKTON

HARMATTAN EAST

CROSSFIELD

ELKTON/SHUNDA CONTACT

0   10
MILES

LEGEND
• WELLS WITH INTRAFORMATIONAL BRECCIA
— — KNOWN OR SUSPECTED FRACTURES
▨▨ OIL FIELDS

CALGARY

Fig. 19.—Dissected Mississippian Elkton landscape north of Calgary, Alberta. Main subsequent valley trends NNW. along east side of Elkton-Shunda contact. Regional dip is toward west-southwest. Oil and gas are trapped in Elkton salients and outliers (gas accumulations not shown). Some WNW.-trending valleys and their tributaries appear to coincide with fault or fracture trends. Intraformational breccia is indicative of underground drainage, which also may follow fracture trends.

the presence of underground drainage during at least one of the erosion periods that affected the southwestern Alberta Mississippian landscape. Similar phenomena are lacking completely in southeastern Saskatchewan, a fact which indicates a change in climate and therefore of erosion pattern after the Triassic and before the Cretaceous.

Another good example of karst topography accompanied by the formation of sinkholes and a significant amount of underground drainage is in the Cambro-Ordovician Arbuckle Limestone hills of the Central Kansas uplift (Walters, 1946; Walters and Price, 1948). The sinkholes and caves in the Arbuckle are filled with green shale, sandstone, and breccia resembling the "Sooy" or basal conglomerate of the Pennsylvanian Des Moines, a formation which also covers most of these hills. Locally, however, a thin layer of Upper Ordovician Simpson shale, sandstone, and dolomite is between the Arbuckle and the "Sooy," indicating that the landscape was first formed in post-Early Ordovician time, and was exhumed again during

the time interval between the Mississippian and the Pennsylvanian transgression. The fact that there was an earlier period of erosion is confirmed by the discovery of Simpson-filled sinkholes in the Arbuckle of eastern Kansas (Merriam and Atkinson, 1956). The Arbuckle landscape of the Central Kansas uplift consists of a paleoplain of low rolling hills, in which a few canyons several hundred feet deep have been cut, whose walls have slopes as steep as 2,100 ft./mi. The bottoms of the caves filled with "Sooy" clastics, which have been found in wells drilled through sections that otherwise consist only of Arbuckle beds, are higher than the bottoms of the valleys, but lower than the bottoms of the sinkholes on the paleoplain. It thus appears that underground rivers flowed from the sinkholes through the caves into the steep-walled canyons. The latter are aligned in straight patterns which suggest a fault or fracture origin.

FOSSIL CLIMATES AND WEATHERING

The difference between the karst and sinkhole type of buried landscape and that described from southeastern Saskatchewan is striking. Therefore, a discussion of the climatic conditions prevalent at the time and locality of landscape development is in order. An important consideration is whether this climate was humid or arid. In humid areas, a vegetation cover develops that restricts erosion and reduces the rate of runoff, resulting in gentle, more-or-less-rounded landscape forms. In an arid climate, on the other hand, steep escarpments can develop freely. The Mississippian and older formations of the Mid-Continent area and of western Canada, even those covered by Pennsylvanian rocks, may well have been exposed to several erosion periods, beginning in Late Mississippian or Early Pennsylvanian time, and separated by times of relative standstill or even submergence. Some of this erosion may have been during humid and some during arid conditions. The overlying transgressive sediments may give some clue to the climate just before they were deposited. However, unless the last erosion cycle has destroyed completely or altered the effects of all preceding ones, knowledge of the climatic conditions at the time of the final stage of erosion will not be sufficient for a full understanding of the buried landscape. The great importance of this question in the case of carbonate rocks is reflected in the following quotation from Thornbury (1954, p. 56): "In humid regions limestone is usually considered a 'weak' rock. Areas underlain by limestone are generally lower than surrounding areas. This is the result not so much of the physical weakness of limestone as of its susceptibility to solution. In arid regions, however, where moisture is deficient and solution insignificant, we frequently find that limestone is a 'strong' rock and commonly is a cliff or ridge former."

King (1953), however, has pointed out that landscapes developed before the middle Tertiary, when the carpet grasses started spreading over the world, generally should have been of the semi-arid (scarp and pediment) type. Russell (1956, p. 454) also considered that the low forms of plants that covered the earth until mid-Mesozoic time could have created only incipient soils in the modern sense, with the result that topographic profiles did not resemble closely those known today in humid regions. Nevertheless, it must be assumed that the distinction between chemical and mechanical weathering existed even with little soil cover present. As pointed out by Peltier (*in* Thornbury, 1954, p. 58–65), chemical weathering is most intense under conditions of high temperature and abundance of water. Mechanical weathering, on the other hand, is most effective in a temperate, relatively dry climate. As already stated, different climatic conditions may have alternated during long or even during relatively short periods of erosion. The best method to determine the type or types of climate which prevailed during a certain period of landscape formation would appear to be to study the factual information available and to see which concept or concepts best fit the data obtained.

The length of time during which many eroded landscapes of the past have been exposed to atmospheric influences would lead one to believe that weathered surfaces must be a very common adjunct of buried erosional landforms. Weathered surfaces nevertheless appear to be absent in many places where they could be expected to occur. Their apparent absence may be the result of removal of the weathering products either by flowing water while the land was still exposed, or by wave and current action during subsequent marine transgression. In either situation, the coarser weathering products would be expected to collect in low places on the old surface and to become mixed with sediments of the transgressive environment. If the in-mixed sediments were clay or carbonate mud, too little porosity would be left in the original weathering products for them to be of potential interest as reservoirs for oil and gas. If, on the other hand, the in-mixed sediment were a sand or gravel, the resultant product most probably would become part of the overlying transgressive reservoir in the form of a "basal conglomerate." Walters (1946, p. 695–699) has given a good description of "non-marine" and "marine" conglomerate bodies derived from the Arbuckle Limestone of the Central Kansas uplift.

Possible conditions under which weathering products would not be removed from their original place of deposition are the following.

1. Absence of (sufficient) flowing water at the surface
   (a) Arid climate
   (b) Karst landscape (underground drainage)
2. Retention of weathering products in the presence of flowing water
   (a) By gravity (eluvium)
   (b) By plant cover.

In an arid climate, winds may remove the finer weathering products, leaving behind a coarse residuum or eluvium as in case 2-a. A similar situation probably also characterizes karst landscapes. The water, as it enters the subsurface through sinkholes, carries with it the finer particles and soluble material and leaves a residuum of the larger chunks of weathered material (see description by Walters, 1946, p. 697). Gravity has the same effect even when water is free to remove the weathering products, and the residual eluvium consists of the heaviest pieces originally present in the exposed sediment, such as rock fragments, silica, concretions, conglomerate pebbles, *etc.* Unless such an eluvium becomes completely infilled or cemented by sediments during the subsequent transgression, such material can make a very good reservoir rock. It appears to be important that the transgressing sea should not alter such a deposit by transporting it, breaking it down into smaller particles, mixing it with other sediments, *etc.* Optimum conditions for the preservation of the reservoir properties of this deposit thus would be provided by a relatively rapid transgression, quickly moving the new shoreline over the original eroded landscape, exposing it for not more than a short period of time to the effects of waves and longshore currents, and covering it with quiet-water sediments.

Among the fields that produce oil from weathered sediments which may have originated in this manner are the Arbuckle Limestone reservoirs of the Central Kansas uplift. Solution erosion of the karst landscape, developed on the Arbuckle surface during both Late Ordovician and Early Pennsylvanian times, formed a mantle of leached residuum 2–30 ft. thick. This is the present Arbuckle reservoir rock (Walters, 1946, p. 700–701). Another example of this type of reservoir occurs on parts of the TXL structure of western Texas, from which the upper part of the Devonian was removed by pre-Permian erosion. Low areas on the Devonian erosion surface are filled with a highly porous, weathered chert known as the "Tripolite Zone," which forms an oil reservoir separate from those encountered in the Devonian (David, 1946). The tripolite is a weathering product of the Devonian, which here consists of an upper and a lower chert (novaculite) section separated by a middle limestone member. A similar residual mantle occurs on the Mississippian buried hills of southeastern Saskatchewan (limestone rubble) and those of southwestern Alberta (chert), but in these areas a reservoir generally is not developed; this is the "Detrital" of Bokman (1963, p. 258).

The probability that plant cover played a less important role in pre-middle Tertiary times has been mentioned (King, 1953; Russell, 1956). Ancient soils certainly are not conspicuous in the stratigraphic column. Even if they were, they most probably would not form reservoir rocks for oil and gas. In contrast to the coarser eluvium and residuum, most ancient soils, being comparatively fine and generally clayey, can be expected to have been severely altered or even removed by subsequent transgressions, to the point of becoming simply a constituent of the transgressive sediments. Only those parts of the soil which occupied rock crevices or were carried down by burrowing animals and plant roots would be preserved.

Weathering also alters the underlying rocks that remain in place. Mechanical weathering may induce fractures. Chemical weathering, especially of carbonate rocks, may enlarge such fractures by solution and form karst landscapes, sinkholes, caverns, and other phenomena related to underground drainage. In the buried Mississippian hills of southeastern Saskatchewan, the porosity of the reservoir rocks, where exposed along an escarpment, probably at first was increased by such chemical processes. However, as the landscape subsided and the "Red Beds" sea encroached on it, anhydrite filled these enlarged pore spaces and led to the formation of a dense "caprock." This caprock appears to be present in all buried hills of this area, and usually extends about ¼ mi. inward from the subcrop of the reservoir bed.

SAND DEPOSITION IN BURIED VALLEYS

The study of buried landscapes is of importance not only to the exploration for oil and gas in buried hills, but also to the search for hydrocarbons trapped in sandstone beds or other clastic reservoirs that accumulated in the valleys between the hills. In the southeastern Saskatchewan area, where the transgressive "Red Beds" contain practically no sandstone of reservoir quality, there is no economic incentive to search for such sandstone bodies. Farther west along the flank of the Canadian shield, however, a vast paleoplain underlain by Devonian, Mississippian, and Jurassic rocks was covered by Lower Cretaceous (locally Upper Jurassic) sediments which include a high percentage of sandstone which could serve as reservoir, particularly in the basal members, such as the "Basal Quartz," Ellerslie, McMurray, Cantuar, *etc.* There are indications, also, that the influence of buried topography, through the medium of compaction, is expressed as high in the section as the Upper Cretaceous Cardium Formation. Some of the producing sandstone bodies (*e.g.*, in the Cantuar field, southwestern Saskatchewan) overlie old highs and were derived from subcropping sandstone beds deposited in an environment otherwise characterized by shale; however, most of the sandstone was deposited in valleys of the old paleoplain.

The first person to make a paleotopographical study of this pre-Cretaceous erosion surface in eastern Alberta and western Saskatchewan was Beltz (1953), who presented an isopachous map of the Mannville Group. On the same map, Beltz also showed the pre-Mannville paleogeology (Devonian, Mississippian, and Jurassic subcrops) and the occurrence of oil and gas in the basal Cretaceous. This map was the first to show the obvious relations between these various factors; the more erosion-resistant subcropping formations formed the topographical ridges, whereas the basal Cretaceous hydrocarbon accumulations were concentrated in the sandstone-filled valleys.

Two parts of this paleoplain have been mapped by the writer, using two different methods. Figure 20 shows the northern part of the Province of Alberta; contour datum is sea-level, and the present elevation of the eroded surface of the Paleozoic above or below sea-level is shown. This map extends toward the west an earlier map published by Martin and Jamin (1963). This method of illustrating the buried topography was chosen because the overlying basal Cretaceous would not lend itself to the construction of an isopachous map of the "cast" of this landscape, as did the

"Red Beds" for the Mississippian surface of southeastern Saskatchewan. This is because the sandstone and shale of the basal Cretaceous McMurray Formation of northeastern Alberta, which covers unconformably the old land surface of the Devonian, are in turn cut deeply by Pleistocene to Recent erosion in much of the "Bituminous Sands" area along the Athabasca River and its tributaries. In places, this erosion cuts even into the Devonian. Furthermore, some of the highest areas of the paleoplain west of the present Athabasca River were not covered by McMurray sediments (see cross section, Fig. 21). This contour method of illustrating the old land surface is useful where the regional post-Cretaceous westward tilt is relatively small or even reversed toward the east. However,

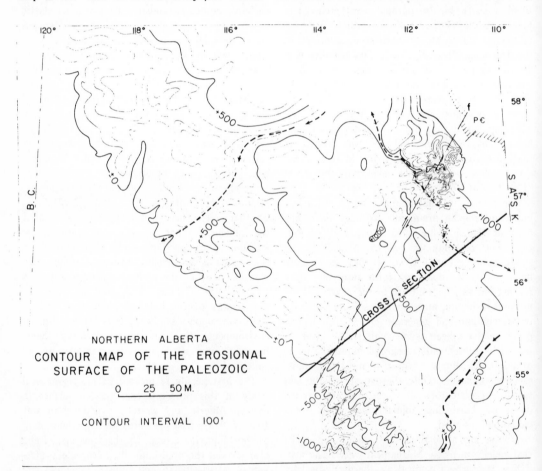

Fig. 20.—Northern Alberta. Contour map of erosional surface on top of Paleozoic. Steepening of dip below zero contour (approximate position of hinge line) caused by subsequent downwarping. Dashed lines with arrows indicate direction of drainage. Athabasca "Bituminous Sands" area occupies northeast quarter of map. Line of cross section indicates location of Figure 21. Contours in feet.

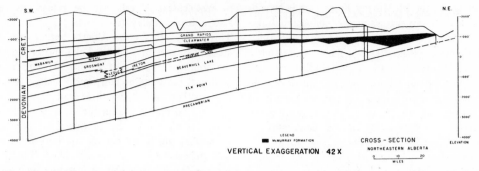

FIG. 21.—Southwest-northeast cross section of Athabasca "Bituminous Sands" area, northeastern Alberta. Note collapse area at right, caused by removal of Elk Point salts by solution. Line of section is indicated on Figure 20.

the drawback is obvious after studying the contours southwest of the basin hinge line, which coincides approximately with the zero (sea-level) elevation of the top of the Paleozoic. Beyond this hinge line, the superimposed steepening of the Cretaceous toward the southwest obliterates most of the original shape of the Paleozoic landscape and thus gives a distorted picture of the paleotopography.

Figure 22 shows the adjoining area of west-central Saskatchewan, by means of an isopachous map, showing the thickness of the "cast," as in southeastern Saskatchewan. The formation used

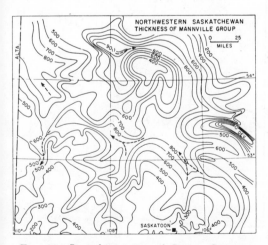

FIG. 22.—Isopachous map of Lower Cretaceous Mannville Group, west-central Saskatchewan, showing post-Paleozoic paleotopography. This figure also illustrates excess thickness of Mannville in center of area (300–400 ft. more than thickness in main consequent valley, at left), which was caused by Devonian salt removal during deposition of Mannville. Contours in feet.

for isopaching is the Lower Cretaceous Mannville Group, which also was used by Beltz (1953). The problem in this instance is that the Mannville was subjected to structural movements during deposition. As a result, the deepest part of the two large subsequent (southeast-northwest-trending) valleys subsided because of salt removal from the underlying Devonian. Selection of a thinner stratigraphic interval (e.g., the basal Lower Cretaceous) might perhaps have given better results. Figure 21 illustrates the thinning that has taken place in the salt-bearing Elk Point Formation near its outcrop; where solution has taken place, the post-Devonian strata have subsided. Nevertheless, the basal Cretaceous sandstone units were deposited by flowing rivers and not in stagnant lakes, as shown by their coarse texture, cross-bedding, and other depositional features. The steeper slopes noted locally on the Devonian paleoplain in the Athabasca "Bituminous Sands" area (360 ft./mi., compared with 233 ft./mi. in southeast Saskatchewan) also appear to be related to later subsidence resulting from salt removal (northeast corner of Fig. 20).

In an earlier paper (Martin and Jamin, 1963), the writer determined the stream-length ratio for the Athabasca "Bituminous Sands" area to be 5.5, which compares well with the ratio of 4.9 established for the Mississippian buried landscape of southeastern Saskatchewan. Applying the principles of quantitative geomorphology, the possible locations of the major consequent valleys that drained this area were calculated. Subsequent work has proved that these valleys are more or less where predicted (Fig. 20). The occurrence in these valleys of bituminous sandstone beds corresponding in age to the McMurray Formation is considered further proof that a properly connected drainage pattern existed during Early Cretaceous time.

Both maps (Figs. 20, 22) show the existence of several northwest-southeast-trending cuestas, which correspond to erosion-resistant layers in the Devonian-Mississippian sequence. Details of the identity of the Devonian cuestas were given by Martin and Jamin (1963), and the cuestas are recognized easily on a cross section (Fig. 21). Extensive gas fields are being developed in some of the buried hills which form these cuestas.

The subsidence which has affected the area east of the Beaverhill Lake subcrop during the Early Cretaceous, especially in west-central Saskatchewan (Fig. 22), makes it difficult to determine from the map alone which was the major consequent valley that drained this area. A massive sandstone section in the basal Cretaceous of the Bellshill Lake oil field and other similar fields

in east-central Alberta can be traced upstream into a broad valley filled with these sediments. This valley crosses the Alberta border into Saskatchewan east of Bellshill Lake and is shown in the southwest corner of Figure 22. This is evidently a third major consequent valley draining this northern area; the three valleys are shown together on Figure 23. On the same map, a major valley is shown in the southwestern corner of Manitoba, which lies outside of the southeastern Saskatchewan area already described. The valley spacing derived from the quantitative geomorphological considerations stated earlier suggests the possible existence of a fifth large consequent stream, the approximate position of which has been indicated by a broken line. Other broken lines illustrate the writer's belief that the origin of some of the largest lakes on the Canadian shield perhaps may be traced to this pre-Cretaceous drainage system. Of special interest is the

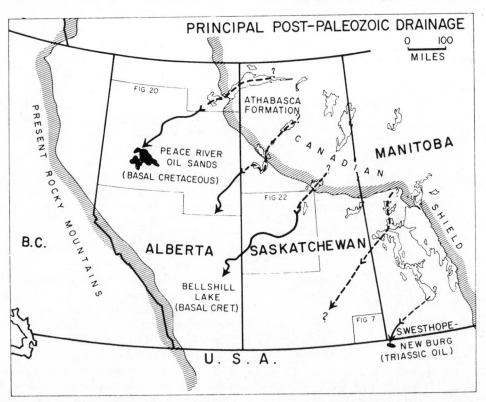

FIG. 23.—Principal post-Paleozoic drainage from Canadian shield. Three western valleys are filled with Cretaceous clastics, the easternmost one with Triassic and younger sediments. Second stream from the east (broken line) is hypothetical. Note outlines of areas covered by Figures 7, 20, and 22. Bend in river above words "Bellshill Lake" is shown in detail on Figure 24. Area of Figure 19 is below word "Alberta." Broken lines on exposed shield area indicate positions of possible pre-Cretaceous streams, now occupied by major lakes.

apparent relation between these major rivers and the occurrence of large oil-bearing sandy deltaic and shoreline deposits of Mesozoic age (Fig. 23).

Busch (1959) published an excellent paper in which he deals in part with sands deposited in buried valleys; he calls them *strike valley sands*. Sandstone bodies of this type are abundant outside of the map areas discussed here, and occur at different levels within the Lower Cretaceous of western Canada. Numerous examples of such sandstone bodies also have been described from United States oil and gas fields.

Less well known is the trapping of hydrocarbons in consequent valleys. Figure 24 shows the details of a small part of the third valley mentioned above, located east of the Bellshill Lake field, Alberta. The outline of a 4–6-mi.-wide valley, indicated by a broken line, is based on the thickness of the Mannville Group (Kb), which is less than 500 ft. on the paleoplain into which the valley has been cut. The cross section (Fig. 25) shows this valley to be filled with a thick section of "Basal Quartz" (Lower Mannville) sandstone beds, which are shaped into several terraces. The terraces (which are not sand bars (Rudolph, 1959) and certainly not "sand reefs," a term used by Bokman (1963)) were formed by the river cutting down into its own sediments, presumably as a result of the subsidence that affected its source area during Mannville time. When the last meandering river course was filled with clay, the possibility was created for hydrocarbons to be trapped against the updip (northeast) side of these meanders. The final position of the river course was interpreted to be that zone where the top of the "Basal Quartz" is more than 400 ft. below the top of the Mannville Group. Using these criteria, it has been possible to trace this valley upstream, as well as downstream, through a large distance. The meanders of the final river course can be reconstructed by combining the known subsurface data with the mathematical

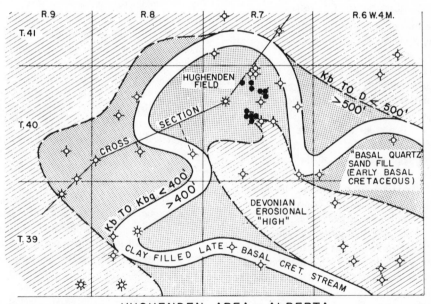

HUGHENDEN AREA – ALBERTA
LOWER MANNVILLE PALEOTOPOGRAPHY

0 _____ 5 M.

Fig. 24.—Lower Cretaceous Mannville Group paleotopography, Hughenden area, Alberta. Meanders of a late basal Cretaceous stream have cut into "Basal Quartz" sandstone beds deposited in broad post-Paleozoic valley. Regional dip is toward southwest. Oil is trapped against a clay-filled meander where it intersects regional strike; gas wells shown correspond to sandstone bodies above "Basal Quartz." Line of cross section indicates location of Figure 25.

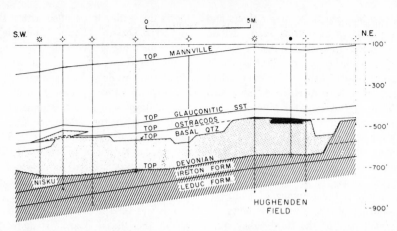

FIG. 25.—Structural cross section, Hughenden area, Alberta, showing incision of late basal Cretaceous stream (now filled with clay) into older "Basal Quartz" sandstone reservoir, creating terraces at two levels. Oil is trapped along updip margin of higher terrace. Line of section is indicated on Figure 24.

formula, derived from quantitative geomorphology, to relate the average dimensions of river meanders.

SUBMARINE CANYONS

Submarine, as opposed to subaerial, erosional topography should be mentioned briefly. Buried submarine canyons occur exclusively in marine sediments originally deposited close to the edge of the continental shelf. Such canyons commonly are several thousand feet deep and many miles wide, and usually are filled with shale. Where a regional tilt is present (e.g., in the area along the Gulf of Mexico), hydrocarbon accumulations may form against the side of the canyon that is regionally downdip (e.g., Yoakum field, Texas; Hoyt, 1959). Other submarine canyons have been described from Louisiana (Bornhauser, 1948) and from California (Frick et al., 1959; B. D. Martin, 1963); the submarine canyons described in the latter two references are filled partly with sandstone which locally is hydrocarbon-bearing. All examples mentioned are Tertiary in age; none is known from the Mid-Continent area or western Canada. The mechanism causing the formation of these channels is very different from that controlling subaerial denudation, and the geomorphological rules governing their occurrence have yet to be formulated in detail.

MAPPING TECHNIQUES

The first step which the geologist usually takes

in studying these old land surfaces is to construct a contour map of the present-day attitude of that surface (e.g., the top of the Mississippian erosion surface in southeastern Saskatchewan). In so doing, the geologist recognizes that this unconformity surface has been tilted and possibly affected by other structural influences since its formation. This first map is needed, however, to study factors, such as fluid contacts, that may be related to present-day elevations. Such a map may be all that is required for the study of landscapes that have been disturbed slightly, or not at all, since their formation (i.e., those buried or partly covered by younger sediments whose present structural attitude is still horizontal or nearly so; Fig. 20). Andresen (1962) has called this the unconformity contour method.

The second step, or method, involves isopachous mapping. An impression of the morphology of the landscape at the time of burial can be obtained from an isopachous map of the transgressive formation overlying the unconformity. Such a map provides a "cast" of the topography (Figs. 7, 18, 22). Depending on the thickness and lithologic character of the interval selected, such isopachous maps will be influenced to some degree by compaction effects. A thick interval may be preferred, for this reason, as long as it is reasonably certain that the interval represents a time of continuous sedimentation, unaffected by later tilting, folding, faulting, or subsidence during deposition, or by a significant amount of post-depositional erosion. On the other hand, if the stratigraphic interval chosen for isopachous mapping is of relatively uniform lithologic character throughout the area studied, the effects of compaction may be disregarded as long as one is primarily interested in relationships (thick or thin isopachous interval, representing low or high topography) rather than in absolute figures. Andresen (1962) has termed this the *datum plane—valley floor isopach method.* The method used to determine the compaction factor was discussed earlier in this paper (Fig. 10). The effects of synsedimentary movements on the validity of such a "cast" isopachous map also have been discussed. Such effects can be minimized by making an isopachous map only of the formation or member just above the unconformity, provided the upper limit of the isopachous interval is known with reasonable certainty to be a time line.

A third step, or method, which the writer has employed with some degree of success to analyze the Mississippian paleotopography of parts of southwestern Alberta (Fig. 19), is to map the thickness of the formation or member just below the erosion surface. This is meaningful only if the strata concerned were subjected to little or no tilting, or other disturbance, during the period of uplift that preceded the formation of the landscape to be studied. The method was used by Siever (1951, Fig. 5), and constitutes a refinement of the *paleogeologic map method* used by the same author (1951, Fig. 4) and described further by Andresen (1962).

Where tilted formations are present, or the strata were otherwise affected by structural movements, the nature of such deformation is important to determine insofar as the deformation preceded or accompanied the formation of the buried landscape to be studied. A map of the present structural attitude of a marker below the unconformity includes the effects of both pre- and post-unconformity movements. The latter may be eliminated by introducing a fourth step. This consists of mapping the thickness of the interval from the pre-unconformity marker to the unconformity and adding to this thickness that of the post-unconformity "cast," restored to its value before compaction. Siever (1951, Fig. 7) used a similar method, but made no allowance for compaction of the Pennsylvanian. For predominantly sandy beds, like the basal Cretaceous of western Canada, the compaction effect is bound to be much less than for clayey formations such as the "Red Beds" of

southeastern Saskatchewan.

A restored pre-unconformity structure map of this type allows one to determine the original spatial position of important reservoir beds as well as of the erosion-resistant, ridge-forming layers, and leads logically to a fifth step. The stratigraphic levels at which the ridge-forming layers occur may be determined from a statistical analysis as outlined in a preceding section (Fig. 14). By means of a summit-level map (Fig. 11), the most likely positions of cuestas and other strike ridges may then be sketched on the restored pre-unconformity structure map. Combining this further with a drainage-pattern map, based on known subsurface data and on the quantitative geomorphological parameters that have been determined for the area of study, a reasonably reliable paleogeomorphic map may be constructed, showing the possible location of unexplored buried hills and valleys in addition to those already known. These prospects may then be defined further by photogeologic and (or) seismic methods of investigation in areas where such methods have proved to be useful in locating buried erosion topography.

CONCLUSIONS

Paleogeomorphology is a useful tool for solving problems connected with the exploration for and development of many hydrocarbon accumulations previously classified as "stratigraphic," in particular those with a marked three-dimensional geometry. Included, in particular, are fossil organic reefs and buried landscapes. Barrier beaches, point bars, submarine canyons, and other predominantly three-dimensional phenomena also are considered to form paleogeomorphological rather than stratigraphic traps, because they lack obvious continuity with the laterally adjoining sediments. Deposits resulting from such factors are not shown easily on simple facies maps and do not lend themselves readily to analysis by means of "layer maps" and similar two-dimensional approaches. A combination of isopachous maps, cross sections, panel diagrams, *etc.* is needed for proper understanding, and the analysis itself must be basically of a geomorphological nature. There is thus a wide scope for paleogeomorphological studies with a correspondingly great variety of applications to petroleum geology.

With regard to buried erosional landforms, it is

important to note that buried landscapes may be analyzed according to quantitative geomorphological principles, making it possible to extrapolate a drainage system and its interfluves (buried hills) from a known, thoroughly drilled area. The geometric pattern of a drainage system is determined further by geological factors such as structure, faulting (and fracturing), and the alternation of erosion-resistant and non-resistant beds in homoclinally dipping areas. Hydrophysical factors to be taken into account in outlining paleotopographical prospects are: the presence and position of a summit level (derived from the elevations of buried hills previously discovered in the area; these elevations are approximately those at which the erosion-resistant levels will tend to form escarpments); the angle of entry of tributary valleys; the alternation of obsequent and resequent interfluve spurs; and others.

Sandstone deposited in the valleys of a buried landscape also contains commercial deposits of oil and gas in many areas. The analysis of the corresponding landscapes must proceed along lines of reasoning similar to that used for buried hills. However, additional attention must be given to such factors as the geometry of river meanders, the formation of point bars and terraces, the influence of valley width on sedimentation, *etc.*

Weathered carbonate surfaces may form good reservoir rocks for hydrocarbons. On the other hand, the same surfaces, where filled with anhydrite or calcite, form a tight caprock over the underlying reservoir. Leaching and consequent widening of fractures increase reservoir volume, particularly in otherwise dense carbonate rocks. Fracture distribution, however, is a problem of structural geology. Underground cavities usually result in a loss of reservoir rock where filled with impermeable sediments, but it is most difficult to detect a systematic distribution pattern of such cavity fills.

Submarine canyons locally are sites of accumulation of small pools of oil or gas, but submarine canyon occurrence is sporadic and the laws governing the distribution of such canyons still need to be formulated.

It is not easy to estimate the hydrocarbon reserves ultimately to be recovered from traps formed by buried landforms, for the simple reason that they commonly are complex traps in which, for example, pre-erosion porosity distribution may play as important a role as trapping

below the unconformity. Similarly, the total influence of paleogeomorphological factors in forming hydrocarbon-bearing reservoir rocks above the unconformity may, in some places, be debatable. As a rough approximation, the writer has estimated that about 9 per cent of the ultimately recoverable oil in western Canada occurs in traps related to buried hills and valleys (Martin, 1960). Bokman (1963) gives figures of 9 per cent of oil and 17 per cent of gas reserves for Alberta alone. In southeastern Saskatchewan, this figure is calculated to be as large as 32.5 per cent of oil reserves (Martin, 1964 a, b). For the "free" world as a whole, on the other hand, Knebel and Rodriguez (1956) estimate that "unconformity traps" contain only 6 per cent of all major oil-field accumulations (6.5 per cent, if the Middle East is excluded). These figures indicate to this writer that a very substantial amount of "unconformity" hydrocarbons still are awaiting discovery in areas where exploration for this type of trap hitherto has not received sufficient attention.

For example, the percentage of paleotopographic traps discovered in southeastern Saskatchewan has risen steadily since the discovery of the Alida field in 1955. It is believed that exploration for such traps may benefit substantially in the future from the application of geomorphological principles to the analysis of buried landscapes, as outlined in this paper for some specific areas. New areas that could be analyzed to advantage in this manner include the Trempealeau trend of Ohio and the Mississippian of Kingfisher and neighboring counties in Oklahoma. Other areas should be sought in the broad belts of Paleozoic rocks that have been subjected to one or more periods of erosion on the North American continent. With the increasing use of computers to solve geological problems, the introduction of trend-surface analysis could help to cut down the time necessary for such regional studies.

REFERENCES CITED

Andresen, M. J., 1962, Paleodrainage patterns: their mapping from subsurface data, and their paleogeographic value: Am. Assoc. Petroleum Geologists Bull., v. 46, no. 3 (Mar.), p. 398–405.

Beltz, E. W., 1953, Topography and geology of eastern Alberta and western Saskatchewan during Early Cretaceous time: Alberta Soc. Petroleum Geologists News Bull., v. 1, nos. 1–4 (May), p. 1–3.

Bokman, J., 1963, Post-Mississippian unconformity in Western Canada basin, *in* Backbone of the Americas: Am. Assoc. Petroleum Geologists Mem. 2, p. 252–263.

Bornhauser, M., 1948, Possible ancient submarine canyon in southwestern Louisiana: Am. Assoc. Petroleum Geologists Bull., v. 32, no. 12 (Dec.), p. 2287–2290.

Busch, D. A., 1959, Prospecting for stratigraphic traps: Am. Assoc. Petroleum Geologists Bull., v. 43, no. 12 (Dec.), p. 2829–2843.

David, M., 1946, Devonian (?) producing zone, TXL pool, Ector County, Texas: Am. Assoc. Petroleum Geologists Bull., v. 30, no. 1 (Jan.), p. 118–119.

Davis, W. M., 1884, Geographic classification, illustrated by a study of plains, plateaus and their derivatives: Proc. Am. Assoc. Adv. Science, v. 33. p. 1–5.

——— 1889, Topographic development of the Triassic formation of the Connecticut Valley: Am. Jour. Science, 3d Ser., v. 37, p. 423–434.

——— 1899a, The peneplain: Am. Geologist, v. 23, p. 207–239; also: 1909, Geographical essays, p. 350–380, New York, Dover Publications (repr. 1954), 777 p.

——— 1899b, The drainage of cuestas: Proc. Geol. Assoc., v. 16, pt. 2 (May), p. 75–93.

Edie, R. W., 1958, Mississippian sedimentation and oil fields in southeastern Saskatchewan: Am. Assoc. Petroleum Geologists Bull., v. 42, no. 1 (Jan.), p. 94–126.

Frick, J. D., T. P. Harding, and A. W. Marianos, 1959, Eocene gorge in northern Sacramento Valley (abs.): Am. Assoc. Petroleum Geologists Bull., v. 43, no. 1 (Jan.), p. 255.

Hack, J. T., 1960, Interpretation of erosional topography in humid temperate regions: Am. Jour. Science, v. 258-A, p. 80–97.

Harris, L. D., 1960, Drowned valley topography at beginning of Middle Ordovician deposition in southwest Virginia and northern Tennessee: U. S. Geol. Survey Prof. Paper 400-B, p. 186–189.

Hill, R. T., 1901, Geography and geology of the Black and Grand Prairies, Texas, with detailed description of the Cretaceous formation and special reference to artesian waters: U. S. Geol. Survey, 21st Ann. Rept., pt. 7, p. 362–380.

Horberg, L., 1952, Interrelations of geomorphology, glacial geology, and Pleistocene geology: Jour. Geology, v. 60, no. 2 (Mar.), p. 187–190.

Horton, R. E., 1945, Erosional development of streams and their drainage basins; hydrophysical approach to quantitative morphology: Geol. Soc. America Bull. v. 56, no. 3 (Mar.), p. 275–370.

Howard, A. D., and L. E. Spock, 1940, Classification of landforms: Jour. Geomorphology, v. 3, no. 4 (Dec.), p. 332–345.

Hoyt, W. V., 1959, Erosional channel in the middle Wilcox near Yoakum, Lavaca County, Texas: Trans. Gulf Coast Assoc. Geol. Soc., v. 9, p. 41–50.

Kay, M., 1945, Paleogeographic and palinspastic maps: Am. Assoc. Petroleum Geologists Bull., v. 29, no. 4 (Apr.), p. 426–460.

Khutorov, A. M., 1958, The formation of secondary oil deposits in the Fergana depression: Petr. Geol. (Geologiya Nefti) v. 2, no. 7-B, p. 643–651.

King, L. C., 1951, South African scenery. A textbook of geomorphology: 2d rev. ed., Edinburgh, Oliver and Boyd, 379 p.

——— 1953, Canons of landscape evolution: Geol. Soc. America Bull., v. 64, no. 7 (July), p. 721–752.

Knebel, G. M., and G. Rodriguez E., 1956, Habitat of some oil: Am. Assoc. Petroleum Geologists Bull., v. 40, no. 4 (Apr.), p. 547–561.

LeGrand, H. E., 1960, Metaphor in geomorphic expression: Jour. Geology, v. 60, no. 5 (Sept.), p. 576–579.

Levorsen, A. I., 1954, Geology of petroleum: San Francisco, W. H. Freeman, 703 p.

——— 1960, Paleogeologic maps: San Francisco, W. H. Freeman, 174 p.

Lobeck, A. K., 1939, Geomorphology: New York, McGraw-Hill, 731 p.

Louis, H., 1964, Über Rumpfflächen- und Talbildung in den wechselfeuchten Tropen besonders nach Studien in Tanganyika: Zeitschr. für Geom., N.F., Bd. 8, Sonderheft, p. 43–70.

Martin, B. D., 1963, Rosedale channel—evidence for late Miocene submarine erosion in Great Valley of California: Am. Assoc. Petroleum Geologists Bull., v. 47, no. 3 (Mar.), p. 441–456.

Martin, R., 1960, Principles of paleogeomorphology: Can. Mining and Metall. Bull., v. 53, no. 579 (July), p. 529–538; also: 1960, Oilweek, v. 11, no. 36 (Oct. 22), p. 84–94; also: 1961, Can. Oil and Gas Industry, v. 14, no. 10 (Oct.), p. 28–40.

——— 1964a, The exploration for Mississippian buried hills in the northeastern Williston Basin (abs): Third Int. Williston Basin Symp., p. 285–286.

——— 1964b, Buried hills hold key to new Mississippian pay in Canada: Oil and Gas Jour., v. 62, no. 42 (Oct. 19), p. 158–162 (complete text).

——— and F. G. S. Jamin, 1963, Paleogeomorphology of the buried Devonian landscape in northeastern Alberta, in K. A. Clark Volume, a collection of papers on the Athabaska oil sands: Res. Council Alta., Information Ser. 45, p. 31–42.

McGee, D. A., and H. D. Jenkins, 1946, West Edmond oil field, central Oklahoma: Am. Assoc. Petroleum Geologists Bull., v. 30, no. 11 (Nov.), p. 1797–1829.

McKee, E. M., 1963, Paleogeomorphology, a practical exploration technique: Oil and Gas Jour., v. 61, no. 42 (Oct. 21), p. 140–143.

Merriam, D. F., and W. R. Atkinson, 1956, Simpson filled sinkholes in eastern Kansas: State Geol. Survey Kans. Bull. 119, pt. 2 (Apr.), p. 61–80.

Meyerhoff, H. A., 1940, Migration of erosional surfaces: Ann. Assoc. Am. Geographers, v. 30, no. 4 (Dec.), p. 247–254.

Penck, W., 1924, Die morphologische Analyse. Ein Kapitel der physikalischen Geologie: Geogr. Abh., 2; Reihe, H. 2; Morphological analysis of land forms, a contribution to physical geology (transl. by H. Czeck and K. C. Boswell), 1953: London, Macmillan, 429 p.

Rich, J. L., 1938, Recognition and significance of multiple erosion surfaces: Geol. Soc. America Bull., v. 49, no. 11 (Nov.), p. 1695–1722.

Rudolph, J. C., 1959, Bellshill Lake field, Alberta: Am. Assoc. Petroleum Geologists Bull., v. 43, no. 4 (Apr.), p. 880–889.

Russell, R. J., 1956, Environmental changes through forces independent of man, in Thomas, W. L., Jr. (ed.), Man's role in changing the face of the earth: Chicago, Univ. Chicago Press, p. 453–470.

Siever, R., 1951, The Mississippian-Pennsylvanian unconformity in southern Illinois: Am. Assoc. Petroleum Geologists Bull., v. 35, no. 3 (Mar.), p. 542–581.

Sloss, L. L., 1963, Sequences in the cratonic interior of North America: Geol. Soc. America Bull., v. 74, no. 2 (Feb.), p. 93–113.

Sparks, B. W., 1960, Geomorphology: London, Longmans, 371 p.

Strahler, A. N., 1950, Equilibrium theory of erosional slopes approached by frequency distribution analysis: Am. Jour. Science, v. 248, no. 10, p. 673–696; no. 11, p. 800–814.

———— 1952a, Quantitative geomorphology of erosional landscapes: C. R. XIX Internatl. Geol. Congr., Algiers, fasc. 15, p. 341–354.

———— 1952b, Dynamic basis of geomorphology: Geol. Soc. America Bull., v. 63, no. 9 (Sept.), p. 923–938.

———— 1960, Physical geography: 2d ed., New York,
Wiley & Sons, 534 p.

Thornbury, W. D., 1954, Principles of geomorphology: New York, Wiley & Sons, 618 p.

Vogt, P. R., 1956, Alida field, southeast Saskatchewan: First Williston Basin Symposium, p. 94–100; also: 1957, Can. Oil and Gas Industries, v. 10, no. 7 (July), p. 97–101.

Walters, R. F., 1946, Buried pre-Cambrian hills in northeastern Barton County, central Kansas: Am. Assoc. Petroleum Geologists Bull., v. 30, no. 5 (May), p. 660–710.

———— and A. S. Price, 1948, Kraft-Prusa oil field, Barton County, Kansas, in Structure of typical American oil fields, v. 3: Am. Assoc. Petroleum Geologists, p. 249–280.

Wheeler, H. E., 1963, Post-Sauk and pre-Absaroka Paleozoic stratigraphic patterns in North America: Am. Assoc. Petroleum Geologists Bull., v. 47, no. 8 (Aug.), p. 1497–1526.

Reprinted from:
BULLETIN OF THE AMERICAN ASSOCIATION OF PETROLEUM GEOLOGISTS
VOL. 51, NO. 1 (JANUARY, 1967), PP. 4-27, 20 FIGS.

UNCONFORMITY ANALYSIS[1]

PHILIP A. CHENOWETH[2]

Tulsa, Oklahoma

ABSTRACT

The shelf or foreland of a geosyncline or basin is an area in which there are commonly many unconformities. As a result of repeated tilting, the strata above and below each unconformity are not strictly parallel. Unconformity analysis is a method of reconstructing the history of each unconformity and determining the relations of each set of strata. Four situations are considered: (1) simple tilt—successive transgressions are essentially parallel, (2) differential tilt—successive transgressions are not parallel, (3) synclinal folding, and (4) anticlinal folding. Southwestern Arkansas Cretaceous and Tertiary strata are examples of (1) and (2), the northeast Texas Cretaceous interval illustrates case (3), and the Pennsylvanian section of the north flank of the Hunton anticline in Oklahoma is an example of case (4). In each of the examples there are three or more important angular unconformities; the distribution of the formations above and below each unconformity constitutes an important phase of the analysis.

The foreland shelf is a region conducive to the migration of petroleum and its accumulation, most especially in stratigraphic traps. Associated with each unconformity are two sets of potential stratigraphic traps—one below and one above. By careful unconformity analysis one can determine the subsurface trend and width of potential producing zones. Without such an analysis one may drill for an objective which is not present beneath the well location.

The conditions presented here in theory and by example are duplicated in many parts of the United States. A list is appended of those areas in which unconformity analysis might be used to find more oil in stratigraphic traps.

INTRODUCTION

Since the early history of geology in the 18th Century, unconformities have been recognized as features of more than usual significance. Hutton (1788, quoted by Adams, 1954, p. 243) stated that unconformities indicate great "revolutions." Because the geologic column indicates that many successive revolutions have occurred, Hutton postulated that periods of quiet sediment deposition have alternated with periods of violent upheaval, the formations above each unconformity being derived from the waste of those below. Thus the geologic record revealed to Hutton a long "succession of former worlds."

It is in some respects unfortunate that the science of stratigraphic geology began in the British Isles. This is true because the rocks of post-Precambrian age, although they are well exposed and neatly arrayed in proper succession, contain several striking angular unconformities. Not only were early English geologists impressed by these surfaces of discontinuity but, imbued as they were with the Huttonian concept of "revolutions," they assigned to them greater significance than they deserved. When geologic researches

spread westward across the north Atlantic, geologists in eastern North America found not only unconformities in apparently the same stratigraphic position as in England but also very similar lithologic types. As a result, the seeds of an erroneous concept were planted, one which continues to affect our modern geologic thinking—a belief that unconformities are of world-wide extent.

For a long time geologists thought that only by means of unconformities could interregional correlations be achieved, and because the "systems" defined in England and Wales were separated from one another by unconformities they must be similarly divisible elsewhere. The ghost of this hypothesis has not yet been laid to rest. An example is the pre-Cretaceous unconformity exposed in much of the North American interior. This unconformity is so prevalent that it becomes natural to think of it as being present everywhere.

To the subsurface geologist it is apparent that unconformities which punctuate the stratigraphic column in outcrop become progressively less distinct as they are traced into subsurface sedimentary basins. Obviously, while erosion is proceeding on land, deposition continues in the sea. Changes in lithologic character and variations in faunal content become less distinct basinward, with the result that in the basin centers uncon-

[1] Manuscript received, March 22, 1966; accepted, April 21, 1966.

[2] Sinclair Oil & Gas Company, Tulsa Research Center. Published with permission of the Sinclair Oil & Gas Company.

4

formities lose their identity. Similar phenomena occur in outcrop, as is well known. A pronounced angular unconformity that is present in one place on the surface may disappear completely within a few miles.

CLASSIFICATION OF UNCONFORMITIES

The classification of unconformities tends to be arbitrary because unconformities change in character from place to place. There is a continuous range from high-angle unconformities to conformity. Numerous attempts have been made to classify unconformities; the most durable of these schemes (Dunbar and Rodgers, 1957, p. 118–119) is: (1) nonconformity, (2) unconformity, (3) disconformity, and (4) paraconformity.

Nonconformity describes the erosion surface between plutonic igneous rocks and metamorphic rocks, below, and stratified sedimentary rocks, above. *Unconformity* (usually modified with the adjective "angular") designates that variety which separates strata of quite different structure. *Disconformity* has strata above and below essentially parallel, yet shows some evidence of erosion between the two sets of strata. *Paraconformity* refers to the situation in which no apparent erosional break exists, but a time hiatus is nevertheless present (key fossil zones *et cetera* are missing.)

UNCONFORMITIES ON BASIN MARGIN

During active tectonism, basins and geosynclines subside irregularly, responding sluggishly to increases in sediment load, to withdrawal of or changes in density of materials in the upper mantle or crust, or to some combination of these and other factors. The basin flanks, including the shelf and the hinge line between the shelf and basin, are repeatedly flexed and tilted. Each withdrawal or shoaling of the sea produces an unconformity. Each deepening of the sea results in the deposition of an overlapping sequence of sediments whose structural attitude is slightly different from that of the strata previously laid down and eroded. Ultimately a coastal plain is produced, characterized by the presence of shallow-marine, gently dipping strata in which there are numerous, repeated, low-angle unconformities. These commonly are difficult to detect, except through regional studies. They form a highly complex and confusing pattern.

IMPORTANCE OF BASIN-FLANK UNCONFORMITIES

An area of repeated unconformities and of gently dipping rocks on the shelf or foreland of a geosyncline or basin is of interest to the petroleum geologist. It is there where the largest and most effective hydrocarbon traps usually are found, and conditions favorable for the generation of oil may have existed.

Repeated tilting causes repeated movements of hydrocarbons; only the most efficient and durable traps still contain the original hydrocarbons. Because of tilting, simple anticlines and domes and most faults are not necessarily the most effective traps, and the stratigraphic trap probably is more efficient than the structural. In this paper, stratigraphic traps include lensing porous rock enclosed in impermeable material, a stratum pinching out above an unconformity, or a reservoir rock truncated and buried beneath an unconformity. No matter how often or in how many directions the stratigraphic trap is tilted, oil may escape only by migrating updip toward the outcrop.

Not only is the stratigraphic trap the most efficient, but theoretically it has the greatest reservoir capacity. A porous stratum thinning landward toward a pinchout between impermeable rocks and merging basinward with finer-grained material (possible "source beds") is tilted down toward the basin. In this way the porous bed receives fluids which move away from the high-pressure region and migrate upward to the wedge-edge. This primary accumulation in itself is not ordinarily of commercial size; secondary arching at an angle to the thin edge permits lateral movement to the locally high areas. Thus in these traps some bending is needed before large oil accumulations can take place. The classic American example of the conditions just described is in East Texas where oil moved eastward (updip) in the sandstone of the Woodbine Group and collected at the wedge-edge of this reservoir. Later arching of the Sabine uplift produced a concentration in the highest part of that uplift. East Texas is only one example; every petroleum province has its counterpart.

In order to view different types of traps in their proper perspective, one should contrast the drainage area available to a structural trap with that available to a major stratigraphic pinchout. Theoretically, an anticline under hydrostatic conditions accumulates only that oil which was gen-

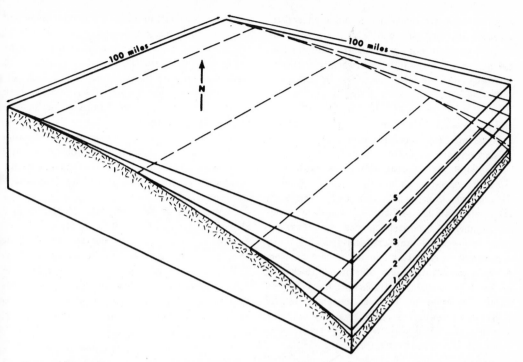

Fig. 1.—Block diagram of overlapping sequence of strata deposited with great uniformity on shelf of gradually subsiding basin. Numbered lines represent key beds which can be traced across facies boundaries and each is assumed to be of same age throughout. Dashed lines are shoreward limits ("strandlines") of key beds. Numbering is in order of relative age of beds, No. 1 being oldest labelled. Block illustrates area scores of miles square; vertical scale is much exaggerated.

erated in the formations in the adjacent syn- clines. Under a hydrodynamic situation the drain- age area may be somewhat larger. Though this may involve a distance of several miles, the area is small compared with that from which a strati- graphic trap may draw hydrocarbons.

Therefore, careful study of basin-flank uncon- formities can be extremely rewarding. Each un- conformity may conceal one or more stratigra- phic traps. To interpret these unconformities cor- rectly one must analyze the sequence of events which led to the present configuration and posi- tion. To exploit them effectively, each must be considered in its relation to the others. The sub- surface strike, width, rate of thinning, and angle of dip of each potential reservoir are important factors to consider in a successful exploration program. The reconstruction of the history of the unconformity, and the determination of the different patterns of outcrop and subcrop related to the unconformity, should be studied. This pro- cedure is termed here, "unconformity analysis."

UNCONFORMITY ANALYSIS—
MODELS AND EXAMPLES

For ease of presentation, and in order to achieve a system which may be applied generally, the descriptions and diagrams which follow are simplified. Several simplifying assumptions have been made which can not be duplicated every- where in nature. The most important of the sim- plifications are the following (Fig. 1).

(1) The areas considered in the examples are measurable in scores of miles.

(2) The strata were deposited in normal shal- low-marine waters.

(3) Numerous key beds, which can be traced across facies boundaries, are regarded as time markers.

(4) During deposition the area tilted uniformly down toward the basin, rotating about a hinge parallel with the shoreline.

(5) Trangression of the coast proceeded uni- formly and gradually as tilting took place. This

resulted in progressive overlap landward and divergence of strata basinward.

(6) Facies are not shown on the simplified diagrams. In the areas illustrated each unconformity is overlain by a basal coarse clastic bed which grades upward and seaward into a finer facies. Many factors can modify this simple relationship, and complex situations can result. For the present it is assumed that basal sandstone is everywhere present on each unconformity surface.

(7) After each retreat of the sea, erosion reduced the region to a plain at or near sea-level.

Case One: Simple tilt (Successive transgressions are parallel).—Figure 1 shows a series of sediments deposited in an uninterrupted transgression. The numbered solid lines are key beds which cross facies boundaries and are contacts between formations. Basinward thickening is shown by a slight divergence of the key beds. After deposition of such a sequence, several things can happen. The entire area may be tilted,

raised or lowered, or warped into anticlines and synclines, or it may remain static.

The simplest case is that involving a slight uplift and downward tilt toward the basin, with dips commonly on the order of ½°–1° (Fig. 2). Erosion reduces the area relatively quickly to a plain at or near sea-level because the sediments are poorly consolidated. On Figure 2, the landward-facing cuestas are shown out of proportion to their true size in order to emphasize the seaward retreat of younger beds because of erosion; landforms in these regions are generally much subdued.

Figure 3 is a plan-view map of the surface shown in Figure 2; it is a combination surface geologic map and a "worm's eye map" of the beds directly above the unconformity. ("Worm's eye" map is used in the sense employed by Levorsen [1960, p. 18]. It depicts the position of strata above an unconformity. The term "lap-out map" is regarded as being less de-

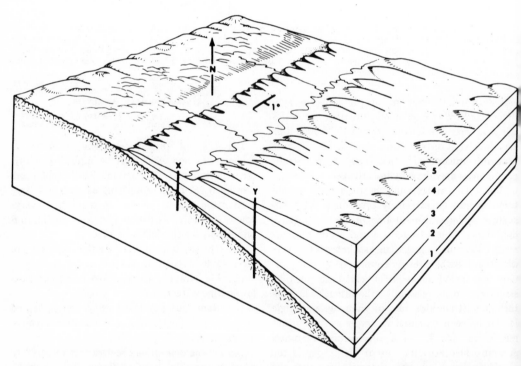

Fig. 2.—Block diagram of area of Figure 1 after slight uplift, sea withdrawal, and erosion. Key beds 3, 4, and 5 are shown as slightly resistant, landward-facing cuestas; their strandlines have been destroyed by erosion. Corresponding strandlines of key beds and 1 and 2 have not been exposed. Small seaward dip established during deposition has been accentuated so that it now has approximate amount shown by dip-and-strike symbol. Wells at X and Y correspond with points X and Y on Figure 3.

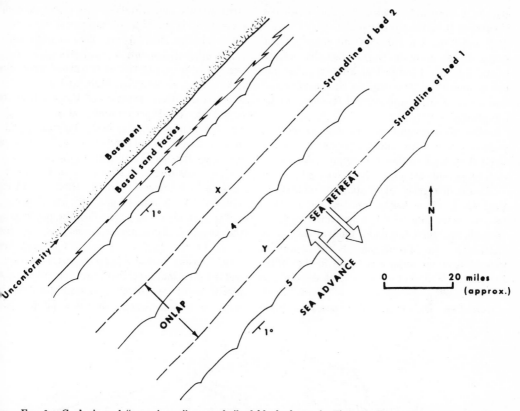

Fig. 3.—Geologic and "worm's eye" map of tilted block shown in Figure 2. Wavy lines represent outcrop of key beds 3, 4, and 5, cuestas of Figure 2. Dashed lines are shoreward limits of key beds 1 and 2 determined by subsurface well control; this part of map is "worm's eye view" of beds above basal unconformity. Bed 2 onlaps bed 1 in amount and direction indicated. Points X and Y are referred to in text as well sites; they are shown in same relative position on Figure 2. Basal sandstone facies everywhere rests on unconformity so that landward edges of all key beds, where preserved, are of same lithologic type.

scriptive, hence less desirable.) The "worm's eye" part is shown by dashed lines labelled "strandline of bed." The subcrop and "worm's eye" parts of his and subsequent maps are determined by sub-surface well control. Also shown are the original direction of sea advance (at right angles to the strandlines) and the direction of sea retreat (the direction toward which the coast was tilted). The true amount of onlap is indicated between the strandlines of beds 1 and 2. It should be noted that the "strandlines," or present shoreward limits (not necessarily the original juncture of sea and land), of beds 3, 4, and 5 have been eroded. A well drilled at point X (see Figs. 2, 3) would reach the lower part of the formation between 3 and 4 and most of the formation between key beds 2 and 3. A well at Y would drill beds 4, 3, and 2 in that order, but not bed 1 because the

well is northwest of the feather-edge of that bed. The lowest facies encountered is a basal sand-stone; it is somewhat older in the well at Y (age 1–2) than at X (age 2–3), and still younger where it crops out. Relative positions of wells X and Y are shown also on Figure 2.

It should be noted that there is no onlap of formations apparent at the outcrop of the uncon-formity. ("Overlap" is used in this paper in the general sense to describe a transgressive relation-ship; "offlap" describes the opposite regressive situation. "Onlap" refers to the relations within an overlapping sequence in which successively younger beds transgress the depositional pinchout of older units. "Overstep" describes the relations between an overlapping succession and the trun-cated strata below the surface of the unconformi-ty [Krumbein and Sloss, 1963, p. 316]). A geologist on the ground could detect the northwest onlap only if he were fortunate enough to have a deep and extensive cut at right angles to the strike of the unconformity. The fact that beds 3, 4, and 5 crop out parallel with one another and the unconformity, however, is a clue to the direction of onlap. They also parallel the zones of facies change.

Under these conditions the most important traps are formed by anticlinal structures which trend at an angle to the unconformity. Associated with these folds are pinchouts resulting from the basinward divergence of strata. Any impervious layer which cuts across the basal porous facies to rest on the unconformity has the same effect; the "strandlines" mark the updip limit of such beds. Relatively minor fluctuations in the shoreward edge of reservoir rocks may result in the formation of traps devoid of structural deformation (Freeman, 1949).

Case Two: Differential tilt (Successive trans-

gressions are not parallel).—A more commo[n] occurrence is warping of a coastal area wit[h] the result that the beds are tilted at some ang[le] to the original shoreline. If the block of Figure were tilted up at the left corner and down at th[e] right corner the situation illustrated in Figure would result. Here, outcrops of the key beds i[n]tersect the basal unconformity at a low angl[e.] The strandlines of the beds, because of th[e] differential tilt, have been truncated. The map [of] the upper surface (Fig. 5) shows the outcrops [of] the key beds diverging slightly as they trend ba[-] sinward from the unconformity. Directions of th[e] advance and retreat of the sea are shown.

The fact should be noted that the outcrops [of] the key beds cross facies boundaries, as illu[s-] trated by the basal sandstone. An outcrop be[lt] which remains in one lithofacies belt after tiltin[g] and erosion indicates that direction of tilt an[d] directions of sea advance and retreat are th[e] same, as in Case One; differential tilt produc[es]

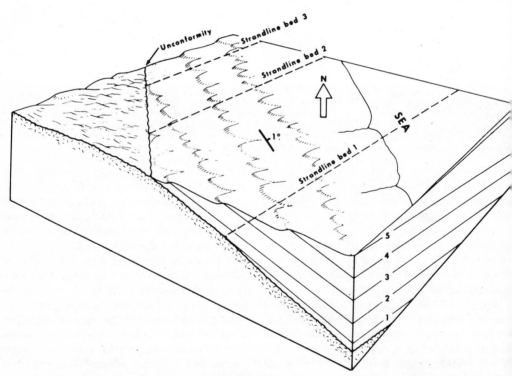

Fig. 4.—This represents block diagram of Figure 1 after a differential tilt, sea withdrawal, and some erosi[on.] Tilting, as shown by dip-and-strike symbol, was toward far right corner of block. Erosion has exposed all k[ey] beds except number 1 but outcrops in this case intersect unconformity at small angle rather than paralleling it [as] in Figure 2. Strandlines of key beds, established during marine transgression shown in Figure 1, *parallel* left a[nd] right edges of block and are shown partly by dashed lines.

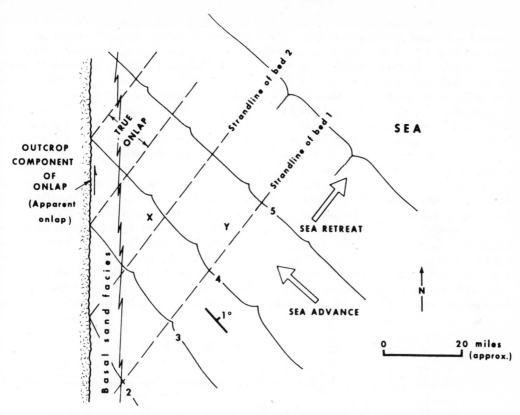

Fig. 5.—Geologic map and "worm's eye" map of block diagram of Figure 4. As in Figure 3, wavy lines indicate outcrop of key beds: dashed lines show shoreward limits at unconformity. Outcrops diverge slightly basinward (toward lower right) and cross facies boundaries landward. True direction and magnitude of onlap as shown in subsurface by double-headed arrow are toward upper left. A component of onlap appears at unconformity where successively younger strata onlap toward top of map. X and Y are well sites described in text.

an outcrop belt which crosses lithofacies belts.

In Case Two it is important to differentiate between true onlap and apparent onlap. At the unconformity, successively younger beds extend farther north along the contact. This "outcrop component" (Melton, 1947, p. 1873) indicated by the half arrow of Figure 5, gives only a suggestion of the true direction of onlap, which is revealed by subsurface information to be toward the northwest. Rate of onlap, i.e., feet of strata onlapped per mile, can be calculated for any stratigraphic unit (Melton, 1947). Wherever a particular porous reservoir bed is involved, rate of onlap becomes one measure of the width of a possible producing trend; angle of dip of the unconformity and inclination of the strata above and below are other measures. Direction of onlap is normal to the trend.

A well drilled at the point X (Fig. 5) would penetrate only key bed 3 before entering the basement. A well at Y would drill beds 4, 3, and 2, but not bed 1.

A considerably more complex set of conditions develops when, after a series has been deposited, tilted, and eroded as previously described, the sea readvances across the area. A new overlapping sequence is deposited on and oversteps the truncated older beds. If this sequence in turn were to be tilted and eroded, the relations shown in Figure 6 might result. This illustration combines a geologic map of the present surface, a subcrop map beneath unconformity No. 2, and "worm's eye" maps above both unconformities. Shown also are the directions of sea advance and retreat. An older series, indicated by the letters L-P, was laid down by a sea which advanced toward the northeast. Tilting toward the southeast exposed this series to erosion with the results that (1) the outcrops developed a northeast strike and (2) an eastward component of onlap appeared at the basal unconformity. After this, the sea invaded the region again, this time advancing toward the northwest and depositing the numbered overlapping series 5–10; bed 5 is the oldest key bed shown of this series. Again tilting and withdrawal of the sea occurred, this time toward the southwest. Subsequent erosion produced the pattern shown. In each of the two series, the outcrops or subcrops diverge away from the unconformity; in each case an outcrop component of onlap at the unconformity only *suggests* direction and rate of true onlap. A well drilled at the location of the dry hole symbol would penetrate the following key beds: 7, 6, unconformity No. 2, N, M, L, and unconformity No. 1. In this hypothetical example, the two main potential reservoirs, the basal sandstones, crop out. Consequently, oil or gas traps in these strata can be present only

where some stratigraphic or structural feature prevents migration of hydrocarbons to the outcrop. There are, however, numerous other possible reservoirs and traps, especially where basinward divergence of strata has produced pinchouts. A thorough understanding of the stratigraphy and regional structure as well as of the areal distribution of the various formations is requisite to efficient exploration. Unconformity analysis is but one major aspect of the over-all effort.

Southwestern Arkansas: example of differential tilting and multiple unconformities.—Relatively simple straight tilting is well illustrated by the Cretaceous and early Tertiary sections of southwestern Arkansas (Dane, 1929; Imlay, 1940). The geologic map (Fig. 7) has been simplified by smoothing of the formation contacts, not illustrating the effects of slight folding and differential erosion, leaving off areas of alluvium,

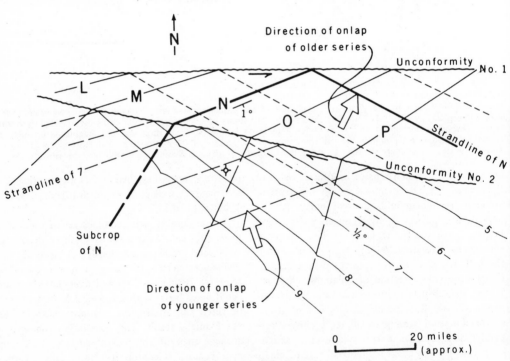

Fig. 6.—Complete analysis of two overlapping series. Older strata, indicated by letter lines L–P, were deposited by sea advancing northeast across area, onlapping unconformity No. 1. These beds were tilted southeast and eroded to sea-level, producing pattern similar to that in Figure 5. A second transgression followed with sea advancing northwest. Numbered beds overstep unconformity No. 2 which was developed on truncated edges of older series. Later tilting southwest and erosion produced present pattern. Shown are: strandlines of each series on two unconformities ("worm's eye" part of the map), subcrops of older lettered series beneath unconformity No. 2 (subcrop part), and outcrops of both series (geologic map part). Dashes are used for all subsurface lines. Shown also are directions of onlap and outcrop components of onlap (half arrows).

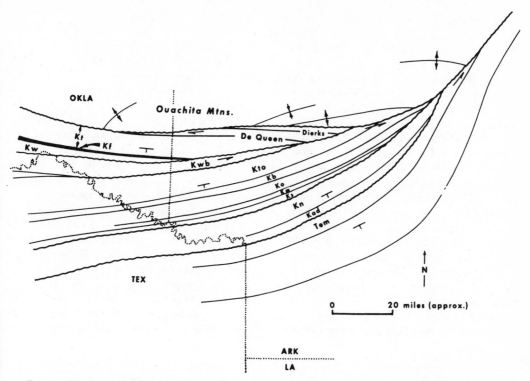

Fig. 7.—Simplified geologic map of Cretaceous and early Tertiary section of part of Arkansas, Oklahoma, and Texas. Formation symbols are those used on Geologic Map of Arkansas (Arkansas Geol. Survey, 1929). Major unconformities are shown by wavy lines. Half arrows indicate outcrop components of overlap; Trinity (Kt) apparently onlaps toward west; Woodbine (Kwb), Tokio (Kto), and younger strata apparently onlap eastward; whereas Nacatoch (Kn), Arkadelphia (Kad), and Midway (Tem) show no outcrop component of onlap in most of area, thus indicating northwestward onlap as illustrated in Figures 2 and 3.

and omitting certain lenticular rock units. Figure 7 is an area of moderately regular beds dipping gently south and southeast. Five unconformities are mapped; the one with the greatest angular divergence is that at the base of the Lower Cretaceous, where strata mapped as Trinity (Kt) rest on the folded Paleozoic sediments of the Ouachita Mountains. The two key beds shown directly above the unconformity are the Dierks and De Queen Limestone Members. The younger of the two (De Queen) extends a considerable distance west of the older, crossing into Oklahoma before being truncated at the unconformity. This outcrop component of onlap suggests that the Trinity seas overlapped in part in a westward direction. To determine the true direction of the onlap, one must ascertain the strike of the wedge-edges of these beds in the subsurface south and west of the point where they converge with the unconformity at the base of the Trinity.

The second largest unconformity shown on the map is at the base of the Upper Cretaceous Woodbine Group (Kwb). Beginning at the west, this unconformity cuts gradually across the Washita (Kw) and Fredericksburg (Kf) Groups at the Oklahoma-Arkansas state line; eastward it cuts out the De Queen and Dierks, with the result that farther east the Tokio (Kto) rests on pre-Cretaceous rocks. Between the Woodbine Group and the Tokio Formation is a relatively minor disconformity developed during a brief withdrawal of the sea during Eagle Ford deposition (Dane, 1929, p. 43). In this paper this disconformity is treated as another key bed. The easternmost pinchout of the Eagle Ford Shale appears (unlabelled) at the western edge of the map. The Tokio overlaps the Woodbine northeastward and is in turn overlapped by the Brownstown (Kb). Thus a definite northeastward outcrop component of onlap in the Upper Cretaceous replaces the westward component of onlap developed in the Lower Cretaceous (shown by the half arrows along the unconformities).

If the Upper Cretaceous and Tertiary are stripped away, something of the history during the Early Cretaceous (Comanchean) is revealed (Fig. 8). In the pre-Late Cretaceous (pre-Gulfian) subsurface the key beds diverge southward, toward the former basin. The strandlines (limits) of the Dierks and De Queen show the direction of the sea advance (slightly west of north). There is no way of determining the updip extent of the overlap, but, based on what is known elsewhere of Comanchean stratigraphy, it is safe to assume that the seas advanced well beyond this area. The Dierks and De Queen Members thicken southward, becoming important in oil-field stratigraphy. The approximately equivalent James Limestone is an oil-producing unit, and the Ferry Lake Anhydrite (equivalent only to the lower 10 feet of De Queen) is the most valuable correlation marker in the Lower Cretaceous

(Crawford, 1951; Imlay, 1940).

Removal of the Upper Cretaceous also exposes an unconformity hitherto unrevealed and a thick series of Triassic(?)-Jurassic rocks (Eagle Mills Group, Smackover Formation, and Cotton Valley Group). Indeed, other unconformities are present too, for Cotton Valley (C.V., Fig. 8) oversteps Smackover and Eagle Mills (Forgotson, 1954, p. 2478) as shown by the fine dashed lines beneath the Lower Cretaceous Hosston et cetera. (this part of the map is actually a pre-Comanchean subcrop map beneath a pre-Gulfian subcrop map). Farther south in Arkansas and north Louisiana there is evidence of additional large unconformities. The Late Triassic (?) Eagle Mills Formation rests on the Pennsylvanian Morehouse Formation (Scott et al., 1961; Hoffmeister and Staplin, 1954). The Jurassic (?) Norphlet Formation, which appears beneath the Smackover, has been found to rest unconformably on the Triassic-Jurassic Louann Salt and Werner Formation, the Eagle Mills Formation, and undifferentiated

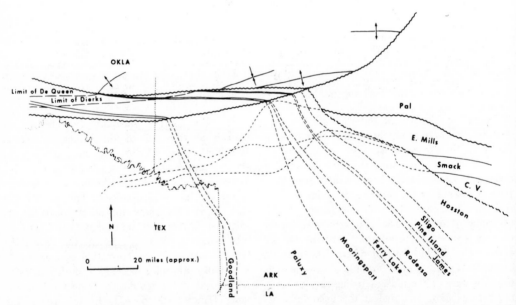

Fig. 8.—Same area as Figure 7 with Upper Cretaceous and Tertiary removed. Pre-Upper Cretaceous subcrops of Trinity and Fredericksburg are shown, with subsurface nomenclature. Map shows pre-Upper Cretaceous subcrop of Triassic(?)-Jurassic strata (Eagle Mills, Smackover, and Cotton Valley) and their pre-Lower Cretaceous subcrop. Indicated also are strandlines (limits) of Dierks and De Queen Limestones; these two lines are "worm's eye" part of map, showing formation traces on pre-Cretaceous unconformity. All formations thicken southward, toward center of basin of deposition. Thus Ferry Lake is nearly 500 feet thick at Arkansas-Louisiana line whereas outcrop equivalent is about 10 feet of gypsum at base of De Queen. Structural attitude of Lower Cretaceous subcrop is similar to that developed at time of uplift and erosion at end of Early Cretaceous time. Later tilting has affected strike and dip to some degree. This is approximately surface across which Upper Cretaceous seas transgressed from southwest to northeast as they returned to flood area.

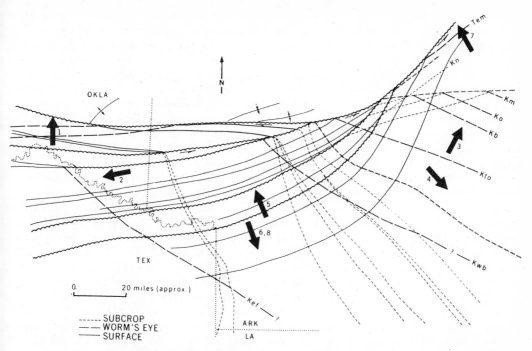

Fig. 9.—Complete analysis, somewhat simplified, omitting Jurassic and older units. Numbered arrows show directions of successive advances and retreats of sea. Post-Woodbine surface is assumed to be disconformity of less importance and is not indicated by arrows. Solid lines represent outcrops of key beds; long dashes are strandlines or shoreward limits; short dashes are subcrops. Eagle Ford (Kef) and Woodbine (Kwb) limits are questioned because they have been altered by pre-Tokio erosion and therefore might be regarded as subcrop lines. A well drilled near arrow number 4 would reach pre-Cretaceous rocks at base of Tokio, whereas one near arrow number 6 would penetrate all Upper Cretaceous except Eagle Ford and almost full section of Lower Cretaceous excepting only Fredericksburg (Goodland of Fig. 8) and Washita. These two hypothetical wells are located about same distance downdip from outcrop of unconformity.

Paleozoic rocks (Jux, 1961; Hazzard *et al.*, 1947).[3] The Smackover, in turn, oversteps the Norphlet and older strata down to the Eagle Mills (Jux, 1961; Andrews, 1960). Relations of these older unconformities are obscure and difficult to interpret and are omitted here for the sake of clarity.

The next younger significant unconformity is that at the base of the Upper Cretaceous Nacatoch (Kn, Fig. 7). Although this unconformity cuts into older beds at the northeastern end of its trace, it appears to be a disconformity through most of the mapped area (Dane, 1929, p. 114). It should be noted that there is no outcrop component of onlap above the unconformity—an in-

[3] Age of the Louann Salt has been the subject of considerable controversy. Jux's (1961) determinations have been seriously challenged but subsequent work (unpublished) by other palynologists has generally confirmed the Jurassic age assignment by Jux.

dication that tilting after deposition was in the direction from which the sea had come. Thus a third direction of onlap (northwestward) is established. This last movement, incidentally, was the first stage in the development of the Late Cretaceous-Tertiary Mississippi embayment.

The youngest unconformity shown on Figure 7 is that at the base of the Midway Group (Tem). Like that at the base of the Nacatoch, this shows no outcrop component of onlap except at the extreme northeastern end. Therefore, the Midway seas advanced from southeast to northwest, and retreated toward the southeast.

Complete analysis of the post-Jurassic unconformities of this area is shown on Figure 9. The apparent jumble of lines can be resolved fairly easily into four groups, each group being related to a single cycle of advance and retreat of the sea. The numbered arrows correspond with each cycle.

Arrow Number 1 (Fig. 9) shows the direction of sea advance and resulting formation onlap during Early Cretaceous time (Trinity, Kt, Fredericksburg, Kf, and Washita, Kw). The amount of onlap is indicated by the subsurface limits of the Dierks and De Queen Members in the northwest part of the map. Erosion has long since removed the shoreward limits of the Fredericksburg and Washita Groups in this region, but it is safe to assume that they once extended much farther updip, because Fredericksburg fossils have been found in northwestern Oklahoma. The small half arrow shows a westward outcrop component of onlap.

After Washita deposition the area was tilted gently west and deeply eroded. Figure 8 is a composite geologic map of the present surface and the pre-Upper Cretaceous surface; it shows the approximate configuration of the Lower Cretaceous strata at the time of submergence beneath the Gulfian seas. It is important to note, however, that only the subsurface part shows the pre-Gulfian surface; the present outcrop area has been subjected to many later periods of tilting and erosion. Furthermore, the subcrop pattern is not precisely the same as during pre-Gulfian time, because several tilts have affected the regional structure since that time.

Arrow Number 2 (Fig. 9) indicates the direction of tilt and sea withdrawal after Washita deposition. The northwest-southeast-striking subcrop patterns were established during erosion after this tilt. Evidence from this area and elsewhere indicates that this was a long period of erosion. At one time many geologists regarded this unconformity as a surface which separated two "systems," Comanchean and Gulfian (a nomenclatural procedure harking back to the Huttonian concepts).

After prolonged erosion, the Upper Cretaceous seas advanced across the area from the southwest (Arrow No. 3, Fig. 9), depositing in onlapping sequence the Woodbine (Kwb), Tokio (Kto), Brownstown (Kb), Ozan (Ko), Marlbrook (Km), and Saratoga (not labelled on Fig. 9). A brief withdrawal of the sea after the deposition of the Woodbine was accompanied by slight erosion of the strandline and concurrent deposition of the Eagle Ford Shale in Texas. To be more precise, the Eagle Ford offlaps the Woodbine, whereas the Tokio and younger units overstep the Eagle Ford and Woodbine. The dashed line

shown as a strandline of the Woodbine probably is close to its original position, but it is also the trace of the slightly eroded edge of the group beneath the Tokio. The outcrop component of this onlapping sequence is shown by the eastward-pointing half arrow at the place on the surface where the Brownstown rests on the Paleozoics.

Arrow Number 4 (Fig. 9) shows the direction of tilt and sea withdrawal after deposition of the Saratoga (Ks, Fig. 7). Erosion produced the strike indicated by the dashed lines in the northeast part of the map and the V-shaped pattern of subcrop and strandline.

The Nacatoch and Arkadelphia Formations (Kn and Kad, Fig. 7) were laid down in a sea which transgressed toward the northwest, shown by Arrow Number 5 (Fig. 9). These formations onlap older beds in the subsurface, but no onlap component appears along the basal unconformity. In the northeastern part of the mapped area the strike of the strandlines of beds in the Nacatoch crosses the subcrops of the Tokio, Brownstown, Ozan, and Marlbrook at an acute angle, parallel with the broken unconformity trace. Lack of an outcrop component of onlap indicates that post-Arkadelphia tilting was at a right angle to the direction of sea advance (Arrow No. 6), as in the hypothetical case shown in Figures 1–3. Prolonged erosion after post-Arkadelphia (post-Cretaceous) tilting was succeeded by invasion by the Paleocene Midway sea. Advance was practically in the same direction (Arrow No. 7) as the last of the Cretaceous transgressions and probably overstepped all of the now-exposed Mesozoic rocks. A remnant of the Midway strandline is indicated in the northeastern corner of the map. There also has been slight differential tilting, as indicated by the fact that younger units onlap the Midway. Retreat began soon after sea advance, in the direction indicated by Arrow Number 8, and has continued with only short interruptions to the present.

The analysis of this region brings out a point of considerable interest. Whereas the early deposits, mainly Cretaceous, bear evidence of repeated deep transgressions with little or no preservation of regressive deposits, the Cenozoic shows the opposite. The outer part of the modern coastal plain, from this area in southwestern Arkansas to the present sea coast, is a sequence of deposits laid down in a retreating sea. It seems

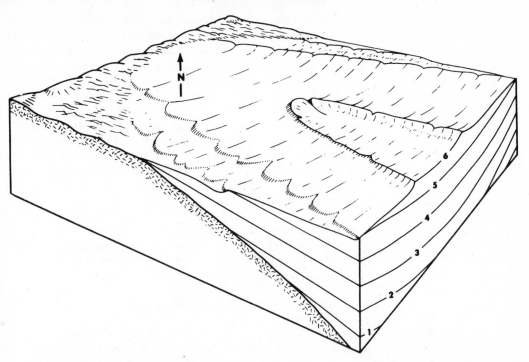

Fig. 10.—Block diagram illustrating pattern of outcrops resulting after overlapping sequence of Figure 1 has been folded into large syncline and partly eroded. For clarity key beds are shown forming cuestas; they converge gradually on outcrop of unconformity, intersecting at angles which increase toward trough.

reasonable to assume that this circumstance is characteristic of all basins and that deposits in any depositional basin ought to contain an increasing proportion of non-marine and coarse sediments upward in the stratigraphic section. The complete life history of most basins includes the final destructive stage during which the deposits are uplifted and eroded. In strongly deformed basins only remnants of the regressive phase are retained, examples being the Dunkard Group of the Appalachian geosyncline and the Whitehorse-Quartermaster Groups of the Anadarko basin.

Case Three: The syncline.—The events which take place if the overlapping sequence of Figure 1 is bent into a syncline are illustrated in Figure 10. Actual downwarping may occur during or after retreat of the sea. Regardless, an individual small segment of each flank could be treated almost as a case of simple or differential tilt as described in the previous sections. The entire outcrop of the unconformity, however, forms a gently curving line. Successively younger key beds extend farther toward the trough of the syncline; the outcrop component of onlap has two converg-

ing directions and the rate of onlap of this component diminishes toward the axis of the syncline. Figure 11 is a map of the surface, combined with a "worm's eye" view, and shows the relations of these beds. True onlap is toward the northwest. Submergence of this sequence results in the formation of a different pattern, as the sea commonly readvances most rapidly along the trough, onlapping toward the flanks of the syncline. Synclinal folding and erosion of this younger series produce an outcrop component of onlap directly opposite that of the older sequence, and the true directions of onlap may be nearly at right angles to one another.

In a situation such as this, trapping of hydrocarbons is effected only where folding, faulting, or other barriers to migration prevent lateral movement toward the outcrop. For the case of the reservoir rock above the unconformity, the Travis Peak of Figure 12, for example, migration toward the wedge-edge is barred by the onlapping impervious Glen Rose Group, but lateral movement to the outcrops in central Texas and southwest Arkansas can be stopped only by local features. In the case of a truncated reservoir rock, accumulation theoretically occupies a crescent-shaped area near the trough of the syncline. If the Maness (Km, Fig. 13), for example, were a porous formation the occurrence of oil pools might be expected near its northern extent.

Northeastern Texas: example of unconformities in a syncline.—The area considered here is contiguous with that previously described, but is considerably larger. The northeastern parts of Figures 12–14 are reductions and simplifications of the maps shown in Figures 7–9.

After Jurassic and Early Cretaceous transgression of the southern border of the continent, northeastern Texas was folded into a broad gentle syncline trending roughly northwest-southeast. A synclinal tendency had been established earlier, because Jurassic and Early Cretaceous seas apparently invaded an arcuate gulf whose shoreline lay about at the position of the present Nacatoch outcrop (Murray, 1952). Downbending in this area, however, had ceased temporarily by Comanchean time, for the Lower Cretaceous Travis Peak and Glen Rose seas advanced northward to-

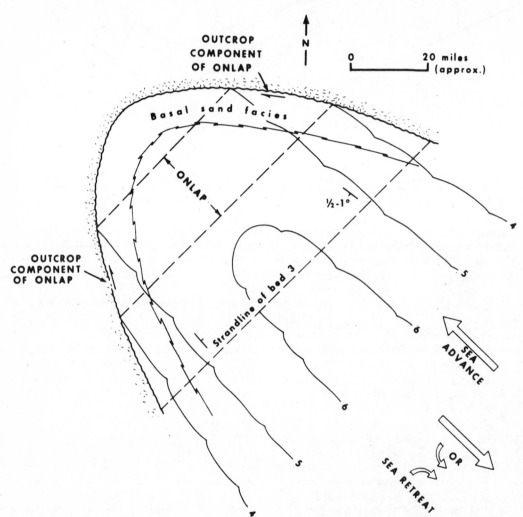

Fig. 11.—Map of block diagram of Figure 10 showing in "worm's eye view" strandlines of key beds 3, 4, and 5. Corresponding limit of bed 6 has been removed by erosion. Overlap was toward northwest. Along outcrop of unconformity there are now two converging, and diminishing, components of onlap. Key beds cross facies boundaries as indicated. There is no indication of time of folding; it could have been during sea retreat or at some later time as shown by large arrows.

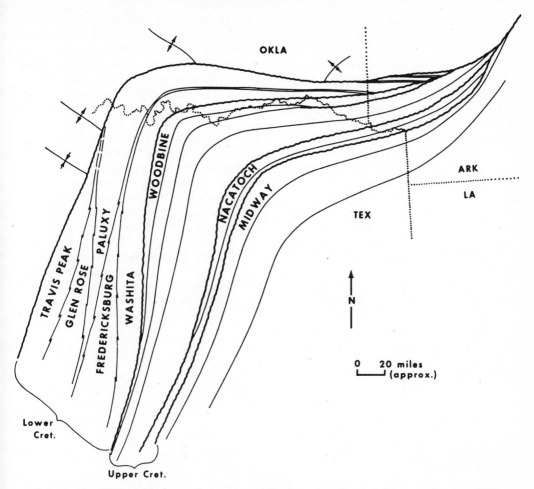

Fig. 12.—Simplified geologic map, Cretaceous and Tertiary of northeastern Texas, southeastern Oklahoma, and southwestern Arkansas. Only major unconformities are indicated. Entire Lower Cretaceous section thickens southward in outcrops even though post-Comanchean erosion has removed upper Washita beds. Southward pinchout of Paluxy sandstone is related to brief regression. Basal sandstone overlies each unconformity; these sandstones merge in Arkansas where unconformities converge.

ward a series of straight, nearly east-west beaches (Arrow No. 1, Fig. 14, the same strandlines as the partly equivalent Dierks and De Queen Members shown on Figs. 8 and 9).

After the syncline formed, erosion produced the pattern essentially as shown in Figure 13, which corresponds with the block diagram of Figure 10. Outcropping and subcropping units form horseshoe-shaped bands concentric with the axis of the syncline. Original direction of onlap is indicated in the subsurface by the dashed limits of members of the Glen Rose (Dierks, De Queen) and is suggested by the outcrop components indicated by the half arrows converging toward the axis. Southward (basinward) thickening causes the

widening outcrops in Texas and similarly widening subcrops in Arkansas. Deep erosion stripped away all Lower Cretaceous rocks in eastern Arkansas, exposing the Paleozoic section and a small part of the Jurassic rocks which underlie much of the region mapped. Subsequent seas advanced up the synclinal trough more readily than across the high areas on either side, probably because of a renewed downbending in the syncline. As a result of later tilt and further erosion, only the lateral strandlines are preserved. On the south flank they indicate a southwestern direction of sea advance; on the opposite side the seas spread northeast (Arrows No. 2, Fig. 14).

There are four important unconformities in the

section here: at the base of the Cretaceous, Woodbine, Nacatoch, and Midway (Fig. 14). Jurassic is omitted for clarity and lack of control. Arrow No. 1 shows direction of Early Cretaceous sea advance, at right angles to the limits of the De Queen and Dierks. Synclinal folding followed Comanchean deposition, the sea retreating in the opposite direction from Arrow No. 1; before retreating, the sea may have withdrawn toward the syncline axis (Arrows, Fig. 11). Areal distribution of the youngest Comanchean unit, the Maness Formation (Km, Fig. 13), favors the latter interpretation (Bailey *et al.*, 1945, Fig. 1).

After post-Comanchean withdrawal, the Upper Cretaceous sea spread into a gulf-like embayment, advancing rapidly up the trough and overstepping the southwestern and northeastern shores (Arrows No. 2, Fig. 14). Subsequent erosion has removed the bayhead record of this overlap, but it is preserved on the flanks and indicated by the diverging outcrop components. The initial deposit

is the sandstone of the Woodbine Group. During younger Nacatoch deposition only the southwestern shore was in this region, because in Arkansas the onlap is northwest (Arrows No. 3); evidently the downwarp of the Mississippi embayment destroyed the eastern flank of this great syncline. Retreat of the Cretaceous seas was southeast in Arkansas and northeast in Texas. By the time of Midway deposition, the gulf, which had begun to lose its character during Nacatoch-Arkadelphia deposition, was no longer in existence. Tertiary seas advanced northwestward toward a relatively straight shoreline (Arrow No. 4). Subsequent offlap, sea retreat, very slight downwarping, and erosion have produced the present-day outcrop pattern.

Case Four: The anticline.—If the sea coast of Figure 1 and its overlapping strata were arched into an anticline, the situation illustrated in Figure 15 might result. Concentric landward-facing cuestas converge on the old land surface. Each

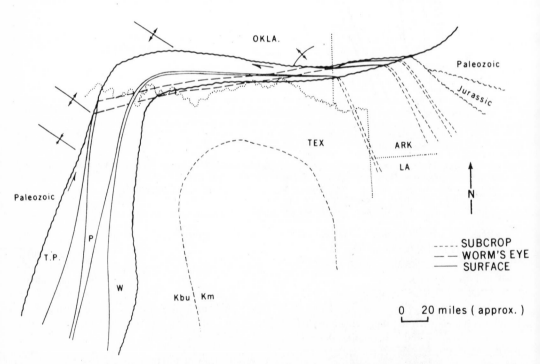

FIG. 13.—Composite outcrop-subcrop-"worm's eye" map, Lower Cretaceous. Same area as Figure 12, but with Upper Cretaceous removed. Heavy dashed lines indicate shoreward limits of Dierks and De Queen Limestone Members in Arkansas; they are correlated with Glen Rose of Texas. Comanchean seas advanced and overstepped at right angles to these strandlines, reaching maximum in Fredericksburg-Washita time. Probably in late Washita time, seas withdrew irregularly, depositing some sandstone in Washita and redbeds in the Maness (Km) near center of syncline. Downblending probably was concurrent with regression though not definitely so. This is approximately same terrane across which Woodbine seas advanced (at first in shape of an arcuate gulf), spreading laterally into Arkansas and central Texas.

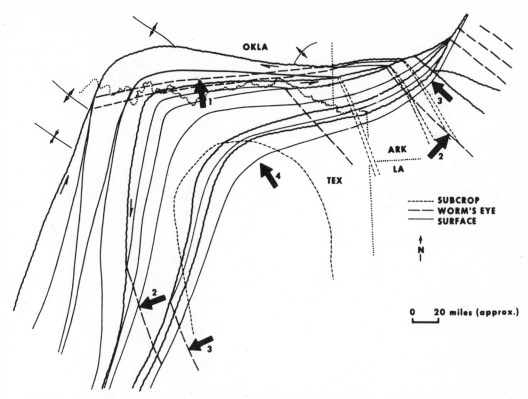

FIG. 14.—Complete analysis. Same map as Figure 13 with Upper Cretaceous and lowermost Tertiary added. Arrows indicate directions of sea advance and resultant overlap: 1, Lower Cretaceous; 2, Upper Cretaceous Woodbine to upper Taylor of Texas, Marlbrook of Arkansas; 3, Nacatoch and Arkadelphia, Navarro of Texas; 4, Tertiary Midway. Woodbine onlap is indicated in two opposing directions, a result of rapid flooding of post-Washita gulf; northwest direction of onlap is not shown but must have taken place. No well-defined embayment preceded Nacatoch transgression, however, as downwarp in Mississippi embayment area destroyed eastern flank. Tertiary overlap is everywhere toward northwest, apparently along nearly straight northeast-southwest coastline. Half arrows show direction of outcrop components of overlap, converging in Lower Cretaceous, diverging in Upper Cretaceous beds. This results from post-Comanchean downbending (converging components) and gulf-like character of shoreline at time of Woodbine transgression (diverging components).

successively younger bed extends somewhat farther away from the axis of the anticline. As in the case of the syncline, an individual small segment of each flank could be treated as a case of simple or differential tilt outlined earlier. Viewed as a whole, however, the differences are considerable.

The geologic map of the eroded anticline shows the key beds crossing facies boundaries and gently converging on the unconformity (Fig. 16). The subsurface limits of each bed, the "strandlines," here trend at right angles to the fold axis, but the true direction can be determined only from subsurface information. Outcrop components of onlap diverge from the axis at an increasing rate. Resubmergence of this area commonly would produce a curving set of strandlines, because the sea would advance more slowly toward the central part of the old uplift. The uplift itself tends to remain a positive area just as the syncline tends to remain essentially negative.

The situation diagrammed produces an extremely favorable environment for the entrapment of oil, where porous strata which thicken basinward and merge in that direction with "source" beds have been arched across a fold. Petroleum, which may have accumulated in small patches at the wedge-edge, is mobilized by the arching and migrates toward the uplifted area where it may accumulate in large pools closed laterally by dip and by facies changes.

Hunton anticline: example of unconformities

103

on an anticline.—The northern slope of the Arbuckle Mountains in Oklahoma is an area with
excellent examples of repeated unconformities in
an arched region (Fig. 17). A large domal uplift,
the Hunton anticline, appeared here during Early
Pennsylvanian (post-Wapanucka, pre-Atoka) time
(Tomlinson and McBee, 1959; Hicks, 1956; Disney, 1960; Rowett, 1963). The seas which
flooded the Atokan basins on both sides overlapped the dome from both flanks but apparently
did not meet across it. Four more uplifts occurred, the first three being succeeded by a readvance of the sea. Erosion prior to each submergence produced a characteristic outcrop pattern
which was later overstepped. The last phase of
sea advance, during which the Ada and Vanoss
Formations were deposited, occurred at a time
when the anticlinal nature of the uplift was obscured temporarily by greater diastrophism in the
Arbuckle anticline on the south. Subsequent
movement produced the present westward tilt
and a very slight rejuvenation of the core of the

dome. Folding and faulting have obliterated the
relationships on the southern flank.

In order to understand the sequence of events
which took place in this region, the unconformities are presented in partial analysis (Figs. 18,
19). Stripping of all younger beds to the pre-
Thurman unconformity (Fig. 18) and the addition of the limits of the older Pennsylvanian
units indicate the nature of the anticline just before Thurman deposition. It must be remem-
bered, however, that numerous subsequent events
have modified the picture, making it difficult in
places to reconstruct these events. For example,
the strandlines of the Atoka, McAlester, and Savanna once extended south of the present outcrop, probably wrapping around the core of the
anticline (Tomlinson and McBee, 1959, Figs. 7,
9). The Boggy Formation may have buried the
whole structure but its strandline has been erased
by erosion.

Some of the strandlines of major key beds as
they now exist are shown in Figure 19. None of

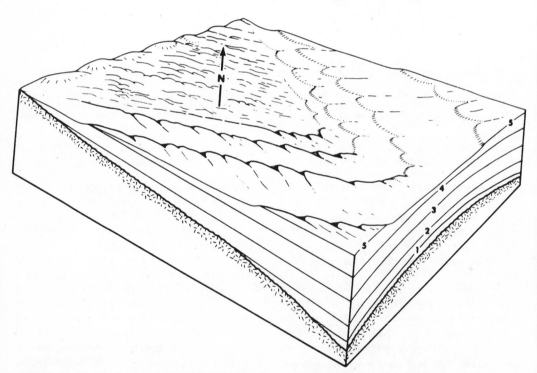

Fig. 15.—Block diagram showing overlapping sequence of Figure 1 arched into broad gentle anticline and
partly eroded. Numbered key beds are shown for clarity as forming arcuate cuestas converging gradually on
basal unconformity. Key bed 1 does not crop out; its strandline forms part of ideal oil and gas trap, provided
proper reservoir and cap rock conditions also exist.

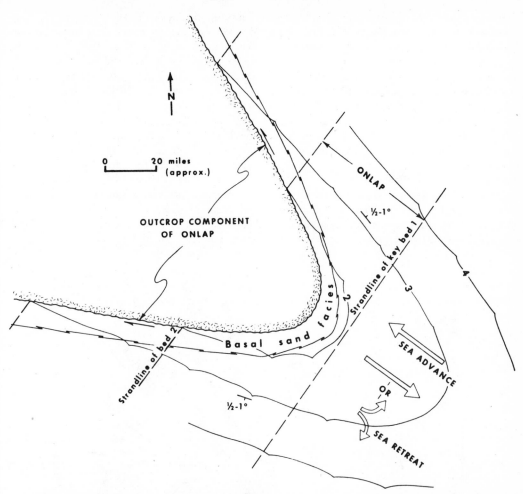

Fig. 16.—Geologic and "worm's eye" map of central part of top of block diagram shown in Figure 15. Irregular solid lines, numbered, correspond with cuestas; oldest is number 2. Successively younger key beds onlap and extend farther west; outcrop components of onlap diverge from axis as shown by half arrows. Large hollow arrows indicate direction of sea advance and alternate possibilities of direction of sea retreat depending on relative time of folding. Key beds cross lines of lithologic change as illustrated by basal sandstone facies. Dashed lines represent strandlines of key beds; except for bed number 1, they have been partly removed. Depending on associated reservoir and cap rock conditions, key bed 1 forms ideal trap for oil and gas. Resubmergence of anticline such as this most commonly proceeds from the flanks toward axis, because fold remains positive area. This is similar, in opposite sense, to flooding of syncline, where outcrop components of overlap are reverse of those shown here (Fig. 11).

the multitude of subcrop lines is indicated but there is a subsurface extension of each bed which is truncated by an unconformity. For example, the Belle City Limestone (shown as B.C., Fig. 20) is overstepped southward by the Ada Formation. Subsurface data show that this important key bed has a subcrop which curves southwest around the buried core of the Hunton anticline (Hicks, 1956, p. 341); this could have been inferred from the general nature of the fold.

Unconformity analysis of this region is not complete because some of the truncated edges and strandlines are not shown (Fig. 20). Arrows indicate the approximate direction of overlap. Each transgressing sea advanced farther toward the crest of the fold, the strandline criss-crossing outcrops of the previous rock units and producing a pattern of great complexity. Arrow Number 1 indicates the direction of sea advance and resultant onlap during deposition of the Atoka and later pre-Thurman formations (Hartshorne, McAlester, Savanna, and Boggy; the first three are not labelled on the map). In this region these shorelines trend nearly north-south; farther northeast they diverge widely, and the Atoka strandline is onlapped for many miles by younger units, particularly by the widespread Boggy. A northwest component of outcrop onlap is indicated by the half arrow on the eastern flank of the uplift and is strikingly apparent on detailed maps of this region.

After Boggy deposition the Hunton anticline again was arched gently and the sea receded toward the east and west. A differential downwarp is indicated by the strandlines of the key beds separating the seven formations from Thurman through Holdenville. Basal conglomerates in the

Thurman and Atoka (Weaver, 1954, p. 35; Rowett, 1963, p. 36), as well as in younger post-unconformity formations (Ham, 1954, Table I), indicate the considerable magnitude of the uplift. During the cycle of deposition represented by the Thurman through Holdenville Formations, the sea overlapped the flanks of the uplift, advancing from the northeast (Arrow No. 2, Fig. 20.) Only the limits of the Thurman and the next two younger units, Stuart and Senora, are shown. Information from the far northwestern plunge of the arch, in the vicinity of Paul's Valley, Oklahoma, suggests that seas advancing from the western basin produced a similar but mirror-image line of deposits. At least one of the several Deese sandstone beds can be traced successfully in an arcuate pattern across the fold and be shown to

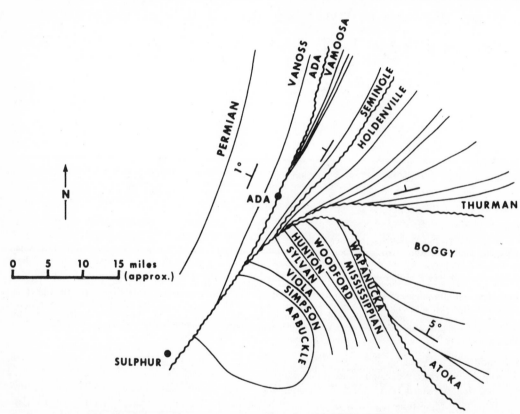

Fig. 17.—Simplified geologic map of northern flank of the Hunton anticline, south-central Oklahoma. Simplification involves straightening of some contact lines, and elimination of folds and faults. Major post-Hunton unconformity is not shown here although arching during that time was similar to Pennsylvanian folding. Smaller unconformities in Pennsylvanian, mostly disconformities, also are disregarded for simplification purposes. Outcrop relations on southern and western flanks of Hunton anticline have been obliterated largely by extensive folding and thrusting which occurred during Late Pennsylvanian orogeny in Arbuckle anticline on south. Pre-Ada erosion was especially deep. Graben-faulting on east side may have begun to develop during Late Pennsylvanian; its effects are omitted here for clarity.

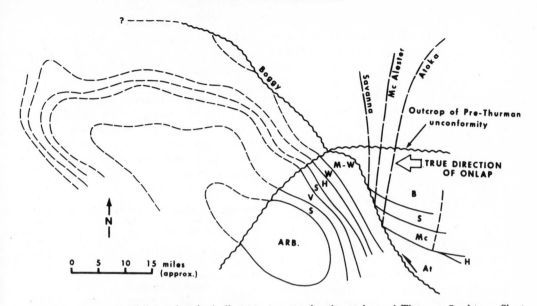

Fig. 18.—Hunton anticline stripped of all strata to unconformity at base of Thurman Sandstone. Short dashed lines indicate subcrop of pre-Atoka formations; long dashes represent strandlines of Atoka, Hartshorne, McAlester, and Savanna Formations. Progressive westward overlap is indicated, with Boggy Formation, whose strandline has been eroded, far onlapping older beds. Because this dome arose rapidly in deeply submerged area where no well-developed overlapping sequence existed, pattern does not duplicate that of Figure 16. Instead, this map illustrates first stage in gradual and intermittent resubmergence of fold, as described in Figure 16 caption.

join the Calvin Sandstone, which rests directly on the Senora Formation. Relations here are difficult to interpret, not just because of facies change but also because the eastern section crops out and has been thoroughly studied and subdivided (on the basis of topographic expression as well as lithology), whereas the western section is entirely in the subsurface, and it has not been possible to subdivide and study it as carefully.

The Holdenville Formation overstepped all older Pennsylvanian units and was deposited well onto the anticlinal core, reaching to the Ordovician Sylvan Shale (Miser et al., 1954). After this deposition a third uplift of the anticline caused the seas to retreat north. Subsequent transgression, beginning with the Seminole conglomerate and sandstone, nearly submerged all of the uplift (Arrows No. 3). Strong uparching at the end of Vamoosa deposition, the Arbuckle orogeny, was followed by deep erosion and a stripping of most of the units from the core. Subsequent burial by sediments assigned to the Ada and Vanoss Formations was largely non-marine and the whole of the anticline probably was buried, or nearly so, by these Late Pennsylvanian rocks and by Per-

mian deltaic and alluvial sediments. A general westward tilt occurred in Late Permian time and erosion reduced the structure to its present pattern. Mesozoic submergence and later epeirogenic uplift had little effect on the general structural outline.

The economic importance of a study such as this is apparent when one considers the number, size, and variety of oil pools within this area. Later folding and faulting have produced new traps while altering the former ones. An analysis of the unconformities is therefore only part of the complete study necessary for an accurate evaluation in this region.

PERSISTENCE OF FOLDING

In each of the four cases described in the preceding section, cycles of uplift, erosion, and resubmergence are the rule. As many as four or five complete cycles may develop in any region. Furthermore, there is a strong tendency for structure to perpetuate itself. For example, if an area is downfolded to form a large synclinal bend, subsequent movements duplicate, or nearly duplicate, the older ones. An area subjected to relatively straight tilting of the shorelines may be bent later into an arch or syncline, but there still appears to be a tendency for former movements to be repeated rather than for new ones to be initiated. It is commonly noted that anticlines, for example, occur in the same general area subjected formerly to arching, even though long periods of time separate the two events. The same tendency has been noted for synclinal folding. In the examples described, the post-Washita East Texas synclinal fold occurred in practically the same area as a much older (Jurassic) downwarp. Similarly, the Hunton anticline (Early Pennsylvanian) formed in approximately the same region as a post-Hunton Group (Devonian) uplift. Evidence from the Nemaha ridge of Kansas, the Forest City basin of Iowa, and elsewhere suggests that this is a common phenomenon. Each period of folding and erosion produces a characteristic subcrop-"worm's eye" pattern. To the oil geologist belongs the task of unravelling the skein.

OTHER AREAS

The entire Atlantic Coastal Plain from Mexico to New England has been subjected to repeated invasions and recessions of the sea resulting in the formation of unconformities such as those described in the preceding sections. Almost any part of the inner margin of the coastal plain can be treated as a straight shoreline which has been tilted directly toward the sea or at some angle, or both. Larger segments are either anticlinal or synclinal. From the Mexican border the major features are the following.

Rio Grande embayment ....synclinal
San Marcos arch, Llano uplift ....anticlinal
Northeast Texas basin ....synclinal
Sabine uplift ....anticlinal
LaSalle arch ....anticlinal
Mississippi embayment ....synclinal
Monroe uplift ....anticlinal
Southern Mississippi platform
 (Wiggins arch) ....anticlinal
Southern end of Appalachian Mtns. ....anticlinal
Ocala uplift ....anticlinal

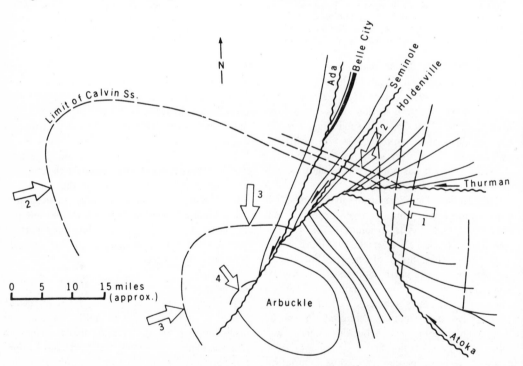

Fig. 19.—Outcrop and "worm's eye" maps, Hunton anticline. Shown are some of strandlines on four major unconformities and outcrops of principal key beds. Large numbered arrows show main directions of overlap for each phase; half arrows represent outcrop components as having been somewhat distorted by later tilting and deep erosion. For simplification only three strandlines are shown for Thurman through Holdenville interval; economically important Calvin Sandstone is traced farthest. On western flank it is known as third Deese sandstone.

----- SUBCROP
— — WORM'S EYE
——— SURFACE

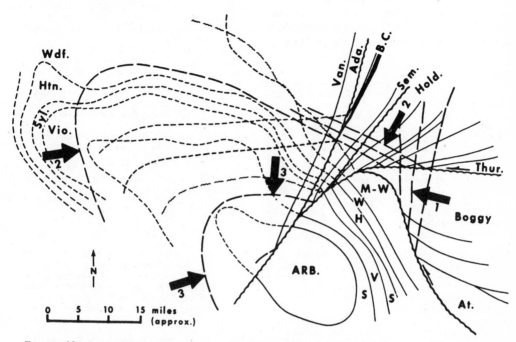

Fig. 20—Nearly complete analysis of Hunton anticline, showing subcrops beneath each major unconformity, "worm's eye" maps above each, and major outcrops. Certain key beds, such as Belle City Limestone shown solid in outcrop, are of great value in reconstructions of this nature. Unfortunate difference in nomenclature between intensely studied surface section on east and subsurface section on west renders precise correlations very difficult. Post-Vanoss tilting toward west and subsequent deep erosion have produced present outcrop pattern.

Peninsular arch ....a ticlinal
Southeast Georgia embayment ....synclinal
Cape Fear arch ....anticlinal
Chesapeake Bay low ....synclinal

The continental margin of almost every basin, or the margins of major uplifts within such basins, also can be analyzed in the manner indicated. Many features have been subjected to more or less severe diastrophism and deep erosion which enormously complicate the task of reconstruction. Furthermore, in order to analyze unconformities and their associated stratigraphic traps it is necessary to have a certain amount of subsurface data. The interpretation and reconstruction of undrilled regions can only be guessed. There are, nevertheless, many areas in the United States where sufficient data are available and where enough is known of the regional stratigraphy and correlations that analyses of this

type might prove to be of value. Following are outstanding examples.

Hugoton embayment and northern flank of the Anadarko basin—Pennsylvanian
Denver-Julesburg basin—Cretaceous
Forest City basin—Pennsylvanian
Eastern shelf of Midland basin—Pennsylvanian, Permian
Bend arch—Pennsylvanian
West-central Wyoming, Montana—Upper Cretaceous
Northwestern shelf of Delaware basin—Permian
Coastal plain of Washington and Oregon—Tertiary
Eastern flank of Cincinnati arch—Devonian

In any or all of the regions listed a careful analysis of unconformities may prove to be a fruitful exercise. Associated with each unconformity are two sets of potential stratigraphic traps, one above and one below. Determination of the width and trend of each possible reservoir and its relation to those stratigraphically higher and lower will yield valuable clues to the location of restricted areas within each region which have the optimum conditions for the accumulation of large quantities of oil. Moreover, if an unconformity analysis has not been made prior to the commencement of drilling operations it is conceivable that one may be drilling for an objective that is not present under the well location.

REFERENCES CITED

Adams, F. D., 1954, The birth and development of the geological sciences: New York, Dover Publications, Inc., 506 p.

Andrews, D. I., 1960, the Louann Salt and its relationship to Gulf Coast salt domes: Gulf Coast Assoc. Geol. Socs. Trans., v. 10, p. 215–240.

Arkansas Geological Survey, 1929, Geologic Map of Arkansas.

Bailey, T. L., F. G. Evans, and W. S. Adkins, 1945, Revision of stratigraphy of part of Cretaceous in Tyler basin, northeast Texas: Am. Assoc. Petroleum Geologists Bull., v. 29, no. 2, p. 170–186.

Crawford, F. C., 1951, The James Limestone and its relationship to the Cow Creek and Dierks Limestones: Gulf Coast Assoc. Geol. Socs. Trans., v. 1, p. 171–181.

Dane, Carle H., 1929, Upper Cretaceous formations of southwestern Arkansas: Ark. Geol. Survey Bull. 1, 215 p.

Disney, R. W., 1960, The subsurface geology of the McAlester basin, Oklahoma: Univ. Okla., Norman, unpub. Ph.D. thesis, 116 p.

Dunbar, C. O., and John Rodgers, 1957, Principles of stratigraphy: New York, John Wiley & Sons, 356 p.

Forgotson, J. M., Jr., 1954, Regional stratigraphic analysis of Cotton Valley Group of upper Gulf Coastal Plain: Am. Assoc. Petroleum Geologists Bull., v. 38, no. 12, p. 2476–2499.

Freeman, J. C., 1949, Strand-line accumulation of petroleum, Jim Hogg County, Texas: Am. Assoc. Petroleum Geologists Bull., v. 33, no. 7, p. 1260–1270.

Ham, W. E., 1954, Collings Ranch Conglomerate, Late Pennsylvanian, in Arbuckle Mountains, Oklahoma: Am. Assoc. Petroleum Geologists Bull., v. 38, no. 9, p. 2035–2045.

Hazzard, R. T., W. C. Spooner, and B. W. Blanpied, 1947, Notes on the stratigraphy of the formations which underlie the Smackover Limestone in south Arkansas, northeast Texas, and north Louisiana, in Shreveport Geological Society, Reference report on certain oil and gas fields of north Louisiana, south Arkansas, Mississippi, and Alabama, Vol. II: Shreveport Geol. Soc., p. 483–503.

Hicks, I. C., 1956, Pauls Valley field, Garvin County, Oklahoma, p. 337–354, in Petroleum geology of southern Oklahoma, Vol. I: Am. Assoc. Petroleum Geologists, 402 p.

Hoffmeister, W. S., and F. L. Staplin, 1954, Pennsylvanian age of Morehouse Formation of northeastern Louisiana: Am. Assoc. Petroleum Geologists Bull., v. 38, no. 1, p. 158–159.

Imlay, Ralph W., 1940, Lower Cretaceous and Jurassic formations of southern Arkansas and their oil and gas possibilities: Ark. Geol. Survey Inf. Circ. 12.

Jux, Ulrich, 1961, The palynological age of diapiric and bedded salt in the Gulf coastal province: La. Geol. Survey Bull. 38, 46 p.

Krumbein, W. C., and L. L. Sloss, 1963, Stratigraphy and sedimentation: 2d ed., San Francisco, W. H. Freeman and Co., 660 p.

Levorsen, A. I., 1960, Paleogeologic maps: San Francisco, W. H. Freeman & Co., 147 p.

Melton, F. A., 1947, Onlap and strike-overlap: Am. Assoc. Petroleum Geologists Bull., v. 31, no. 10, p. 1868–1878.

Miser, H. D., et al., 1954, Geologic map of Oklahoma: U.S. Geol. Survey and Okla. Geol. Survey.

Murray, G. E., 1952, Volume of Mesozoic and Cenozoic sediments in central Gulf Coastal Plain of United States, Part III of Sedimentary volumes in Gulf Coastal Plain of the United States and Mexico: Geol. Soc. America Bull., v. 63, no. 12, pt. I, p. 1117–1192.

Rowett, C. L., 1963, Wapanucka-Atoka contact in the eastern and northeastern Arbuckle Mountains, Oklahoma: Okla. Geol. Survey, Oklahoma Geology Notes, v. 23, no. 2, p. 30–48.

Scott, K. R., W. E. Hayes, and R. P. Fietz, 1961, Geology of the Eagle Mills Formation: Gulf Coast Assoc. Geol. Socs. Trans., v. 11, p. 1–14.

Tomlinson, C. W. and William McBee, Jr., 1959, Pennsylvanian sediments and orogenies of Ardmore district, Oklahoma, p. 3–52, in Petroleum geology of southern Oklahoma, Vol. II: Am. Assoc. Petroleum Geologists, 341 p.

Weaver, O. D., 1954, Geology and mineral resources of Hughes County, Oklahoma: Okla. Geol. Survey Bull. 70, 150 p.

UNCONFORMITY ANALYSIS: ADDENDUM[1]

In the article, "Unconformity Analysis," printed in the January, 1967, *Bulletin,* the writer made insufficient acknowledgment to the unpublished work of Frank A. Melton of Norman, Oklahoma. This note is added to the paper in the sincere desire to correct that oversight. Professor Melton has for many years included a similar treatment of the

geometry of low-angle unconformities in his course in stratigraphy, using some of the same examples and the same general approach to their analysis. During his years as co-teacher of this course the writer benefited from many stimulating hours of discussion and exchange of ideas with Melton. Though the writer's research was independent and his conclusions likewise independent, he has perhaps been remiss in not acknowledging this source of information and counsel.

[1] Manuscript received, February 8, 1967; accepted, March 10, 1967.

Reprinted from:
BULLETIN OF THE AMERICAN ASSOCIATION OF PETROLEUM GEOLOGISTS
VOL. 52, NO. 2 (FEBRUARY, 1968), PP. 313-321, 6 FIGS.

REGIONAL UNCONFORMITIES OF FLATLANDS[1]

FRANK A. MELTON[2]

Norman, Oklahoma 73069

ABSTRACT

This paper is a discussion of (1) the preparation of simplified regional stratigraphic map-diagrams and (2) the usefulness of such map-diagrams in the study of the onlap and strike-overlap at known regional unconformities both on the surface and in the subsurface. Such map-diagrams are simpler than the maps from which they are made, but they may contribute greatly to a clear understanding of verifiable geologic history. They are especially useful in studying the position, trend, and rates of truncation and onlap of significant stratigraphic units of regional extent. Their proper field of use is the coastal plain and low-dip "strata-benchlands," especially in marginal cratonic and shelf areas. If they have any usefulness in highly disturbed tectonic zones, it is yet to be demonstrated.

These regional stratigraphic map-diagrams are distortions of the regional geologic-contact maps, because contacts in the diagrams do not have their correct geographical positions. They do have, however, a logical and meaningful geologic position, which I believe contributes to their usefulness in tracing pinch outs into the subsurface.

Certain basic simplifying assumptions are necessary and are presented. A time-distance regional unconformity-diagram illustrates the immutable nature of the elapsed-time value of regional unconformities. Diagrams are presented and discussed, illustrating subsurface trends which are classified as (1) parallel, (2) nonparallel, and (3) complex.

INTRODUCTION

The purpose of this paper is to document further the simplified regional analysis of known regional unconformities.[3] This is a subject which I have explored for nearly 30 years in the development and presentation of an advanced course in regional stratigraphy and geologic history at the University of Oklahoma. During that period I enjoyed the help of several people who, in the order of their tenure at the University of Oklahoma, include: (1) John H. Speer, professor of geology, East Texas College, Commerce, Texas; (2) Carl Branson, professor of geology, University of Oklahoma, Norman, Oklahoma; and (3) Philip A. Chenoweth, from 1954 to 1960 on the staff of the geology department, University of Oklahoma. William Arper, professor of geology, Texas Technological College, Lubbock, Texas, was interested in this subject while a student at the University of Oklahoma. Later, in 1958, he collaborated with me in the writing of "Onlap and Strike-Overlap Analysis," an unpublished work with at least 30 simplified regional stratigraphic-structural diagrams of unconformity relations in the low-dip "flatlands." The library of the School of Geology and Geophysics at the University of Oklahoma

has a supply of this manuscript for consultation, as does the Oklahoma Geological Survey at Norman.

Chenoweth's paper (1967) is well written, and clearly illustrated; the perspective drawings show considerable artistry. However, certain definitions, concepts, implied assumptions, and basic axioms should be clearly stated and understood before the unconformity analysis can be effectively and confidently used.

The low-dip "coastal plains" with outcropping bedrock truncated by erosion are the focus of attention in this article. A short, adequate geomorphic name is "low strata-benchlands." Any strata-benchland of relatively low dip may be represented by these diagrams and serve as a *situs* for their use in searching for local or regional stratigraphic traps (*e.g.*, the Permo-Pennsylvanian lowlands of the Osage strata-benchlands of Kansas, Oklahoma, and north Texas, or the upper coastal region of the Gulf Coast strata-benchlands).

Regional "stratigraphic map-diagrams" used in studies of this type are much simpler and far more regular than the true situations which they represent. It is at once evident that they may have no application at all in highly disturbed tectonic zones such as the Coast Ranges of western North America or elsewhere. Regardless, if such diagrams are usable there, the associated problems must be considered separately from those of the shelf or marginal shelf areas.

[1] Manuscript received, March 14, 1967; accepted, June 27, 1967

[2] Professor of geology emeritus, University of Oklahoma, and geological consultant.

[3] Chenoweth, P. A., 1967, Unconformity analysis: Am. Assoc. Petroleum Geologists Bull., v. 51, no. 1, p. 4–27.

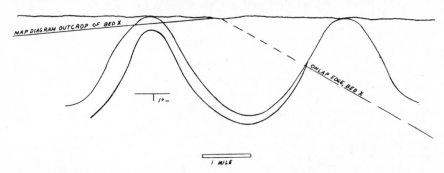

Fig. 1.—Illustrates method employed to simplify regional stratigraphic maps. Contact lines and associated pinchouts are straightened in manner which eliminates effects of late geologic erosion. Stratigraphic continuity of a region may be more readily seen on such a map. Though lines are not in true geographical position, they are in a meaningful geologic position.

The simplified stratigraphic map-diagram of a particular region such as the Arkansas-Oklahoma Mesozoic strata-benchland may contribute greatly to a clear analysis and understanding of geologic history. It facilitates the study of a complex regional situation (Chenoweth, 1967, Figs. 7–9, 12–14). The procedure which I have followed with my classes, in my 1947 publication, and which Arper and I used in the 1958 manuscript and diagrams, is to smooth out the contact lines of a regional map by bridging the valley re-entrants from divide to divide, restoring, in a sense, the rocks that have been eroded from these valleys. This converts the map into a stratigraphic diagram in which no contact line has either its true shape or true position, except on the ridge-tops. This procedure may be thought of as eliminating most of the effects of late geologic erosion (probably Quaternary and/or late Tertiary), depending on the geologic character of the terrane and the location. The procedure also eliminates much of the confusing complexity of eroded bedrock outcrops in low-dip strata-benchlands or in low-dip "plateaus." One can thus focus his attention more clearly on regional relations than without this aid.

To undertake such a distortion of regional geologic maps one must assume, at least temporarily, (1) that the published regional geologic maps are accurate, (2) that the regional unconformities are verifiable, (3) that these unconformities represent true episodes of emergence, sea retreat, and erosion, followed by readvance of a sea and its deposits, (4) that the entire undertaking may be worthwhile as a preliminary probe of regional relations, and (5) that the study ultimately may

lead to a better understanding of stratigraphic origins and of mineral and hydrocarbon occurrences. One must be continually aware that contact lines are not in their true geographical position, but merely in a relatively correct geological position (Fig. 1). Formation pinchouts are especially important and critical because, as a contact line is moved in the updip direction into a position approximately parallel with the regional strike, a point of truncation must be moved laterally as well as updip according to the restoration of the subsurface trend.

One must be cautious in generalizing from known unconformities. In very few areas are these well known, and some have characteristics out of harmony with the simple conditions which are assumed to be the basis for the historical analyses given below.

BASIC SIMPLIFYING ASSUMPTIONS

Simplifying assumptions are useful only if they help keep the ideas clear. Carefully planned simplification is one of the most widely used logical methods of the basic sciences. Following is a brief statement of basic simplifying assumptions which correspond roughly to conditions today, or at a stated time in the past. They are presented, with slight changes, from Melton and Arper (1958).

A. The simplified stratigraphic diagrams represent regions of considerable size, such as stratigraphic or geomorphic provinces, and only the gross geometrical relationships are considered.

B. All rocks considered are of shallow-marine origin.

C. All rocks considered were deposited with essentially flat formational dips. This follows partly from B.

D. All rocks considered have been regionally tilted to an angle several times as steep as the regional dip at which they were deposited—say to ½°–1°.

E. The present surface of the ground in the flatland regions is close to sea level and was eroded to this position since the last uplift. (This is tolerable as an assumption *only* in a regional sense.)

F. Regional unconformities as originally eroded were flatter and closer to sea level than the present *Recent* eroded surface. This of course does not apply to mountainous regions, but seems to be indicated by subsurface data in the flatlands of the world, insofar as they have been adequately explored.

G. Key beds (*i.e.*, a thin gypsum, evaporitic dolomite, or bentonite layer) represent true time horizons. They are assumed to be lithologically recognizable throughout their extent.

H. All key beds are assumed to have been deposited parallel with each other, as shown in the simpler diagrams. Lithologic facies changes of local extent are ignored, except for an assumed continuous basal onlapping marginal facies.

I. The regions shown are on the flank of a permanent continent and at the edge of a permanent sea.

J. The relative positions of the permanent continent and the sea are assumed never to have been reversed, nor have they been changed beyond a relatively minor degree, except in charts involving a geosyncline and a geanticline.

K. The crustal movements are assumed to have been uncomplicated and of relatively short duration.

L. The outcrop pattern has been simplified by ignoring local structural anomalies as well as erosional irregularities (*i.e.*, the diagrams show only the regional strike of contact lines).

M. All unconformities (assumed to be regional) show onlapping and/or truncating relations in these three-dimensional studies. This follows partly from the preceding assumptions.

Such basic assumptions conform sufficiently well, considering their proposed geometric regional use, with the coastal strata-benchlands of southern and eastern North America. It is possible that

some of these assumptions will be changed or abandoned after further trial use. I believe, however, that this *quantitative regional method of unconformity analysis,* which I began to use in my classes at least 25 years ago, and which requires the use of simplifying regional assumptions, will prove very useful in the accurate charting of regional and subcontinental geologic events. This method probably will be of most benefit in charting shelf and marginal-shelf regions, whether marginal craton or marginal geosyncline.

Time-Distance Unconformity Diagram

Simplified regional map-diagrams are easier to comprehend if a time-distance diagram is considered which will help one visualize the true elapsed-time value of regional unconformities such as those at the bases of the Upper Cretaceous Woodbine-Eagle Ford Groups and Upper Cretaceous Tokio-Brownstown Formations in the three-state area of Arkansas, Oklahoma, and Texas (Fig. 2). Many stratigraphers are no doubt aware of these time relations. Nevertheless, they are seldom mentioned in print.

The time-distance unconformity diagram of Figure 2 has two perpendicular axes—a horizontal axis with an arbitrary but appropriate *distance* unit, for example 100 mi, and a vertical axis with an arbitrary and suitable *time* unit, for example 1 million years. Eliminating unnecessary complications, two unconformities, each caused by a sea retreat and contemporaneous erosion on the land, followed by a sea advance, should have the appearance shown in Figure 2. The sea withdrawals produce an offlap of sediments and the advances produce onlap. All events of the same time value are on the same horizontal line in the diagram, such as time "J." The position of the point "J" on the margin of a "permanent" continent and sea is shown on the lower axis at position "J." Time "J" and position "J" establish the extreme landward limit of series or group "IJK" in the diagram. The wedge-shaped area, disappearing seaward at K, represents the elapsed time (the "hiatus") of unconformity JKL in a regional sense.

It is evident, if this diagram has any true resemblance to the geological reality, that all unconformities formed in similar marginal situations must resemble this diagram insofar as the elapsed time represented by true regional unconformities

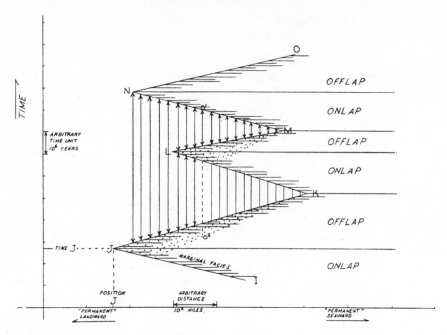

Fig. 2.—Time-distance regional diagram illustrating elapsed-time value of true regional unconformities of a continental margin. Because vertical coordinate is *time, and not thickness of beds*, it is apparent that elapsed time, represented by double arrows disappearing seaward, is immutable and indestructible. Even though some deposits (shown by large dots) may be destroyed by erosion during subsequent sea retreat, they cannot be erased from diagram nor from geologic history.

is concerned. For example, all similar unconformities *must* disappear seaward. All must expand regularly in elapsed time value toward the permanent land up to a certain maximum, beyond which the elapsed time suddenly increases to a much greater value as two adjoining unconformities merge into one. Hence, unconformity JKL merges with LMN into an unconformity of much longer time duration. The elapsed time that this diagram represents cannot be tampered with once it has passed. Nothing can be erased from this diagram if it truly represents geologic time.

One of the most useful qualities of this diagram is its *immutability*. For example, one may suppose that the sediments in series IJK and KLM which are indicated by large dots are destined to be destroyed by erosion during the subsequent sea retreat. This is a difficult concept unless the fact is kept in mind that vertical distance in the diagram represents time and not thickness. The dotted part of IJK would be eroded away during time interval JKL; likewise a part of series KLM would be eroded.

Because of erosion, the pebble O[1] at the base of series MNO may be in contact across the uncon-

formities with pebble O[2] in series IJK. Nevertheless, the elapsed time of the two erosional unconformities separates them, and is as shown in the diagram.

Figure 6 illustrates an unconformity with a double hiatus between the Upper Cretaceous-Tokio and the Lower Cretaceous-Trinity near Murfreesboro, Arkansas. The elapsed time of the unconformities shown in the other diagrams given below should be interpreted in the light of Figure 2.

PARALLEL SUBSURFACE PINCHOUT TRENDS

Following is a minimal list of the necessary events in geologic history indicated by Figure 3. They are numbered in the reverse order of time and, for accuracy of understanding, are considered in the same order. A restatement of these simple events in correct time order is always desirable. However, in solving the puzzle of geologic time, unless one removes the latest geologic events first and works toward the past, he may miss essential steps in the history.

I. (7) Deep erosion approximately to sea level.

II. (6) Downward tilting toward S. 15° E. (with

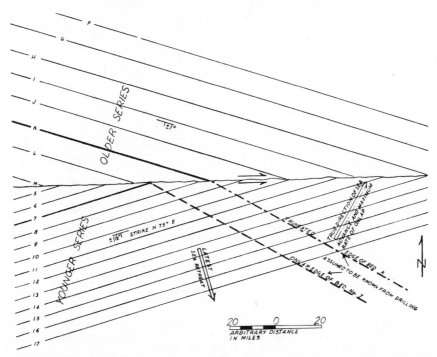

FIG. 3.—Simplified regional stratigraphic map-diagram of a coastal plain unconformity between an older truncated series and a younger onlapping series. Seven necessary minimum geologic events that could explain (in a step-by-step sequence) origin of this stratigraphic relationship are described in text.

associated uplift toward N. 15° W.), sea retreat southeastward and beginning deep erosion. (If the tilting were 50 ft/mi, the uplift near the unconformity at the western edge of the diagram would be of the order of 3,000 ft.)

III. (5) Warping which changed the direction of the coastline, such that an uplift occurred at or near the west side of the area, and/or subsidence permitting farther advance of the sea at or near the east side. (This movement in the case of a regional unconformity may well be a major tectonic episode of a subcontinent.)

IV. (4) Advance of the sea across the deeply eroded surface (see V below) toward N. 30° E. to an unknown but presumably great distance beyond the areas of this map, and deposition of the strata of the "Younger Series" by onlap. The outcrop component of onlap of stratigraphic interval in feet per mile of horizontal distance is measured due eastward above the unconformity. The half arrow indicates the direction of outcrop component of onlap. The true direction of onlap, and its greatest rate in terms of feet of strata per mile are toward N. 30° E. This should always be a greater number of feet per mile than the outcrop component in the coastal, and for the most part in the interior, Paleozoic strata-benchlands of the world.

V. (3) Geologically significant erosion, approximately to sea level, of the entire region.

VI. (2) Downward tilting toward S. 30° W. with associated uplift toward N. 30° E., sea retreat south-westward, and beginning deep erosion of the "Older Series." If the tilting were 50 ft/mi the uplift of bed "K" near the east edge of the map, a distance of 40 mi from the present subcrop of "K," would have been of the order of 2,000 ft above the K subcrop.

VII. (1) Inasmuch as the unconformity and the onlap relations at the base of the "Older Series" are not given, no definite statement can be made regarding direction or distance of sea advance to deposit this series. However, this is not necessary to illustrate the principles under consideration.

This list of seven *necessary minimum* geologic events should be re-examined carefully in the correct time order (using the Arabic numbers) and compared again with the map-diagram in Figure 3.

The illustration here of clear-cut separate tiltings, warpings, and erosional truncations are an intentional simplification of geological movements, which, in reality, may be very complex. For example, the warping referred to in III (above) could be a continuous wave-like change of elevation proceeding through the continent or its margin. Such a complex movement could be referred to as a migrating crustal wave or flexure, and possibly could be caused by an advancing phase change (polymorphic transition) in the lower crust or at a deeper level. The term "dif-

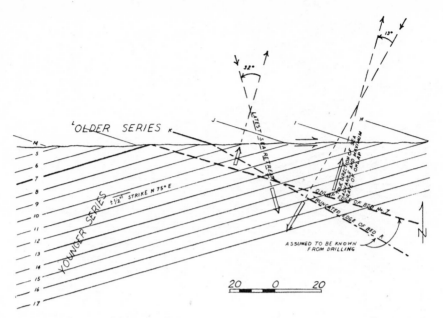

Fig. 4.—Simplified regional diagram which is identical with that of Figure 3 in respect to beds below unconformity. In this figure, however, a complicated relationship above unconformity requires at least one additional geologic event in necessary minimum explanation of geologic history

ferential tilting" might be used if the unconformity is of local extent. In any particular case, there may have been many movements and some possible reversals of tiltings or uplifts. One can only interpret the net result, a simple compartmentalized statement of which is given above.

I infer that this case of subsurface parallelism between overlying onlap shores and underlying truncated subcrops in Figure 3 must be a very common situation, especially on local unconformities of slight extent and small elapsed-time value. This should be true where a readvance of the sea, following uplift, tiltings, and erosion, was caused by a eustatic rise of sea level.

Nonparallel Subsurface Trends

Figure 4 illustrates a situation identical to that of Figure 3 except for the onlapping edge of bed 7 which crosses the subcrop of bed "K" of the Older Series.

Items 1 through 3 of Figure 3, taken in the order of time, apply equally well to Figure 4. Event 4, in Figure 4, however, was an upwarp in the northwest and/or a downwarp in the southeast with the result that the strike of the coastline was changed 13° in a counterclockwise sense. Significant geologic erosion is assumed to have accompanied and/or followed this warping.

5. The sea advanced N. 17° E. and deposition of the "Younger Series" took place by onlap in that direction to an unknown distance beyond the limits of this map-diagram.

6. An upwarp occurred in the west and/or a downwarp occurred in the east so that the sea invaded the continent even farther in the east, and withdrew from the continent in the west. The change of coastal alignment was 32° in a counterclockwise sense.

7. A downward tilting toward S. 15° E., with associated uplift toward the N. 15° W., caused sea retreat southeastward and the beginning of deep erosion.

8. Deep erosion brought the entire region approximately to sea level.

Figure 4 has a more complicated history than Figure 3 because of one necessary simple event. The fact that the two coastal warpings rotated the coastline in the same counterclockwise sense suggests that it be called a "normal" epeiric series of warpings, in contrast to those represented by the map-diagram of Figure 5 which could be called "abnormal." In Figure 5 the two warpings are in the opposite sense of coastal rotation and thus form a less uniform series of epeiric tectogenic events.

Events 1 through 3 of Figures 3 and 4, taken in the order of time, apply equally to Figure 5. Event 4, in Figure 5, however, was an upwarp in the southeast and/or a downwarp in the northwest with the result

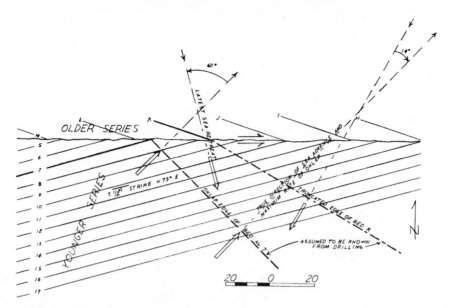

FIG. 5.—Diagram illustrating regional coastal plain geology, inferentially more complicated than that of Figure 4, though necessary minimum number of geologic events is same. Because two coastline changes (warpings) produced rotations in opposite sense, this is termed an "abnormal" case compared with Figure 4, which may be considered as "normal." Figure 4 may be a more common occurrence.

that the coastline orientation was changed 14° in a clockwise sense. Significant geologic erosion is assumed to have accompanied and/or followed this warping as in Figure 4.

5. The sea advanced N. 45° E. and deposition of the "Younger Series" took place by onlap in that direction to an unknown distance beyond the limits of this map-diagram.

6. An upwarp occurred in the west and/or a downwarp occurred in the east so that the sea invaded the continent even farther in the east and withdrew from the continent in the west. The change of coastal alignment was 60° in a counterclockwise sense. (This contrasts with 32° in the same sense in Figure 4.)

7, 8. These events are the same as in the discussion of Figure 4.

Figure 5 is thus more complicated by implication than Figure 4, though the necessary minimum number of events is the same.

COMPLEX SUBSURFACE TRENDS

Chenoweth's Figure 6 (1967, p. 11) is a possible geologic example but seems to be unnecessarily complicated, involving the following sequence of necessary minimum events in the true time order in which they occurred:

1. Sea advance in an approximately N. 25° E. direction, and deposition of the older of two series by onlap on a deeply eroded surface.

2. Great warping, changing the coastline orientation approximately 85° by upwarp in the west and/or downwarp in the east.

3. Downward tilting (using bed N for measurement) toward S. 57° E. (with associated uplift toward N. 57° W.), sea retreat in that southeasterly direction, and beginning deep erosion.

4. Deep erosion approximately to sea level. This could have followed 5.

5. Change in coastline orientation approximately 40° in the sense of upwarp in the east and/or downwarp and sea invasion in the west.

6. Sea advance in an approximately N. 20° W. direction and deposition of the upper series by onlap.

7. A third great warping of the coastline by approximately 50° in the sense of additional upwarp in the east and/or downwarp in the west.

8. Downward tilting in an approximately S. 35° W. direction (using bed 7 for the measurement), accompanied by uplift of the land on the northeast, sea retreat toward S. 35° W., and erosion of the latest lowland coastal plain surface truncating beds 5–9.

Such a case as this involves three very great changes in coastline direction, one by nearly 90° in a counterclockwise direction, and two approximating 45° and 50° in the clockwise direction. Such great changes probably will prove to have been infrequent in the homoclinal low strata-benchlands and are not *normal* in the sense of the

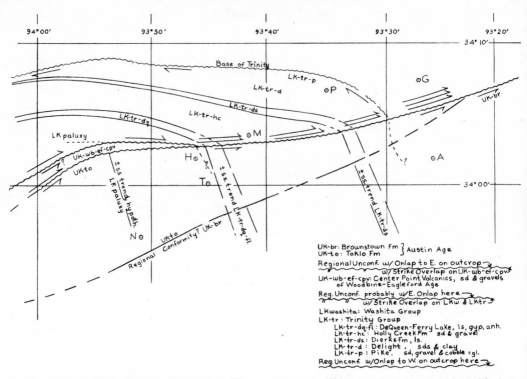

Fig. 6.—Simplified diagram redrawn from Miser *et al.* (1928). This part of map centers about Murfrees-boro, Arkansas, the rocks being of Comanchean and Gulfian ages. Regional onlap is shown in two places—probably three. Regional strike-overlap (slow truncation) is shown in three places. Regarding elapsed geologic time and feet of stratigraphic truncation per mile of outcrop, basal Tokio (UK-to) unconformity changes abruptly to double and possibly triple unconformity as underlying groups of strata disappear beneath it. With sufficient subsurface information about trends of truncated contacts, it would be possible to compile a verifiable statement of detailed geologic history.

normality shown by the Upper Cretaceous progressive downwarp or crustal flexure in the upper Mississippi embayment. This downwarp or flexure changed the coastline orientation of that time in a counterclockwise sense, apparently without significant interruption in deposition of the Tokio to Nacatoch (Late Cretaceous) and perhaps even Tokio to Midway (Paleocene).

Many other stratigraphic-structural situations, such as those in geanticlines and geosynclines, and various complications of pinchout by onlap or by strike-overlap, and repeated pinchouts at nearby unconformities, could be illustrated.

The word "overlap," used by most geologists without further explanation, is almost devoid of any exact meaning. The term "strike-overlap" is, in effect, a brief description of the strike differences between a regional unconformity and the truncated beds below, not to mention the strike difference between the lower beds and the onlapping upper beds, if any onlap is known. It is almost a misuse of the terms to speak of "truncation" where the loss of section beneath the unconformity is of the order of 5 ft/mi or less. The term "strike overlap" for this "truncation" should not meet with objection from surface geologists who can *see* no truncation at the outcrop. For those who wish to explore the means by which outcrop components of onlap and truncation can be measured, the reader is referred to Melton (1947).

Southwest Arkansas

One stratigraphic and tectonic situation, illustrated by Figure 6 of this paper, is worth special attention because of its small area, its complexity, and its central location in the southwestern United States. Figure 6 is a simplified stratigraphic diagram of a part of the area of Lower and Upper Cretaceous outcrop in southwestern Arkansas. It was taken from a 1928 1:250,000 geologic map (Miser et al., 1928).

The lower Trinity (LK-tr-d) shows an outcrop component of onlap toward the west. It is part of a N. 10°–15° W. Trinity onlap, judging from information gained from the area farther west, and illustrated by Chenoweth (1967, Fig. 13). Manley (1965) has confirmed this direction by pebble-orientation studies in the basal Trinity.

The Center Point "volcanics" wedge of the Woodbine-Eagle Ford Groups (UK-wb-ef-cpv) is believed by many to onlap eastward at the outcrop and probably northeastward in the subsurface. Critical information seems to be lacking on the true three-dimensional onlap of this group of strata.

If the map by Miser et al. (1928) is correct in placing an unconformity between the Tokio and the Woodbine-Eagle Ford, there must be easterly strike-overlap at the top of this UK-wb-ef-cpv wedge, as well as onlap at the base of the same beds in the area of this map. Inasmuch as this section apparently has never been subdivided, it is not now possible to determine the amount of onlap and/or overlap of each bed in feet per mile or in any other unit. Whether or not the strike-overlap in the UK-wb-ef-cpv at this locality exceeds the amount of onlap at the base is not known at present.

I believe that the basal Tokio unconformity truly exists above the UK-wb-ef-cpv. It is a very significant unconformity for approximately 16 mi from near Highland (H on the map) to southwest of Graysonia (G on the map). In this 16 mi, the elapsed time of this unconformity is certainly that of a double hiatus. If the full Trinity section below the top of the DeQueen Formation in this locality totals 785 ft (Shreveport Geol. Soc., 1961), then the outcrop component of strike-overlap is 785 ft/16 mi = 49 ft/mi. The part of this thinning which was accomplished during the pre-Woodbine-Eagle Ford truncation, and that accomplished during the pre-Tokio truncation is not at once evident from data which I have been able to find. Where the Tokio strike-overlaps the base of the Trinity and lies on the Paleozoic, the double

hiatus no doubt becomes a triple hiatus, combining those of the basal Trinity, basal Woodbine-Eagle Ford, and basal Tokio-Brownstown. If the Jurassic section ever extended this far north before being eroded downdip to its present subsurface edge, the basal Tokio-Brownstown may overlie a quadruple hiatus.

The outcrop component of Tokio-Brownstown basal onlap toward the northeast may continue through the Ozan, Marlbrook, and Saratoga formations (Late Cretaceous age, Taylor) which succeed the Brownstown upward in the section. I have never seen sufficient local subsurface information either to establish this possibility, or to disprove it.

SELECTED REFERENCES

Caplan, Wm. M., 1954, Subsurface geology and related oil and gas possibilities of northeastern Arkansas: Arkansas Geol. Survey Bull. 20, 124 p.

Chenoweth, P. A., 1967, Unconformity analysis: Am. Assoc. Petroleum Geologists Bull., v. 51, no. 1, p. 4–27.

Dane, C. H., 1929, Upper Cretaceous formations of southwestern Arkansas: Arkansas Geol. Survey Bull. 1, 215 p.

Imlay, Ralph W., 1940, Lower Cretaceous and Jurassic formations of southern Arkansas and their oil and gas possibilities: Arkansas Geol. Survey Inf. Circ. 12, 64 p.

Manley, F. H., 1965, Clay mineralogy and clay mineral facies of the Lower Cretaceous Trinity Group, southern Oklahoma: unpub. Ph.D. thesis, Oklahoma Univ., 116 p.

Melton, F. A., 1947, Onlap and strike-overlap: Am. Assoc. Petroleum Geologists Bull., v. 31, no. 10, p. 1868–1878.

—— and Wm. B. Arper, Jr., 1958, Onlap and strike-overlap analysis: privately reproduced, Norman, Oklahoma, and Lubbock, Texas, 42 p. (Copies may be consulted at the library of the School of Geology and Geophysics, University of Oklahoma, Norman.)

Miser, H. D., and A. H. Purdue, 1929, Geology of the DeQueen and Caddo Gap quadrangles, Arkansas: U. S. Geol. Survey Bull. 808, 195 p.

—— et al., 1928, Geologic map of the Cretaceous formations in southwestern Arkansas: U. S. Geol. Survey, scale = 1:250,000.

Renfroe, Chas. A., 1949, Petroleum exploration in eastern Arkansas, with selected well logs: Arkansas Geol. Survey Bull. 14, 159 p.

Shreveport Geological Society, 1961, Spring field trip guidebook, Cretaceous of SW. Arkansas and SE. Oklahoma: 89 p.

Spooner, W. C., 1935, Oil and gas geology of the Gulf coastal plain in Arkansas: Arkansas Geol. Survey Bull. 2, 474 p.

Reprinted from:
BULLETIN OF THE AMERICAN ASSOCIATION OF PETROLEUM GEOLOGISTS
VOL. 57, NO. 5 (MAY, 1973), PP. 825-840, 10 FIGS., 2 TABLES

Radar Geology—Petroleum Exploration Technique, Eastern Panama and Northwestern Colombia[1]

RICHARD S. WING[2] and HAROLD C. MACDONALD[3]

Princeton, New Jersey 08540, and Fayetteville, Arkansas 72701

Abstract Petroleum exploration in eastern Panama and northwestern Colombia has gained impetus by recent side-looking-radar, geologic reconnaissance mapping. Radar-derived geologic information is now available for approximately 40,000 sq km where previous reconnaissance investigations have been extremely limited because of inaccessibility and almost perpetual cloud cover.

With radar imagery as the sole source of remote-sensing data, the distribution, continuity, and structural grain of key strata provide evidence that the eastern Panamanian Isthmus can be divided into three main physiographic-structural parts: two composite coastal mountain ranges separated by the taphrogenic Medial basin, which trends southeastward from the mouth of the Bayano River to the Atrato River valley of northwestern Colombia. Within the Medial basin, most of the clearly exposed surface structures are not particularly attractive petroleum prospects because prime reservoir strata have been stripped from their crests. However, several large geomorphic anomalies which have been mapped in the Medial basin may be reflections of subsurface structures having a complete stratigraphic section. Possibilities for gravity-type hydrocarbon accumulations in fractured organic shales, siltstones, and carbonate rocks are suggested within several synclinal elements along the axis of the Medial basin. The southwestward extension of the Medial basin trend, coincident with the western Gulf of Panama, may have potential as a future petroleum-producing province. A relatively thick marine stratigraphic section should be present here, with associated paralic and deltaic clastic rocks derived from acidic San Blas terrace since mid-Miocene time. The occurrence of active shell bars in the Bay of San Miguel and present reef trends on the northern Caribbean coast suggest possible offshore sites for geophysical surveying.

Introduction

Petroleum exploration techniques, which in previous years were highly dependent on exploitation of visible-spectrum sensors, have been augmented by a new family of other-wavelength remote sensors. No longer is the petroleum geologist restricted to the analysis of conventional aerial photography for reconnaissance studies, although cameras *per se* have by no means been displaced or abandoned. Aerial photography, although it provides valuable data for the compilation of highly accurate maps, is severely limited by light and weather conditions. Thus, many regions of the world, especially the tropics, are poorly mapped because adequate photographs cannot be obtained. Radar imaging systems provide the capability of obtaining terrain data independent of most weather conditions, and this information may be sufficient to provide a data base for geologic reconnaissance map construction.

Radar has been used extensively in aerospace, astronomy, meteorology, and military studies, but its use for oil and gas exploration has not received wide publicity. Most of the radar imagery used in past and present geologic investigations has been collected over areas in the U.S. where temperate climatic conditions prevail and where adequate geologic field data are generally available. The extension of our experience and knowledge to a climatic environment where geologic data are extremely limited has been made possible through the availability of extensive radar imagery covering approximately 40,000 sq km in eastern Panama and northwestern Colombia. This part of Central America serves adequately as an area typical of tropical, cloudy environments with dense vegetal cover, characteristic of most countries in equatorial latitudes where adequate geologic mapping is missing.

The primary objective of this study was to determine the utility of radar in the compilation of geologic reconnaissance maps for petroleum exploration. That radar is useful in certain terrain studies was known, but the degree of geologic reconnaissance information that radar will provide the petroleum geologist in the tropical environment had not been well documented. To determine this, a geologic reconnaissance map of eastern Panama and northwestern Colombia was constructed and evaluated. This region is characterized by areas where existing geologic maps have been compiled from such limited and scat-

[1] Manuscript received, September 12, 1972; accepted, November 15, 1972. Paper presented before the Association at the 57th Annual Meeting, Denver, Colorado, April 17, 1972.

[2] Advance Geology Group, Continental Oil Company.

[3] Department of Geology, University of Arkansas.

This study was conducted at the Center for Research in Engineering Science, University of Kansas. Supported by NASA contract No. NAS-9-7175, and USAETL contract No. DAAK 02-68-C-0089. Monitored by U.S. Army Engineer Topographic Laboratories Geographic Information Systems Branch, Geographic System Division, Ft. Belvoir, Virginia.

tered field data that they usually fail to delineate regional continuity of geologic formations and structure.

RADAR IMAGERY ACQUISITION

The radar imagery, covering all of Darien Province, Panama, part of northwestern Colombia, and much of east-central Panama, was obtained in 1967 and 1969 by Westinghouse for the U.S. Army Engineer Topographic Laboratory (USAETL) during less than 20 hours of actual imaging time. Prior to this, during the preceding 30-year period, there had been minimal success in obtaining conventional aerial photographic coverage because of almost perpetual cloud cover, especially over the mountains and foothills.

The 1967 and 1969 radar imagery is the product of a high resolution AN/APQ-97 K-band Side Looking Airborne Radar (SLAR) system. The 1967 imagery provided a slant-range display that progressively incorporates compression in the near range, *i.e.*, a geometric distortion just the opposite of that inherent in conventional oblique photographs (MacDonald and Waite, 1971). In contrast to the 1967 data, the 1969 imagery consists of a ground-range display, which means that a hyperbolic sweep correction was applied to radar signals returned from the ground. Theoretically, this would result in a planimetric display of the terrain (*i.e.*, no compression in the near range), if the terrain were perfectly flat. In areas having rugged topography, however, considerable radar imagery distortion can occur (MacDonald and Waite, 1971). Much of the ground-range 1969 imagery, of same-look and with up to 50 percent overlap, lent itself to stereoscopic viewing. Individual radar strips of approximately 18 km width have been joined to provide the uncontrolled radar mosaic in Figure 1.

The operation of side-looking radar can be summarized by examining Figure 2, where a microwave antenna (*A*) is repositioned laterally by the velocity of an aircraft (*Va*) carrying the SLAR system. Each radar-frequency pulse transmitted (*B*) returns signals from the targets within the beam width. These target returns are converted to a time/amplitude video signal (*C*), which is imaged as a single line (*E*) on photographic film (*F*). Returns from subsequently transmitted pulses are displayed on the cathode ray tube (CRT) at the same position (*D*) as the previous scan lines. By moving the photographic film past the CRT at a velocity (*Vf*) proportional to the velocity of the aircraft, an image of the terrain is recorded on the film (*F*) as a continuous strip of radar imagery.

HISTORY OF GEOLOGIC MAPPING—EAST-CENTRAL PANAMA AND NORTHWEST COLOMBIA

East-central Panama has been the least mapped, geologically, of any part of eastern Panama. Terry (1956) provided a geologic reconnaissance map which was lacking in all but the grossest structural aspects of this part of Panama, and included only speculative versions of even supposed major faults. This fact is mentioned, not to detract from his epic work, but only to illustrate the state of geologic knowledge prior to the recently available radar imagery, which has made far more detailed and more reliable reconnaissance geologic mapping possible. The climate is tropical, with vegetation canopy continuous and terrain marked by extremes of inaccessibility. Roads do not exist in most of the region; even the Pan American Highway has not been constructed through this part of Panama. Outcrops commonly are deeply weathered and poorly exposed. Only during the 3 or 4 months of the so-called dry season, beginning around late January, is field work feasible; however, the term "dry season" is misleading, because in some years up to 30 cm of precipitation have been recorded during this period.

Terry's (1956) geologic observations are based on field work and overflights in the Republic of Panama between the years 1920 and 1949. In 1966, a geologist with the Panama Canal Commission, published a report on the general geology of Bayano basin, based on field work through the period from 1955 to 1965 (Stewart, 1966). More recent ground reconnaissance studies have been completed in easternmost Darien Province, Panama, and northwestern Colombia by the Corps of Engineers (Interoceanic Canal Study Comm., 1968, 1969).

Prior to radar imaging, Darien Province was better mapped geologically than other provinces in eastern Panama, however, this previous mapping was quite general and in many places incorrect. For example, entire landforms were off-shape or misoriented. Nevertheless, various gold mining operations and some petroleum exploration had resulted in a smattering of publications and mapping of gross elements as summarized on Terry's (1956) map. Shelton (1952) provided geologic data for the central part of the Tuira-Chucunaque subbasin. Some additional data have accrued from limited wildcat drilling (Johnson and Headington, 1971); however, the Interoceanic Canal Study Comm. (1968) study represented the most recent geologic data available for Darien prior to radar geologic reconnaissance mapping (MacDonald, 1969). Since completion of the 1969 radar mapping a new geologic map

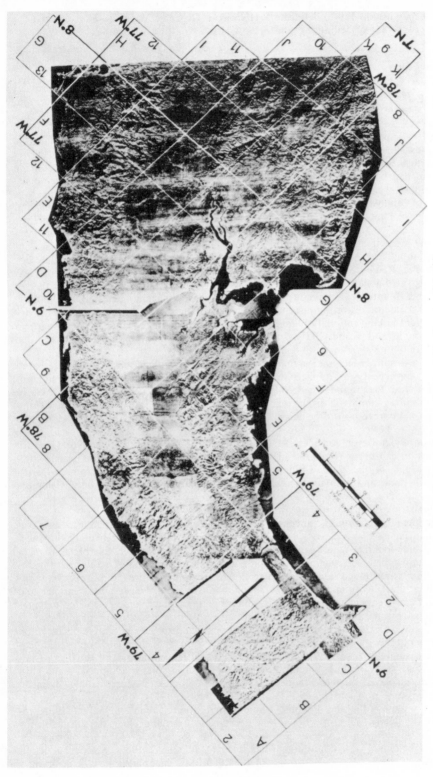

FIG. 1.—Radar mosaic, eastern Panama and northwestern Colombia.

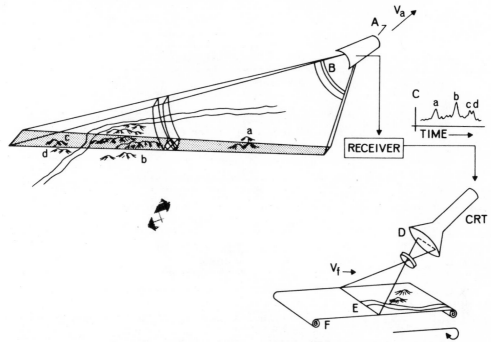

Fig. 2—Sketch diagram, typical side-looking airborne radar system.

has been issued which covers a small part of eastern Panama (Rió Pito, 1:40,000, Direccion General de Recursos Minerales, República de Panamá, 1972).

IMAGERY ANALYSIS

Radar imagery has been shown to be a most valuable supplementary tool in mapping surface-expressed geologic phenomena in many diverse physiographic and structural environments (Wing and Dellwig, 1970; Wing et al., 1970; Gillerman, 1970). In the tropical environment, where field work is at best difficult and sometimes impossible, radar may be the only adjunct to field surveys. Radar imagery is often superior to vertical air photos for display and detection of surface fractures—i.e., faults and megajoint zones (MacDonald et al., 1969). Although vertical air photos may reveal more clearly the smaller details of structural elements and patterns, radar imagery can show, as well or better, the true nature and extent of structural patterns—the proverbial forest as well as the trees. Radar imagery is also superior to conventional photography for detection of land-water boundaries, because smooth water is a specular reflector (for radar wavelengths) and consequently returns little of the incident signal to the aircraft antenna.

There are, of course, limitations. For example, most earth-science interpreters have found that radar imagery favorably and preferentially displays those topographically expressed linears most nearly parallel with the flight track of the aircraft (orthogonal to radar-look direction). Conversely, linears most nearly parallel with the look direction are minimized for lack of shadowing, and some to such an extent that a small percentage of linears are not detectable at all (MacDonald, 1969). For areas of extremely rugged terrain, where extensive radar shadowing is encountered, trade-offs between loss of geologic data and radar geometry must be taken into account during mission planning sessions (MacDonald and Waite, 1971).

There is essentially no penetration of vegetation by the SLAR imaging system when operating at K-band radar frequencies (MacDonald, 1969), and consequently, in the jungle environment, only the canopy surface of the dense foliage is imaged. Whether depicted on air photos or displayed on radar imagery, the canopy surface mirrors topography, reflecting outcrops which have been weathered and eroded differentially according to varying resistance of diverse rock types, structural attitudes, and fractures. Although conventional vertical air photos tend to show a welter of canopy irregularities, the inte-

Table 1. Summary of Key Stratigraphic Data, Eastern Panama

Age	Formation	Thickness	Description
			TUIRA-CHUCUNAQUE SUBBASIN
Early Pliocene-late Miocene	Chucunaque		Sandstone and shale containing foraminiferal fauna; occupies narrow inner trough of SD-VII
Middle Miocene	Pucro	610 m	Calcareous sandstone and thin limestone beds
	Gatun	1,070 m	Shale, underlain by conglomerate, sandstone, and thin limestone lenses
Early Miocene	Aquauqa	610-1,370 m	Bentonitic and bituminous shale, limestone, and sandstone
Late Oligocene	Arusa	460 m	Massive, brownish, tuffaceous marlstone
		310 m	White-weathering, tuffaceous, massive mudstone
Middle early Ologocene	Clarita	300 m	Tuffaceous carbonate rocks (mostly limestone) and sandy limestone
Late Eocene	Corcona	350-400 m	Shale, sandy, marine, calcareous; massive agglomerate, conglomerate, limestone (includes volcanic material)
Middle early Eocene			Chert (possibly equivalent to Quayaquil chert of Eduador), limited distribution (Pacific Hills area)
Pre-late Eocene	Basement		Andesite flows; crystalline rocks (somewhat silicic on Caribbean side)
			BAYANO SUBBASIN
Late Miocene		915-1,220 m	Soft, argillaceous siltstone, Foraminifera
Middle Miocene		460 m	Reef-type nearshore limestone beds, 15 km long; present on south side of SD-III syncline
Early Miocene			Sandstone and siltstone
Late Oligocene			Tuffs, conglomerates, calcareous and tuffaceous sandstone, and tuffaceous siltstone
Middle early Oligocene			Limestone (calcareous algae, Foraminifera)
Late Eocene			Sandy shale, agglomerate, conglomerate, tuffaceous and calcareius sandstone, tuffaceous siltstone
Pre-late Eocene	Basement		Andesite flows; crystalline rocks (silicic on Caribbean side; mafic on Pacific side)

gration inherent in the relatively coarse SLAR resolution cell has a smoothing effect which facilitates discrimination of large-scale structural patterns and major elements thereof. Contrasting types of vegetation grow selectively on the most favorable types of strata or on regolith derived from different types of underlying bedrock. Thus, some vegetal patterns and types can be differentiated on SLAR imagery and lithic inferences can be made (Lewis and MacDonald, in press).

Radar terrain return depends on: (1) surface roughness, (2) incidence angle of the slanted microwave beam, (3) polarization, (4) wavelength of the electromagnetic signal used to "illuminate" the landscape, and (5) the complex dielectric constant. As with aerial photography, interpretation of the radar display involves considerations of tone, texture, shape, and pattern. Tone is the intensity of signal backscatter recorded on film as shades of gray, from black to white. Texture, in simplest terms, is the degree of erosional dissection, and variously textured drainages can be used to infer lithologic type. For example, igneous outcrops generally have a massive appearance on radar imagery and are distinguished by "rugged and peaked divides." Shape alludes to telltale outlines of surface-expressed features such as those of alluvial fans, dikes, igneous plugs, volcanos, and folds. Pattern is the arrangement of geologic, topographic, or vegetal features such that one may distinguish linears—e.g., fractures (faults and megajoints), bedding-trace grains, en échelon folds, etc. Thus, it is possible to extract considerable geologic data through careful interpretation of this imagery (Wing, 1971).

Probably the most important interpretive phase of geologic reconnaissance mapping with radar imagery is selecting rock units that have sufficient areal extent to be significant and are still distinctive enough on the imagery to be easily mapped. One is generally restricted by the scale of the imagery, but the primary limitation for the differentiation of rock units (either on aerial photographs or radar imagery) is topographic relief. Depending on the rock type involved, climatic conditions, thickness of the regolith and vegetal canopy, radar imagery may reveal abundant interpretive data, or it may show almost none. The geologic data interpreted from radar imagery for this study were derived from surrogates rather than through direct identification. Similarly, the time-stratigraphic assignments used for geologic map units must be inferred from corroborative data, and are obviously not interpretable from radar imagery. A summary of key stratigraphic data from the previously mentioned geologic reports is presented in Table 1.

Geologic reconnaissance mapping of the area shown in Figure 1 was presented in a study by Wing (1971) which incorporated prior radar mapping of Darién Province by MacDonald (1969).

PETROLEUM PROVINCES

Through examination of the radar mosaic (Fig. 1), eastern Panama and northwestern Colombia can be divided into three main physiographic-structural parts: two composite coastal ranges separated by a taphrogenic Medial basin (Wing, 1971), which extends southeastward the length of the eastern Isthmus, from the Gulf of Panama on the Pacific, to the Atrato River Valley of northwestern Colombia. A more detailed subdivision of major structural and physiographic elements has been derived from the interpretation of individual imagery strips (Wing, 1971). The resultant classification has been outlined in Figure 3 and indexed in Table 2.

Medial Basin

The Medial basin can be divided into three main segments. The transverse Canazas platform (SD-V and VI of Fig. 3) structurally separates the Bayano subbasin (west) from the Tuira-Chucunaque subbasin (east). The Bayano structural subbasin includes Medial basin subdivisions SD-I, II, III, and IV. The Tuira-Chucunaque structural subbasin includes Medial basin subdivisions SD-VII and VIII. The Bayano subbasin is a great half graben (e.g., SD-III), whereas the Tuira-Chucunaque subbasin (e.g., SD-VII) is an asymmetric synclinorium. The basin contains up to 4,600 m of strata, ranging in age from late Eocene to Holocene (Stewart, 1966). A great belt of en échelon folds evidences substantial past wrench movements between the Medial basin and the Pacific coastal ranges (Wing, 1971).

Bayano Subbasin—The oldest strata within the Bayano subbasin are of late Eocene age and the youngest, widely distributed, are of Miocene age (Terry, 1956). Pliocene-Pleistocene strata are reported in the axial parts of the subbasin, especially in the SD-IV syncline. Holocene sediments are notably present in the SD-I area.

Medial basin subdivision SD-I is a synclinal trough which plunges south-southwest and is continuous with a shallow trough submerged beneath the western Gulf of Panama. It is flanked by the Pearl Islands on the east and the Azuero Peninsula on the west. SD-I includes the mouth of the Bayano estuary and related Holocene beach-ridges, marshlands, and low-lying jungle (Fig. 4).

The unique wide band of beach ridges near the mouth of the Bayano estuary (Fig. 4) indirectly suggests the possible presence of quartz sands derived from silicic rocks of the San Blas Range, which may long have contributed debris for the periodic formation of reservoir strata in the western part of the Gulf of Panama. SD-I and this western part of the Gulf of Panama are actually a south-southwest-trending extension of the Medial basin, and this particular area may have considerable potential for hydrocarbon entrapments. A thick, prospective, marine, stratigraphic succession of paralic and deltaic clastic rocks, is anticipated here in water generally less than 50 m deep. The reservoirs are very likely to be found in quartz sandstone deposited in a high-energy environment. The incipient Bayano delta just south of the Upper Maje fault, in the S 1/2, S1/2, C-4 (Fig. 1), may be especially prospective. At this same site, deltas and offshore bars probably have been present sporadically from the Miocene to present time.

Medial basin subdivision SD-II is a projected northwestward subsurface extension of the Maje Range anticlinorium. The relatively thin stratigraphic section (610-915 m) makes this part of the Medial basin unattractive as a petroleum prospect. Medial basin subdivision SD-III is an east-trending synclinal segment of the Bayano subbasin. Stewart (1966) estimated that the Bayano subbasin deepens to approximately 2,135 m near the Maje River in SD-III. The radar imagery displays an inner closed structural low in the NE 1/4, NW 1/4, SE 1/4, C-5 (Fig. 1), expressed in outcropping Miocene strata that outline the structurally lowest part of the SD-III syncline.

The arcuate Igandi anticline (Fig. 5) trends approximately N60°W through the SW 1/4, NW 1/4, C-6, separating the SD-III and SD-IV synclines. The anticline is asymmetric and its steeply-dipping northeast flank probably is bounded by a reverse fault. It is less than 3 km north of a northward bulge (Pavo prominence) in the adjacent front of the Maje Range (Fig. 5), to which it is almost certainly genetically related.

Medial basin subdivision SD-IV is a northwest-trending synclinal segment of the Bayano subbasin, which extends from the Igandi anticline (east) to the west side of the Canazas platform (SD-V, Fig. 3). Stewart (1966) estimated that the Bayano subbasin stratigraphic section is 3,050 m deep in the vicinity of the Ipeti River in SD-IV.

Canazas structural platform—The Canazas platform is composed of Medial basin subdivisions SD-V and SD-VI. Along the northwest side of the Canazas platform is the distinctive outcrop of lower-middle Oligocene carbonate rocks (Fig. 6). The Canazas anticline (W 1/2, SE 1/4, SW 1/4, C-7; also "a" Fig. 6) trends north-northwest, subparallel with the Congo-Torti fault zone, to which it may be genetically related (Wing, 1971).

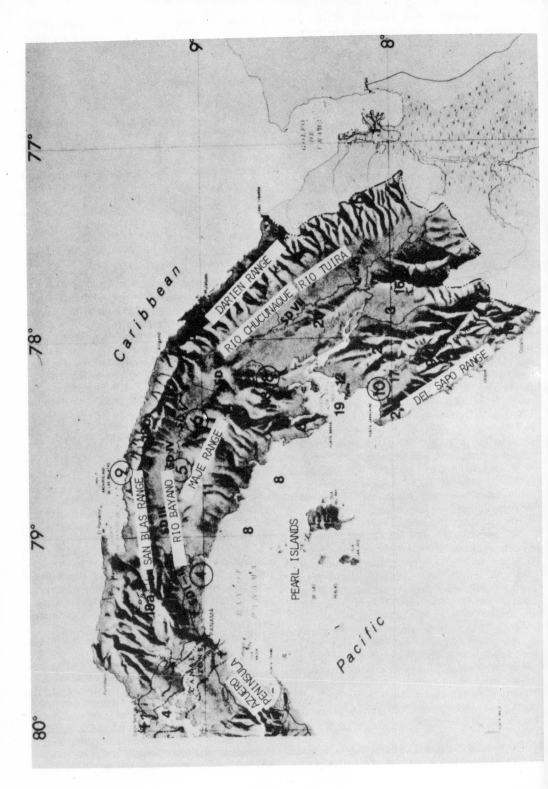

Table 2. Major Structural and Physiographic Elements,
Eastern Panama and Northwestern Colombia

Nos. on Fig. 3		Locations on Fig. 1
1	Atrato delta	G-13
2	Bagre Range	G-8
3	Balsas embayment	G-9
–	Bayano subbasin (Medial basin)	C-5
4	Canal Zone	C-1
–	Canazas platform (Medial basin)	C-7
5	Congo basin	E-7
6	Congo Peninsula basement block	F-7
7	Darien Range	D-9
7a	Darien Range-Morti segment	C-9
7b	Darien Range-Limon segment	G-12
8	Gulf of Panama basin	E-5
9	Jungurudo Range	I-8
10	Jurado Range	J-9
11	Maje Range	D-6
12	Mogue embayment	F-8
13	Pacific Hills anticlinorium	E-7
–	Medial basin	C-5
14	Pan American Hills	D-7
15	Pearl Islands	F-5
16	Pirre (anticline) Range	H-10
17	Sambu basin	G-7
18	San Blas Range	B-4
18a	San Blas Range-Azucar segment	B-6
18b	San Blas-Diablo segment	B-7
19	San Miguel Bay	G-7
20	Sanson Hills fold belt	F-9
21	Sapo Range	H-7
–	Tuira-Chucunaque subbasin (Medial basin)	E-9

SD-I of Bayano subbasin		C-3
SD-II of Bayano subbasin		C-4
SD-III of Bayano subbasin		C-5
SD-IV of Bayano subbasin		C-6
SD-V of Canazas platform		C-7
SD-VI of Canazas platform		D-8
SD-VII of Tuira-Chucunaque subbasin		E-9
SD-VIII of Tuira-Chucunaque subbasin		G-10

A similar anticline, which culminates in a prominent dome ("b" of Fig. 6), also has been delineated on the radar imagery. Between these two prominent anticlines, a more subtle northeasterly-trending arcuate fold belt has been mapped (NE 1/4, SW 1/4, C-7, Fig. 1) which must be related to the northeast-trending shear zone, expressed in the basement core of the San Blas Range on the north. The Medial basin subdivision SD-VI is a transverse, northeast-trending topographic divide.

Tuira-Chucunaque subbasin—The Tuira-Chucunaque subbasin is composed of Medial basin subdivisions SD-VII and VIII. Paralleling the continental divide along the northern part of eastern Panama, moderately dipping strata eventually flatten out to form the broad structural and topographic depression containing the Tuira-Chucunaque basin. The boundary between SD-VII and SD-VIII, within the Tuira-Chucunaque basin, is the transverse Chico River fault (N 1/2, G-10) upthrown on the southeast. This fault is part of a trans-isthmian disturbance that includes the Pirre anticline (H-10) and the anomalously uplifted southeastern segment of the Darien Range, shown by a widened basement outcrop (Fig. 7, starting in the S 1/2, F-11). Stewart (1966) estimated the Tuira-Chucunaque subbasin

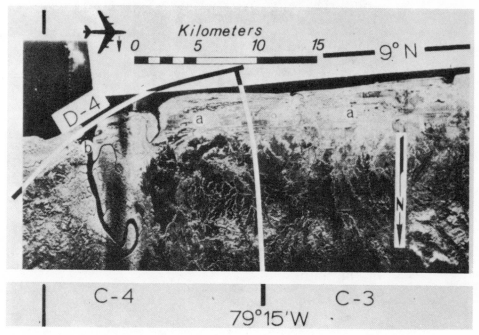

Fig. 4—Radar imagery of Gulf of Panama Coast, showing beach ridges (*a*) in proximity to mouth of Bayano Estuary (*b*).

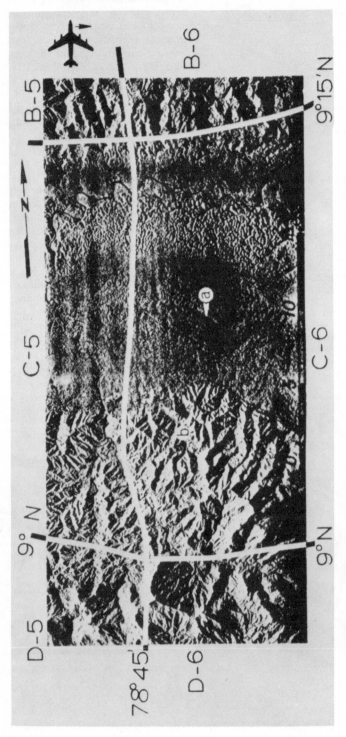

Fig. 5—Radar imagery, part of eastern Bayano subbasin, and adjacent central Maje Range, showing Igandi anticline (a) and Pavo prominence (b).

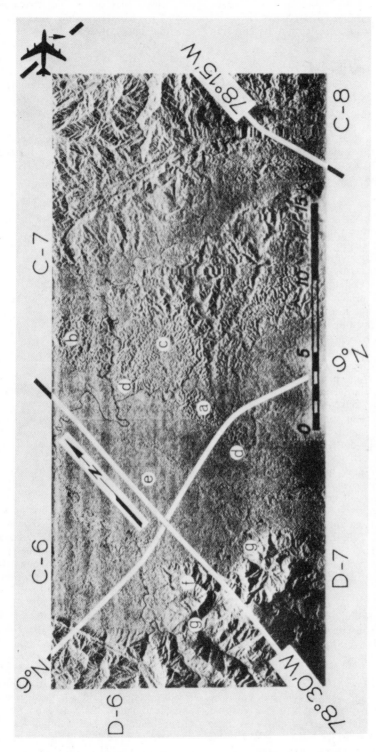

Fig. 6.—Radar imagery, western Canazas platform (SD-V) showing Canazas anticline (*a*), prominent dome (*b*), and karst topography in Clarita limestone (*c*). Also shown are Congo-Torti fault (*d*), part of SD-IV syncline (*e*), fault block (*f*), and east-west trending fault (*g*).

to be approximately 4,575 m deep in the Darien area near the Membrillo River (center E-9), and this is probably the greatest depth to basement in the Medial basin.

Sanson Hills Fold Belt

The Sanson Hills fold belt is comprised of a series of genetically related en échelon anticlines which, individually, trend an average N21°W (Fig. 8). Most of these anticlines are asymmetric and bounded by reverse faults on their southwest sides. The youngest strata exposed in the anticlinal cores are probably of Oligocene or early Miocene age (Interoceanic Canal Study Comm., 1968).

PETROLEUM PROSPECTS

There have been no oil or gas discoveries in Panama to date. Most of the drilling thus far has been in Darien Province, where significant shows were found in the Arusa and Gatun Formations (upper Oligocene and middle Miocene, respectively, Table 1), which include some organic shales. The Aquaqua Formation (lower Miocene) contains some possibly prospective sandstone reservoirs. However, reservoirs are sparse, inasmuch as most of the subsurface stratigraphic succession is quite tuffaceous, and most clastic material was derived from quartz-poor basic igneous rock types. The Clarita carbonate rocks (lower-middle Oligocene) have been considered to be potential pay zones by some, but there is little to substantiate their merit. They are exposed over broad areas on the north side of the Medial basin; also, there is a change of facies to shale south and west of the Tuira-Chucunaque subbasin. On the south side of the basin, the anticlines of the Sanson Hills fold belt generally are stripped at least to Oligocene strata on their crests, except for the Upper Sabana and northwestern Santa Fe closures (Fig. 8). This leaves but a very narrow part of the Medial basin which could be prospective, and it is not particularly well protected against flushing, especially from north flank exposures.

Thus, the outlook for eventual terrestrial Medial basin hydrocarbon production does not seem bright—at least not with respect to conventional entrapments. However, there is the possibility of gravity-type entrapments in fractured organic shales, siltstones, or carbonate rocks, within closed structural lows, such as the SD-III syncline. The SD-IV and SD-VII synclines could be similarly prospective, but depths probably would be greater. The radar imagery displays an inner closed structural low, in the NE 1/4, NW 1/4, SE 1/4, C-5 (Fig. 1), expressed in outcrops of Mio-

cene strata that outline the structurally lowest part of the SD-III syncline. It is of considerable potential importance as the possible locus of a Canon City embayment type hydrocarbon accumulation. What is envisaged is the possibility of gravity entrapment within the closed structural low in fractured Oligocene carbonate or siltstone strata, the source being overlying late Oligocene carbonaceous shales. Stewart (1966) described a major carbonate section that might be prospective for gravity hydrocarbon accumulations in fractures where it is present in the subsurface of the inner synclinal deeps of SD-III, or SD-IV, or both.

Five small anticlines were mapped by Shelton (1952) within the confines of the Tuira-Chucunaque basin; however, because of the lack of topographic expression, only two of them could be inferred from radar imagery interpretation. The anticlines mapped by Shelton are located as follows:

Yape anticline 8°8'N, 77°34'W—NW 1/4, SE 1/4, G-10
Quebrada Sucia dome 8°15'N, 77°35'W—SE 1/4, SE 1/4, F-10
Capete anticline 8°4'N, 77°34'W—NW 1/4, SE 1/4, G-10
Rancho Ahogado anticline 8°35'N, 77°52'W—NE 1/4, SW 1/4, E-9
Tuira anticline 8°08'N, 77°34'W—NW 1/4, SE 1/4, G-10

Geologic exploration for oil in the Tuira-Chucunaque basin was initiated by Sinclair Panama in 1925, when three shallow test holes were drilled to 332, 1,100, and 1,200 m. The deepest of these three tests was drilled on Yape anticline, NW 1/4, SE 1/4, G-10. A basal 110-ft oil-stained sandstone (Aquaqua Formation) was tested by Sinclair and recovered 1,500 bbl/day of salt water, with good shows of gas (Johnson and Headington, 1971).

Shelton (1952), in field work associated with petroleum exploration, provided geologic data for the central part of the Tuira-Chucunaque basin, and Terry (1956) reported on the regional geology of eastern Panama. Subsequent to the work of Shelton and Terry, the most recent and deepest drilling was completed (dry hole) by the Delhi-Taylor Corporation in 1959 at a total depth of 3,498 m. This test hole was on the Rancho Ahogado anticline centered at the NE 1/4, SW 1/4, E-9. Sufficient gas was found in the thin sandstones of this well to cause considerable drilling problems. Well-log and sample analysis provides evidence that the total stratigraphic section is still untested over this large structure (Johnson and Headington, 1971). A geomorphic anomaly of substantial size, embracing at least 400 sq km, has been mapped from the radar imagery, and is centered at 8°22'N, 77°41'W, F-10. This anomaly

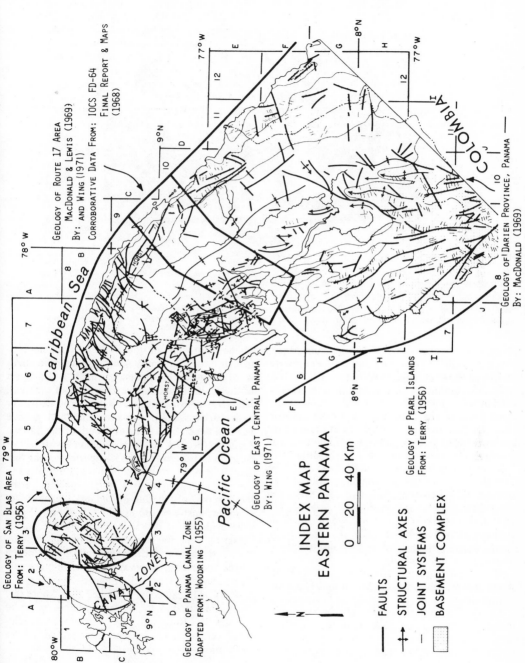

Fig. 7—Generalized geologic reconnaissance map, eastern Panama and northwestern Colombia.

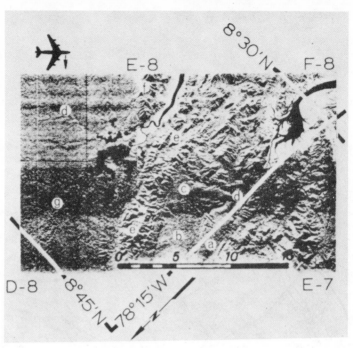

FIG. 8—Radar imagery, part of Sanson Hills fold belt showing Cucunati anticline (a) and syncline (b), Trans-Cucunati anticline (c), unnamed fault (d), Santa Fe anticline (e), Arreti anticline (f), and unnamed fold (g). Sabana Estuary can be seen in upper right.

appears to be reflecting a subtle, subsurface, structural high which is the result of the northerly plunge of Pirre anticline (SW 1/4, G-10; W 1/2, H-9). The outline of this anomaly is especially evident when one examines the drainage of the Rios Tuquesa and Tupisa on the radar mosaic (Fig. 1). Not only does this structure provide excessive size, but an almost complete stratigraphic section should be found by the drill.

In the Sambu basin (location SE 1/4, G-7; NW 1/4, H-8; Fig. 1) Gulf Oil Company drilled three shallow wells between 1922 and 1927. Two of these tests were in the vicinity of 8°04'N, 78° 15'W (NE 1/4, SE 1/4, G-7) and the third well at approximately 8°03'N, 78°18'W (C, SE 1/4, G-7). Oil shows were present in the Gatun beds of Miocene age, and the anticlinal trends located on land by field mapping have been projected offshore in San Miguel Bay (Johnson and Headington, 1971).

The Gulf of Panama is generally less than 100 m deep. Terry (1956) illustrated the extensive area of the shallow Gulf, which is several thousands of meters higher than the adjacent bottom of the Pacific basin. He stated: "If the ocean were withdrawn to the 100 fathom (183 meters) isobath, the area of Panama would be increased by about one-third, most of the addition being on the Pacific side."

The west side of the Gulf of Panama is a trough, plunging gently south-southwest, a shelfward extension of Medial basin SD-I. That part of the Gulf of Panama which is west of the Pearl Islands is considered to be highly prospective for hydrocarbons. As previously mentioned, a wide band of beach bars near the mouth of the Bayano Estuary (N 1/4, D-4) is indirect evidence of quartz sands derived from silicic rocks of the San Blas Range, which may long have contributed to the periodic formation of reservoir strata in the western part of the Gulf of Panama. Beach ridges (trending east-west along area "a" of Fig. 4) provide evidence of progradation along the Pacific Coast between the Rio Bayano and the Panama Canal. Back-swamp drainage north of the beach ridges and mangrove swamps (area "b") are easily delineated on the radar imagery.

The Caribbean coast is flanked by offshore reef chains manifested by the archipelagos north of the San Blas Range (Fig. 9). In areas where Holocene reef chains are present, there probably have been middle and late Tertiary reef chains as well, though not necessarily in exactly the same places. Thus, geophysical exploration for buried reefs and carbonate bands would likely be successful offshore in the Caribbean, north of the San Blas Range, where the water is less than 100 m deep.

FIG. 9—Radar imagery of Caribbean coast north of San Blas Range, showing windward part of active reef zones (a). Coral sand islands are on leeward slope of barrier reef north of (b).

The shelf involved has a fairly irregular bottom, probably reflecting the intense folding in evidence on shore along the north side of the Caribbean coastal range; and some such irregularities may even mark prospective structures offshore where the stratigraphic cover has not been stripped away. Edgar (1968) cited considerable structural disturbance in deep water on the north side of the eastern Panamanian Isthmus.

One other phenomenon of note is the presence of shell bars in the Gulf of Panama (C, SE 1/4, G-7), especially near the mouths of some rivers (Fig. 10) and around the Pearl Islands (Wing, 1971). Transgressions and regressions of middle and late Tertiary time should have resulted in migrations of the shell bars, such that their distribution *in toto*, in the subsurface, may be in the form of long, linear, porous carbonate belts generally transverse to the land.

CONCLUSIONS

The term "reconnaissance" is applied to incomplete or generalized mapping which usually precedes more detailed or localized studies. Similarly, reconnaissance mapping can enlarge the scope of local studies by providing a general geologic picture of the surrounding region. From a practical standpoint, reconnaissance mapping may be the only feasible method for geologic exploration because of the limitations of time, funds, adequate base maps, and accessibility.

The interpretation of radar imagery facilitates physiographic differentiation and geologic reconnaissance mapping on a regional scale. At the very minimum, a ready subdivision generally can be made between igneous and sedimentary rocks. Large-scale structural units can be synoptically studied, and the single-strip imagery format used in conjunction with a radar mosaic enables the petroleum geologist to become quickly familiar with the essential features of structural provinces. On a regional scale, gross lithologic and structural subdivisions can be interpreted at levels of detail which commonly exceed those available on existing small-scale geologic maps in many parts of the world.

Stereoscopic viewing, using same-look flight strips of 50 percent overlap, can provide a wealth of detail, including semiquantitative dip components and relative fault offsets. This method is currently the primary interpretive technique; some such stereo work was part of the subject Panama mapping effort.

On a more detailed scale, a relative stratigraphic sequence can be determined by using radar imagery, but only if the lithic units are

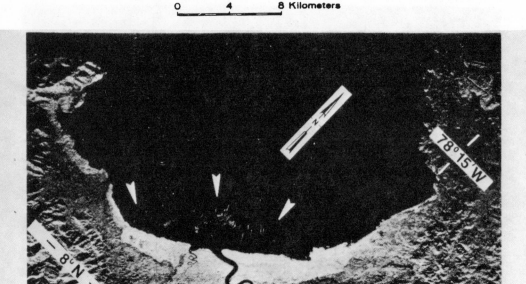

FIG. 10—Radar imagery near mouth of Rio Sambu in San Miguel Bay. Imagery taken at low tide, exposing shell bars growing perpendicular to shoreline (outlined by white arrows).

expressed in the terrain configuration and the structure is not too complicated. In this regard, stereo viewing of large-scale opaque prints is a great help. Where collateral field data are available and are used in conjunction with radar imagery interpretation, the problem becomes less complicated.

With the exception of those data provided by field investigation, the geologic information interpretable from the radar imagery of eastern Panama far exceeds those data previously available through conventional reconnaissance methods. Certainly, radar remote sensing offers the only practical technique for reconnaissance mapping in the wet tropics; however, even where conventional aerial photographic coverage can be obtained, radar imagery can be a valuable supplement because of its unique data content. Radar geologic reconnaissance—preferably with, but even without, air-photo support—can serve the petroleum geologist as an important exploration tool because of its substantial physiographic-geologic data content.

REFERENCES CITED

Edgar, N., 1968, Seismic refraction and reflection in the Caribbean Sea: Ph.D. dissert., Columbia Univ., 159 p.

Gillerman, E., 1970, Roselle lineament of southeast Missouri: Geol. Soc. America Bull., v. 81, p. 975-982.

Interoceanic Canal Study Commission, 1968, Geology final report, route 17, v. 1: IOCS Memo FD-64, Field Director, Office Interoceanic Canal Studies, Panama, 32 p.

—— 1969, Geology final report, route 25, v. 1: IOCS Memo FD-80, Field Director, Office Interoceanic Canal Studies, Panama, 110 p.

Johnson, M. S., and E. Headington, 1971, Panama—Exploration history and petroleum potential: Oil and Gas Jour., April 12, p. 96-100.

Lewis, A. J., and H. C. MacDonald, in press, Radar mapping of mangrove zones and shell reefs in southeastern Panama: Photogrammetria.

MacDonald, H. C., 1969, Geologic evaluation of radar imagery from Darien Province, Panama: Modern Geology, v. 1, p. 1-63.

—— and A. J. Lewis, 1969, Terrain analysis with radar—a preliminary study: Interim Tech. Progress Rept., Proj. THEMIS, Univ. Kansas, Lawrence, Kansas, p. F1-F12, October 1969.

—— and W. P. Waite, 1971, Optimum radar depression angles for geological analysis: Modern Geology, v. 2, p. 179-193.

—— J. N. Kirk, L. F. Dellwig, and A. J. Lewis, 1969, The influence of radar look-direction on the detection of selected geological features: 6th Symp. Remote Sensing of Environment Proc., v. 1, Univ. Michigan, Ann Arbor, Michigan, p. 637-650.

—— A. J. Lewis, and R. S. Wing, 1971, Mapping and landform analysis of coastal regions with radar: Geol. Soc. America Bull., v. 82, no. 2, p. 345-358.

Shelton, B. J., 1952, Geology and petroleum prospects of Darien, southeastern Panama: Master's thesis, Oregon State Univ., 61 p.

Stewart, R., 1966, The Bayano basin, a geological report: Panama Canal Commission Rept. (IOCS Memo PCC-r, File PCC-200.02), 17 p.

Terry, R. A., 1956, A geological reconnaissance of Panama: California Acad. Sci. Occasional Paper 23, 91 p.

Wing, R. S., 1971, Structural analysis from radar imagery of the eastern Panama Isthmus: Modern Geology, v. 2, p. 1-21, 75-127.

—— and L. F. Dellwig, 1970, Radar expression of Virginia Dale Precambrian ring-dike complex, Wyoming/Colorado: Geol. Soc. America Bull., v. 81, no. 1, p. 293-298.

—— W. K. Overbey, Jr., and L. F. Dellwig, 1970, Radar lineament analysis, Burning Springs area, West Virginia—an aid in the definition of Appalachian Plateau thrusts: Geol. Soc. America Bull., v. 81, no. 11, p. 3437-3444.

Woodring, W. P., 1955, Geology and paleontology of Canal Zone and adjoining parts of Panama: U.S. Geol. Survey Prof. Paper 306-A, 145 p.

Reprinted from:
BULLETIN OF THE AMERICAN ASSOCIATION OF PETROLEUM GEOLOGISTS
VOL. 49, NO. 12 (DECEMBER, 1965), PP. 2246-2268, 6 FIGS., 1 TABLE

RELATION BETWEEN PETROLEUM AND SOURCE ROCK[1]

DIETRICH H. WELTE[2]
Würzburg, Germany

ABSTRACT

The relations between petroleum and source rocks can be understood only if we know the important geochemical and geological data of both systems. Therefore, in the first two sections, the latest results of oil and sediment research are discussed. The third section deals with migration, which may be interpreted as the connecting link between crude oils and their source rocks. In the light of the data presented, it is then shown that oil genesis and migration are very closely related to the development of a basin and that one source rock can deliver a whole series of different crude oils. At the beginning of this series are the heavy petroleums and at the end the light ones. This development is basically the result of the progressive thermal degradation of the organic material, which was finely disseminated throughout the source rocks.

INTRODUCTION

The question of the origin of petroleum is one of the most fascinating problems of geoscience. Despite long and intensive research, it has not yet been answered. Recent progress in techniques, especially analytical methods, opens entirely new possibilities for the study of oil generation and migration. Today, for instance, it is possible to obtain almost any desired sample of crude oil or rock from deep wells and subject it to study; the outstanding methods of investigation are gas chromatography and mass spectrometry.

New results in this field were reported during the 6th World Petroleum Congress in Frankfurt/Main. In the light of these new results, it appeared appropriate to study anew the relations between petroleum and source rock, which are linked together through migration.

[1] Manuscript received, March 31, 1965. Originally published as "Über die Beziehungen zwischen Erdölen und Erdölmuttergesteinen" in Erdöl und Kohle, Erdgas, Petrochemie, 17 Jahrgang (1964), p. 417–429. Translated from the German and arrangements made for publication of English version by Robert E. King, American Overseas Petroleum Limited, and Martin Forrer, Texaco Inc., with the consent of the author and of Erdöl und Kohle, Erdgas, Petrochemie. English translation checked and some revisions and additions made by the author.

[2] Geologisch-Paläontologisches Institut der Universität, Pleichertorstrasse 34, Würzburg. The writer expresses his thanks to W. Philipp, Gewerkschaft Elwerath, Hannover, for reviewing the manuscript and for many stimulating discussions. Also the author wishes to thank Robert E. King and Martin Forrer for their interest in this paper and their excellent translation work. Thanks are also due to Donald R. Baker, Marathon Oil Company, Denver Research Center, for reviewing and correcting the translation. The writer is grateful to C. B. Koons and S. R. Silverman for their valuable criticisms, remarks, and clarification of the English text.

CRUDE OIL ANALYSES

The goal of our investigations is an exact knowledge of the chemical composition, qualitatively as well as quantitatively, of as many crude oils as possible; the more we know about the composition of petroleum the better we can form an opinion concerning its origin and history of migration. In order to understand the problem better we very briefly sketch the chemical range of variation of the most important hydrocarbons.

Among cyclic compounds and also among iso-paraffins, the side chains are very short. They are mostly methyl groups. Polycyclic aromatic-cyclo-paraffinic compounds have to be included with the aromatic because of the method of analytical separation.

Analyzing crude oils from different depths in different oil basins, one finds a general decrease in specific gravity of the crude oils with depth; however, this relationship is only relative and not related to absolute depth. The change in specific gravity is caused by a change in chemical composition. Whereas crude oils from reservoirs under thin sedimentary cover generally are very rich in hetero-compounds (that is, molecules that contain other elements such as O, N, or S besides carbon and hydrogen) in addition to actual hydrocarbons, the deeper-zone oils usually are relatively lean in such components.

Barton drew attention to this fact in 1934, based on investigation of crude oils from the Gulf Coast, and he suggested an evolution of crude oils in which either increase in age or in depth of burial may be interchangeable to a degree, because in principle both lead to the same result.

If the fact that deep oils are lighter than shallow oils is really based on an evolution, then this change must be mirrored to a large degree in their chemical composition, in a sort of "chemical evolutionary series." However, too few analytical data on crude oils from well known basins have been published to permit the study of such an evolution.

1. N-Paraffins

Martin *et al.* (1963) analyzed by gas chromatography 18 crude oils from the United States. The most important result was the identification of certain saturated hydrocarbons, that is, the homologous series of paraffins and certain iso-paraffins. Interestingly enough, they found an n-paraffin distribution curve strongly divergent from the usual one (Fig. 2).

Whereas most oils show a maximum at the beginning of the n-paraffin series, more or less from n-C_5 to n-C_{15} (exemplified by Darius oil), the oil from the Uinta basin exhibited a maximum toward the end of the n-paraffin series, that is, between n-C_{25} and n-C_{30}. Martin *et al.* explain this somewhat unusual n-paraffin distribution by the

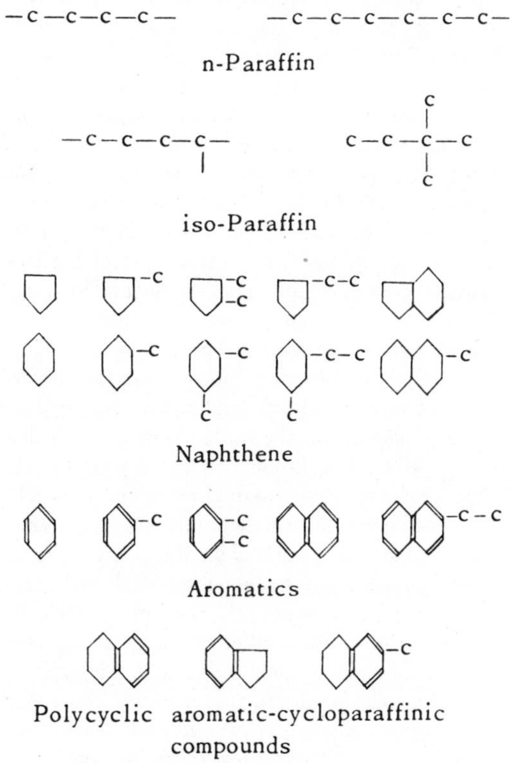

$$-c-c-c-c- \qquad -c-c-c-c-c-c-$$

n-Paraffin

$$-c-c-c-c- \qquad c-c-\overset{\overset{c}{|}}{c}-c$$

iso-Paraffin

Naphthene

Aromatics

Polycyclic aromatic-cycloparaffinic compounds

FIG. 1.—Most important types of hydrocarbons in crude oils.

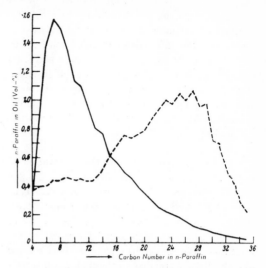

FIG. 2.—N-paraffin distribution curve of Darius oil (Iranian offshore field) and Uinta basin oil (after Martin *et al.*, 1964).

predominance of waxy compounds in the source rock, because waxes in general have hydrocarbon chains with more than 20 carbon atoms. The authors also point out that this explanation fits the non-marine origin of Uinta basin oils, waxes being mostly concentrated in terrestrial plants. A similar n-paraffin distribution curve was found by Martin *et al.* (1964) in the crude oil from the State Line field (Wyoming), which also appears to be of non-marine origin.

Dean and Whitehead (1964) found a similar n-paraffin distribution curve with a maximum toward the end of the homologous series in a Nigerian oil (Imo River). In all of these studies the oils are very young, that is, of Tertiary age.

Furthermore, Martin *et al.* found that certain oils had a predominance of n-paraffins with odd carbon numbers between n-C_{10} and n-C_{20} as well as between n-C_{22} and n-C_{35}. This poses new problems, especially the predominance of such n-paraffins in the high molecular region (n-C_{22} to n-C_{35}) of the Uinta basin oil (Fig. 2). Bray and Evans (1961) had investigated the n-paraffin distribution curve, which they considered to be a clue to the recognition of source rocks, of 39 crude oils and found a regular distribution between even and odd carbon numbers among the higher n-paraffins. N-paraffins extracted from Recent or sub-Recent sediments showed a strong preponderance of odd carbon numbers, whereas fossil sediments had a well-balanced, or nearly so, n-paraffin distribution curve. Bray and Evans concluded that the only sediments which can be considered as source rocks are those whose extracted n-paraffins show a well-balanced distribution curve. In rocks with a preponderance of n-paraffins containing odd carbon numbers, not enough n-paraffins had been formed to dilute the original predominance of homologues with odd carbon numbers.

Somewhat complementary to the observation of Bray and Evans are the findings of Cooper (1962), who observed that oil-field waters contained fatty acids with 14-30 carbon atoms with well-balanced distribution between even and odd acid molecules. Fossil sediments showed a distribution curve similar to that of oil-field waters, whereas Recent sediments exhibited a strong predominance of fatty acids with even carbon numbers.

Cooper and Bray (1963) suggested a mechanism by which a change from a preference for even carbon-numbered fatty acids in Recent sediments to no preference may be possible in petroleum reservoir waters. According to these authors, each fatty acid can lose CO_2 to form an intermediate which reacts to give two products, an n-paraffin and a fatty acid. Each product will have one less carbon atom than the original acid. However, it is questionable whether this process alone can account also for the disappearance of an odd-predominance above 29 carbon atoms, as by the proposed mechanism the freshly formed n-paraffins tend to have shorter chain lengths than their precursors and the amount of natural fatty acids above 30 carbon atoms is relatively small compared with those having around 20 carbon atoms. In this connection it is interesting to know, as we shall see later, that mild thermal cracking of a fatty acid apparently can also produce n-paraffins with more carbon numbers than contained in the original acid (Jurg and Eisma, 1964).

2. ISOPARAFFINS

In gas chromatographic analyses of saturated hydrocarbons, many oils, after removal of n-paraffins by molecular sieves, exhibit marked elution peaks whose most frequent representatives have retention times near $n-C_{15}$, $n-C_{16}$, $n-C_{17}$, and $n-C_{18}$. These elution peaks commonly are caused by so-called isoprenoid (terpenoid) hydrocarbons.

These are isoparaffins which at regular intervals, that is, at every fourth carbon atom, exhibit a methyl branch. Isoprenoid hydrocarbons are typical of biosynthetic processes. The representatives found in crude oil probably are products of the destruction or alteration of natural isoprenoid substances such as various pigments or terpenes. The most prominent isoprenoid hydrocarbon found in crude oil is pristane, a 2, 6, 10, 14-tetramethylpentadecane with the following structure:

$$C-C-C-C-C-C-C-C-C-C-C-C-C-C-C$$
$$| \quad\quad | \quad\quad | \quad\quad |$$
$$C \quad\quad C \quad\quad C \quad\quad C$$

In petroleums this single component may amount to as much as 0.5 per cent of the total crude oil (Bendoraitis et al., 1962).

Bendoraitis et al. (1964) isolated and identified 7 such isoprenoid hydrocarbons in an East Texas crude oil. Five of the components that they described had a methyl (e.g., pristane) or n-propyl, or n-pentyl group instead of the terminal ethyl group. The authors concluded from this that the molecules which originally had been formed through biosynthesis and which certainly had a terminal ethyl group, had been altered by destruction in the source rock. They point out that many of the smaller isoparaffin molecules could well be fragments of larger isoprenoid components. Mild thermal cracking in the source rock could explain both the change in the terminal ethyl group and the disintegration of the isoprenoid molecule into smaller isoparaffin fragments.

Based on a comparison between natural light oil fractions on the one hand and either thermally cracked, catalytically cracked, or hydrocracked light oil on the other, Martin et al. also arrived at the conclusion that maturation of oils must be the result of thermal cracking. They left open the question whether this cracking takes place in the source rock or in the reservoir rock.

The absolute content of isoprenoid hydrocarbons in different crude oils changes appreciably and the quantitative relations between these components show no obvious connection. This apparently irregular occurrence of various isoprenoid hydrocarbons in most oils moved Dean and Whitehead to suggest using them as identifying characters, or "fingerprints" for crude oils. This recommendation is well worth following up.

As mentioned in earlier literature (Smith and Rall, 1953; Rossini et al., 1953), the distribution of isomers in crude oils does not correspond with a thermodynamic equilibrium. This is the reason Martin et al. largely exclude catalytic cracking in the formation of oil. In catalytic cracking carbonium ion reactions occur; as experiments show, they lead to a distribution of isomers close to thermodynamic equilibrium.

According to Dean and Whitehead (1964) the maximum molecular weight of crude oil components is about 2,000, and all homologous series reach this level, even if only in very small amounts. It is, therefore, not surprising that with more analytical data the impression is gained that most crude oils are similar in their general composition and only differ in relative content of certain components or of homologous series and their distribution curves.

Thus Colombo and Sironi (1961) found great similarity among Italian oils and asphalts even though they came from different basins. Geochemical investigation, furthermore, showed that transitions exist from the light mixed paraffinic to the heavy naphthenic oils and finally to asphalt.

Here, we come again upon the evolutionary series of crude oils mentioned at the beginning. Whereas qualitatively the differences are relatively small, the quantity of the higher molecular-weight, specifically heavier, components may be large at one end of a possible evolutionary series and small at the other. In such a series we would merely be dealing with a shift in maxima without addition of new components. Nevertheless, no such evolutionary series have heretofore been described. The reason may lie in the strong scatter of analyzed oils with regard to regional and age provenance.

Available data, however, do suggest that a certain relation exists between age—or, better, state of evolution—of oils and their chemistry. Smith and Rall (1953) found from an analysis of 32 crude oils of different ages that the ratio isohexane/n-hexane decreased with increasing geological age; that is, the normal component became relatively enriched. Martin et al. reported similar conditions. In younger oils they found more isoparaffins than n-paraffins. Furthermore, younger oils contained less volatile components and more cyclohexane, and the distribution of hydrocarbons was generally less regular.

3. CYCLIC COMPOUNDS AND HETERO-COMPONENTS

Compared with n- and isoparaffins, the saturated cyclic compounds have been less thoroughly investigated. The reason lies largely in increased analytical difficulties. The few available data do not help to clarify the problem of oil generation because of lack of data for comparison. However, it seems to be true that young, immature oils are especially rich in saturated cyclic components.

Just as little is known about aromatics and their distribution as about saturated cyclic compounds. Analyzing alkyl-substituted benzenes of different crude oils, Martin et al. observed that multiple methyl substitution (C_6 to C_{10}) at the benzene molecule occurs much more commonly than single substitution of larger groups. The impression generally prevails that the aromatics are concentrated in the more mature oils.

Among hetero-compounds the oxygen-bearing components are of special interest. Above all, acids have been described, among which 5- or 6-cyclic naphthenic acids are predominant. Also to be mentioned are straight-chain and branched-chain fatty acids. Here also belongs the group of high molecular-weight asphalt compounds.

Porphyrins, however, have commanded most of the attention. Ever since Treibs (1936) isolated and identified metal porphyrins from many crude oils, he not only solved the dispute about the origin of petroleum in favor of those believing in an organic origin but he also laid the groundwork for many further investigations in this direction. The main representatives of these components which are found in all crude oils are the porphyrin pigments containing a central nickel or vanadium atom. It is no longer doubted that the porphyrins originate from chlorophyll, which is widely distributed in nature.

Several authors, above all Blumer (1950), Blumer and Omenn (1961), Dunning et al. (1954), Dunning and Moore (1957), Hodgson et al. (1960), and Hodgson and Peake (1961), have tried to reconstruct the conditions reigning during the alteration of chlorophyll to metal porphyrin complexes, conditions which must have been those under which crude oil was generated.

The most important processes are hydrogenation and decarboxylation. Of special interest is one of the most recent publications by Hodgson et al. (1963). Based on an investigation of Japanese oils and sediments, the authors conclude that decarboxylation may also take place as late as in the reservoir. As the porphyrin components would not have lost their acid groups during migration, they would be surface-active and could play an important role in migration. As the porphyrins are of high molecular weight, they occur in the specifically heavier crude oils in larger quantities than in the light oils; therefore, the porphyrin content very commonly decreases with depth in oil basins. Park and Dunning (1961) have shown from C^{13}/C^{12} isotope analyses that the porphyrins are primary components and were not formed at a later date in crude oil, for example, by bacterial activity.

DISTRIBUTION OF CARBON ISOTOPES IN CRUDE OIL

Isotope analysis has become more and more important in petroleum geochemistry. Eckelmann et al. (1962) give a compilation of the distribution of the carbon isotopes C^{13} and C^{12} in 128 crude oils (Fig. 3).[3] In this compilation it is remarkable that most oils lie in the region between -26 per mil and -30 per mil.[4] If there has been no further isotope fractionation, this would mean that most oils originated from organic material formed in terrestrial, limnic, or at least

[3] Carbon isotope ratios are expressed as per mil deviation (δ) from a standard material as follows:

$$\delta\%_0 = \left[\frac{C^{13}/C^{12} \text{ sample}}{C^{13}/C^{12} \text{ standard}} - 1 \right] \cdot 1000$$

As a standard directly or indirectly a Cretaceous belemnite from the Peedee Formation of South Carolina is used.

[4] S. R. Silverman has been kind enough to make the following comment: "Citation of Eckelmann's isotope data for 128 oils is misleading in the context used. Practically all of the values in the non-marine range (-30 per mil or lower) are for early Paleozoic oils. Such oils were reported to be consistently low in C^{13}/C^{12} ratio (in spite of their marine origin) by S. R. Silverman in his paper 'Evidence for an age effect in the composition of natural organic materials' (Abstract, Annual Meeting of Geological Society of America, Nov. 4, 1961). This presentation covered isotope ratios for about 1,200 crude oils and indicated that the ratios of post-Paleozoic petroleums of marine origin were consistently greater than -27 per mil."

The present writer agrees with this comment, as he was not aware of the fact that practically all of the values in the non-marine range of Eckelmann's presentation were for early Paleozoic oils. It is, however, felt that the following arguments against the hypothesis that "most oils are derived from organic materials formed in the atmosphere or in fresh (or brackish) waters" may still be put forward, as they help to clarify this very problematical matter.

brackish environments. From a compilation of C^{13}/C^{12} ratios of various materials (Fig. 4) we can see that the isotope ratio of the overwhelming majority of crude oils coincides with the isotope ratio for organic material of terrestrial origin. This result contradicts both geological observations and geochemical considerations, according to which the origin of oil is predominantly marine or at least marine to brackish.

Silverman (1964) cut several petroleums in narrow distillation fractions and determined their C^{13}/C^{12} ratios. He found a continuous steep increase of this ratio upward from methane until, in the boiling range between 90°C. and 130°C., a maximum was reached. Above 130°C. there was a gradual decrease in the C^{13}/C^{12} ratio with increasing boiling range. This finding in connection with the observation that $C^{12}-C^{12}$ bonds are broken about 8 per cent more frequently than are $C^{13}-C^{12}$ bonds suggests, according to Silverman, that the hydrocarbons of the gas and gasoline fractions arise through decomposition of more complex compounds. Thus we see that here again a mild thermal-cracking process seems to be the best explanation of the foregoing findings.

SEDIMENT ANALYSES

There now remains no doubt that petroleum originated in the organic material of fine-grained sediments, which in many instances are also thin-bedded. Any fine-grained sedimentary rock with a certain amount of finely distributed organic material may, therefore, be a potential source rock. To reduce this large choice or, better yet, to define a source rock accurately has been the goal of many investigations. Most authors have begun with a comparison between the chemical composition of crude oils and the chemistry of soluble organic material in sedimentary rocks. The rocks were dried, pulverized, and later treated with one or more organic solvents. The extracted material and the crude oil were compared. This method has given very many and very good results and will continue to do so, but we must always remember that extraction is something entirely different from migration under natural conditions. Comparison of extract and related crude oil is, therefore, of limited value.

It is not our intention to discuss the enrichment and preservation of organic material in sediments in detail, but only insofar as this may contribute to an understanding of our problem.

Bitterli (1963) concluded from a comprehensive study of primary bituminous rocks of Cambrian to Pliocene age from western Europe that the formation of bituminous sediments is especially favored at certain paleogeographic key points, such as transgressions or regressions. This results from the fact that most sediments of this type contain both autochthonous and allochthonous organic material. The autochthonous material consists mainly of marine plankton whereas the allochthonous is mainly vegetal material of terrestrial origin. The findings of Bitterli are consistent with the results of Ronov (1958). The latter investigated 26,000 samples of different ages and environments from oil and non-oil provinces. The highest amount of organic matter in both kinds of provinces was found in marine nearshore, epineritic clay sediments. Both authors come to the conclusion that about 0.5 per cent organic matter is the lower limit for a potential oil source rock. However, it is clear that such a figure can give only an idea about the order of magnitude and nothing more.

We are especially interested in the diagenetic and post-diagenetic alterations of organic material which led to the genesis of the components

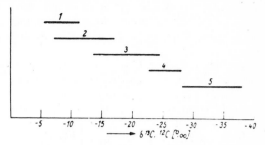

FIG. 4.—Distribution of carbon isotopes in different media (after Park and Dunning, 1961). 1. Atmospheric CO_2. 2. Marine plants. 3. Marine lipids. 4. Terrestrial plants. 5. Terrestrial lipids.

of petroleum. The lipid fraction of the dead organic substances is of particular importance because, structurally, it is closest to the long, straight-chain, and the cyclic, saturated petroleum hydrocarbons.

1. LIPIDS AND KEROGEN

Abelson and Parker (1962) have investigated the fatty acid content of Recent sediments. They found only traces of unsaturated fatty acids and only a very low content of saturated fatty acids, although living algal material that provided the largest portion of the organic detritus contains 5-25 per cent of saturated fatty acids. The combined results of Cooper (1962) and Emery (1960) confirm these observations. Abelson concludes that the low fatty acid content in the sediment is insufficient for the generation of hydrocarbons and that immediately after the decay of the algae most fatty acids are incorporated in the kerogen.

Kerogen is that high molecular residue insoluble in organic solvents which gives off hydrocarbons during pyrolysis. The total composition of kerogen is variable. Hunt and Jamieson (1956) report H/C ratios of 0.90, 0.81, and 0.68 for the kerogen of two shales and a dolomite. Breger and Brown (1962) interpret the variable hydrogen content of kerogen in shales with wide regional distribution to be a consequence of varying terrestrial influence. As terrestrial organic material contains less hydrogen (5.5 per cent) than aquatic (10-12 per cent), a decrease in hydrogen content would indicate an increase in terrestrial organic components in the total kerogen.

It would be most interesting to supplement these investigations by C^{13}/C^{12} isotope analyses.

The formation of kerogen must take place

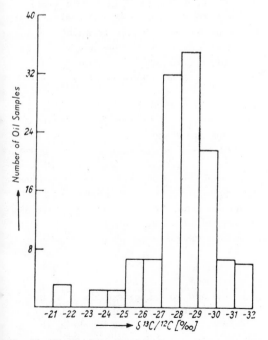

FIG. 3.—Distribution of carbon isotopes in 128 crude oils (after Eckelmann et al., 1962).

very early, shortly after the decay of the organ-- isms in the surface sedimentary layer, and as long as highly unsaturated compounds are still available.

Abelson (1964) compares the formation of kerogen with the reactions taking place in drying of linseed oil. As long as oxygen exists in the interstitial waters, peroxide-like oxygen bridges would form and cause strong cross-linking.

Indispensable for the formation of kerogen are reactive organic substances; these reactive molecules may be increased by the bacterial action that occurs in a fresh sediment. These reactive components cause the cross-linking which may involve even less reactive components and entire "molecular blocks," that is, not fully destroyed relics of organisms. Inside and with such giant molecules there may later occur a reorganization of groups (*e.g.*, migration of hydrogen) by which the kerogen adjusts itself to the new environmental conditions caused by diagenesis. A clarification of these relationships is most desirable. Pyrolytic experiments on kerogens, such as those carried out in the study of the constitution of coals, would certainly be helpful.

2. DISTRIBUTION OF HYDROCARBONS IN RECENT AND FOSSIL SEDIMENTS

We are much better informed on the composition of extractable organic material than on the insoluble kerogen. Smith (1954) described paraffins, naphthenes, and aromatic hydrocarbons from Recent sediments. The distribution of molecular weights of these components, however, did not correspond with that of crude oils. Low molecular-weight components (from 2-14 carbon atoms) are lacking, whereas they occur in large amounts in crude oils. This was emphasized strongly by Sokolov (1959) and Erdman (1961).

Dunton and Hunt (1962) compared low molecular paraffinic and naphthenic hydrocarbons from Recent sediments in five different areas with those of fossil sediments. The Recent samples contained no hydrocarbons between C_4 and C_8; all fossil samples, on the other hand, showed a whole series of petroleum-like paraffinic and naphthenic components of this molecular-weight range.

If one would draw a distribution curve for n-paraffins, the Recent sediments would have a maximum in the high molecular weight range

similar to the curve in Figure 2 (Uinta basin oil). However, the predominance of odd carbon numbers in the higher n-paraffins would seem to be somewhat stronger. On the other hand, fossil sediments would exhibit an n-paraffin distribution curve more like that of the Darius oil in Figure 2, that is, like the majority of crude oils. This implies that the volatile saturated hydrocarbons are newly formed in fossil sediments in the course of time. The question is when and at what depth. Erdman (1961) observed similar conditions among the aromatic hydrocarbons in comparing Recent and fossil marine sediments. The fossil samples had a higher content throughout in aromatic components. Here, too, aromatic, low molecular hydrocarbons formed later in the sediment.

We have already mentioned the preponderance of higher n-paraffins with odd carbon numbers in Recent and sub-Recent sediments. This preponderance decreases with increasing age and greater depth until a well-balanced n-paraffin distribution curve is reached.

Bray and Evans (1961) ascribe this phenomenon to the later formation of higher paraffins which would cause a dilution of the original composition.

Isoprenoid hydrocarbons have been found in various fossil sediments, just as in crude oils. Of special interest may be the occurrence of the isoprenoid substances pristane (2, 6, 10, 14-tetramethylpentadecane) and phytane (2, 6, 10, 14-tetramethylhexadecane) in a Precambrian shale (Eglington *et al.*, 1964; Meinschein, 1964). Blumer *et al.* (1963) found that pristane is a major component of the body fat of certain marine planktonic microorganisms. It is thus to be expected that it occurs also in Recent marine sediments. Interestingly enough, especially in view of oil generation, pristane has been found many times in the marine environment and in all these cases it may be traced back to microplankton. On the other hand, only two occurrences in continental organisms are reported, in wool wax (Mold *et al.*, 1963) and in the fruit of the anise plant (Brieskorn and Zimmermann, 1965).

It has been shown that generally after a certain time and with increasing depth of burial, new extractable hydrocarbons are formed in the sediments. Their source lies largely in the insoluble kerogen and in the extractable organic substance. Thus the ratio of the extractable amount of organic carbon to the residual organic carbon is indicative for hydrocarbon generation.[5] This means that the fresh formation of extractable components from insoluble kerogen must cause an increase in the extract/carbon ratio. This increase may sometimes be difficult to detect, especially if the sediment already contained much insoluble, coaly material.

3. GENESIS OF HYDROCARBONS

A valuable contribution to the problem of the genesis of hydrocarbons was made by Hunt (1962), who subjected kerogen for a short time to high temperatures, with exclusion of oxygen. After only 10 minutes at 400°C., he found that the product of pyrolysis contained more than 1,000 ppm of saturated hydrocarbons as well as the whole series of aromatic hydrocarbons with only extremely small amounts of olefins. Among the saturated components were methane and ethane, and furthermore propane up to octane. Among aromatics were benzene, toluene, xylenes, ethylbenzene, and isopropyl benzene.

Apart from decarboxylation, which certainly occurs during oil generation (e.g., during the alteration of fatty acids into paraffins or during the formation of porphyrins when carbon-carboxyl-carbon bonds are split), there must also occur destruction of simple carbon-carbon bonds. There would thus, above all, occur simple thermal disintegration reactions of the kerogen or the lipid fraction as reactions of first order. With the help of the Arrhenius equation (which permits the calculation of first-order reactions), Abelson (1964) tried to determine approximately the time needed for the formation of hydrocarbons under the afore-described conditions. From the calculated temperature/time curve, it appears that under the expected temperatures between 50° and 180°C. in source rocks, an interval of one million years is sufficient for decarboxylation to a large extent. Cracking, which demands a higher activation energy, would need a temperature of 100°C. for 100 million years to obtain a similar amount. Fairly large amounts of low molecular-weight hydrocarbons could have been formed in a 100-million-year-old source rock under an average temperature of 160°C.

[5] It is believed that this parameter was introduced by Philippi (1957).

The temperature factor can be compensated to a certain degree by the time factor. Both increase in temperature or greater age would, therefore, result in more advanced hydrocarbon genesis.

It is not yet known whether and to what extent the catalytic action of clay minerals, heavy metals, etc. is a necessary proviso for the genesis of crude oil. Jurg and Eisma (1964) report that during their experiments behenic acid yielded hydrocarbons after heating only if there was bentonite present as a catalyst. However, the problem is not completely understood, because in the process of oil generation in source rocks water is present, and experience has shown that catalytic reactions normally do not proceed if the active surface of the catalyst is blocked by water molecules. Furthermore, it seems to be true that in catalytic reactions the reaction product has a tendency to show a thermodynamic equilibrium, which, however, can not be observed in crude oils with respect to the iso- and n-paraffin ratio.

The decay products of chlorophyll which are considered to be the primary source of crude oil porphyrins occur in many Recent and fossil sediments. The important steps which lead to an alteration of the porphyrins are hydrogenation and decarboxylation.

According to Hodgson et al. (1963), hydrogenation occurs either in the Recent sediment, or even before sedimentation in the living organism. The loss of the acid groups—the decarboxylation—may occur either in the plant or in the sediment, perhaps as part of the migration process. Hodgson et al. (1960) observed a change of the decayed products of chlorophyll to true crude oil porphyrins under reducing conditions below the surface in Recent limnic deposits.

4. DISTRIBUTION OF CARBON ISOTOPES IN ORGANIC MATERIAL OF SEDIMENTS

In discussing crude oil analyses, we have already mentioned that Eckelmann et al. (1962) favored derivation of crude oils from organic material of terrestrial origin, provided that the equivalence of the carbon isotope ratios in crude oils and their corresponding source shales is confirmed by further analyses. Because the conclusion of the authors that crude oil was generated largely from terrestrial organic material has far-reaching consequences, a critical investigation is in order.

The first problem lies in assigning a shale source rock to a corresponding crude oil. As we shall see, this is an extremely delicate matter. For the moment we shall assume that the assignment has been made correctly.

Krejci-Graf and Wickmann (1960) investigated 58 Lias-α cores and some corresponding Lias-α crude oils from northwestern Germany. The carbon isotope ratio of the cores ranged from δ $C^{13}/C^{12} = -22.5$ to -28 per mil, that of the crudes from -25 to -30 per mil. Whereas the difference between rock and crude oil samples hardly exceeds 2 per mil (according to Eckelmann et al., 1962, this is still considered equivalent), we must not overlook the fact that the crudes show a somewhat more "terrestrial" C^{13}/C^{12} isotope composition than the Lias-α clay-shales (Fig. 4). In the same publication (Krejci-Graf and Wickmann), a marine algal gyttja,[6] the Ordovician Kuckersit, is shown to have a δ C^{13}/C^{12} ratio ranging from -30.2 to -30.7 per mil. This value is entirely "terrestrial." Yet there is no doubt that the Kuckersit is marine and that terrestrial plant detritus can hardly be expected, as terrestrial plant life was not very strongly developed during the Ordovician.

According to Hoering (1962) the organic substance in five Precambrian rocks with algal structures showed the following isotope composition: $-18.8, -23.5, -29.1, -29.9,$ and -31.7 C^{13}/C^{12} per mil. The three last-named values are within the range of terrestrial organic material according to the usual classification. But where would terrestrial organisms have come from in the Precambrian?

There are only two explanations possible for these examples. Either the C^{13}/C^{12} isotope ratio of sea water changed during the course of geologic history, or reactions took place in the organic material of these sediments which changed the original isotope ratios through a relative loss of heavy C^{13} isotopes. The first alternative is unlikely, even though not impossible; it has been discussed by authors better qualified than the

writer. The second alternative is discussed briefly.

Silverman and Epstein (1958) report a continuous change of δ C^{13}/C^{12} from -20.3 to -26.0 per mil with only small increases in depth (about 50 cm.) in Recent samples from the Florida Keys. Landergren (1954) also found a relative impoverishment in C^{13} with increasing sedimentary thickness in the organic material from deep-sea sediments in the Pacific. In one sample he found a change of 2 per mil for a 20-cm. depth difference; in another the C^{13}/C^{12} ratio changed from -19.5 to -23.3 per mil, that is, 3.8 per mil, in a depth difference of 160 cm.

Sackett and Thompson (1963) investigated plankton from the mouth of the Mississippi and from the open Gulf of Mexico. They also collected bottom sediment. The carbon from the bottom sediments was lower throughout in C^{13} than the plankton floating above it.

These data indicate that relatively shortly after sedimentation, probably at the same time that conversion of organic matter begins, an impoverishment in C^{13} takes place. This loss of heavy C^{13} isotopes can not be explained by a selective preservation only of the lipid fraction, which with regard to isotopes is lighter than the remaining substance of the organisms. This could happen only if lipids changed into non-lipid material which forms a noticeable part of the organic material in the sediments; this, however, is not very likely according to our present state of knowledge.[7]

A partial explanation might lie in an observation by Abelson and Hoering (1961). They found that the protein fraction of lower organisms is heavier with regard to isotopes than the total carbon of the organisms. The proteins, therefore, in a way make up for the lighter lipid fraction. They also found that the carboxyl group of amino acids, on the other hand, is again heavier than the rest of the molecule. Decarboxylation, which plays an important role in the alteration of the organic material in sediments, would therefore cause a relative lowering in C^{13} in the protein fraction,

---

[6] "A sapropelic black mud in which the organic matter is more or less determinable, characteristic of eutrophic and oligotrophic lakes": A. G. I., 1957, Glossary of geology and related sciences, p. 133.

[7] A critic of this paper has written: "Selective preservation of lipid components accompanied by complete destruction or loss of non-lipid organic matter would explain adequately the isotope ratio differences between modern organisms and ancient organic deposits." The author disagrees with this, as it is well known that there is no complete destruction or loss of non-lipid organic matter. One can find it easily in any sediment in question.

the latter certainly contributing to the constitution of kerogen. Thus, the organic substance in the sediment would become more "terrestrial."

As these examples indicate, it seems that the dead organic material laid down in the sediments undergoes an isotope fractionation which makes it lighter, and therefore causes it to appear more "terrestrial" than the original substance. It would appear that this fractionation takes place largely in the fresh sediment.[8] It is not known whether bacteria play a significant role in this connection.

To this fractionation effect which occurs upon the death or the alteration of the organic substance is added the effect of a mixing of marine (autochthonous) organic material with terrestrial detritus (allochthonous), especially near the mouths of large rivers.

Sackett and Thompson (1963) were able to show this very clearly for the Gulf of Mexico area. The C^{13}/C^{12} ratio of the organic material in the Recent sediment changes continuously from about -27 per mil in the terrestrial water system down to -21 per mil in the open Gulf.

As can be seen, Eckelmann's hypothesis on the terrestrial origin of crude oils appears to go too far.

MIGRATION

The migration of crude oil occupies a key position in the relation between petroleum and its source rock. Migration is not only essential for the formation of petroleum deposits but is, in the strict sense, also the decisive process which makes a true oil source rock out of a potential source rock. This process, viewed as a whole, is found to be an only partly understood complex of a multiplicity of specific problems. We shall try to isolate those problems with a bearing on this

---

[8] After this manuscript had been completed a paper was published by W. M. Sackett, which deals especially with this matter (Sackett, W. M., 1964, The depositional history and isotopic organic carbon composition of marine sediments: Marine geology, v. 2, no. 3, p. 173–185).

At the annual meeting of DGMK, October 6-9, 1965, in Hanover, Germany, a paper was presented by W. R. Eckelmann entitled: "The C^{13} organic carbon composition of marine plankton and its relationship to marine sediments." It was shown that the C^{13} composition of marine organic sediments is controlled not only by the amount of terrestrially derived organic carbon, but also by the water temperature in which the organisms grew. Low water temperature causes a depletion of C^{13} compared with that in plankton which grew in warm water.

---

discussion and to treat them in more detail insofar as possible.

We have shown that oil and gas have formed out of widely distributed organic material in sediments during the course of time and with increasing depth. The first droplets of oil and bubbles of gas (insofar as the gas is not dissolved in the oil) are disseminated throughout the sedimentary body in relation to the distribution of the organic material, and isolated from each other by the carbon-free mineral constituents. The newly formed oil-like components permeate or surround the organic material, probably as cloudy concentrations. At least a part of this oil must be able to migrate from these island-like "oil nests" (Wassojewitsch (1960) calls the finely distributed oil in the sediment "micronaphtha"). The movement of oil within the finely porous source rock, before it reaches the coarsely porous reservoir rocks, is called primary migration. Migration within the permeable reservoirs into the trap is called secondary migration.

1. PRIMARY MIGRATION

Potential petroleum source rocks as a rule are fine-grained sediments rich in clay minerals. A characteristic of these rocks is that they originally had high porosity, and with increasing age and greater depth of burial this was gradually reduced. Through compaction the pore volume shrank considerably, and the water and other substances capable of fluid movement in the pore spaces were pressed out. Compaction is therefore the cause of primary migration. This phenomenon was recognized and thoroughly documented (among others) by Hobson (1954), Gussow (1954), and Levorsen (1954).

Hedberg (1926, 1936) and Athy (1930) published the first detailed data on the connection between the porosity of argillaceous sediments and overburden pressure, i.e., depth of burial. Hedberg (1936) distinguished in these four different phases of compaction:

I. Mechanical adjustment of the mineral components and decrease in porosity from 90 to 75 per cent. This occurs with a sediment cover of 0 to 0.1 m. Free water is expelled.

II. As porosity is reduced from 75 to 35 per cent the water content is strongly reduced. This continues as a sequence of Phase I to a depth of 200-300 m. It ends when the clay minerals are

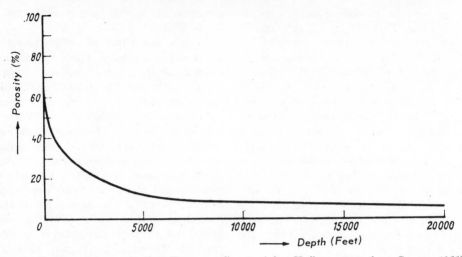

FIG. 5.—Depth-porosity curve of argillaceous sediments (after Hedberg, 1936; from Gussow, 1955).

directly in contact, and as a result only a relatively small amount of free water remains.

III. With mechanical deformation of the mineral components porosity decreases from 35 to 10 per cent. The sediment sinks during this phase from 300 to 2,000 m. Fluids are further expelled from the decreasing pore spaces.

IV. Recrystallization processes take place within the rock. Decrease in porosity below 10 per cent is accomplished very slowly, so that at 3,000 m. depth figures of 8 per cent are still measured. Only absorbed water remains.

Obviously the conclusions of Hedberg, based on the study of shale core samples from Venezuelan oil wells, can be applied to other regions only after further verification. Recent studies have confirmed the general applicability of the basic factors of his results (Engelhardt and Tunn, 1954; Engelhardt, 1960; and Engelhardt and Gaida, 1963).

We shall attempt to determine what takes place during the progressive compaction of an oil source rock. The "oil nests" previously mentioned are first influenced by compaction processes when the clay minerals come into direct contact and mechanical deformation begins. At earlier stages probably no noteworthy quantities of oil-like components are in motion, with the exception of those constituents soluble in the pore waters (e.g., certain organic acids and phenols) and from the beginning in direct contact with them. It seems highly unlikely that primary migration could begin at a much shallower depth. Aside from the question of whether enough oil had been formed at such a shallow depth, it still has to be explained how oily matter can be driven out of the insoluble kerogen in which it has been generated. We would expect that the oily matter, because of its chemical structure and its partly polar character, will be absorbed in the sedimentary organic material and also adsorbed to clay minerals. Therefore it is thought that a considerable input of energy is necessary to squeeze oily matter out of the sedimentary organic material and to break bonds (e.g., functional groups, hydrogen bridges) in order to free it from clay minerals. However, if there are still porosities of about 35-40 per cent and temperatures of 30°C. or even lower, as are prevalent in sediments at 500 m. and less, it is not evident how primary migration can occur on a larger scale.

We must therefore agree with Gussow (1955) that primary migration, from the standpoint of the compaction phenomena described, can begin at a depth of about 500 m., but only if enough hydrocarbons capable of migration are present. The end of primary migration is reached when the depth-porosity curve (Fig. 5) flattens out and the porosity reaches a range from 5 per cent to 0. It is difficult to fix a lower depth limit, as different sedimentary basins have different conditions, but it must be in the range from 3,000 to 6,000 m. Migration of gas is certainly governed by different conditions.

Weller (1959) estimated the temperatures in the depth range in which the porosity of argillaceous sediments through compaction has reached the minimum as about 200°C. It is certainly no coincidence that, on the basis of porphyrin occurrence, temperatures of a maximum of 200°C. for the formation of petroleum are likewise indicated.

The question next arises, how much time, geologically, does oil movement take in the source rock, and specifically how is it effected? The cause of primary migration is compaction, which at least continues as long as the respective sedimentary basin continues to deepen and the water in the compressed argillaceous rocks is able to escape. During this period of time, oil can be expelled from the "oil nests" into the water-filled pores when the capillary pressure and the specific absorption of the insoluble organic material are overcome. During the sinking of the basin the pore diameters also continually decrease, and in the course of time, with increasing depth and temperature, new hydrocarbons are formed from kerogen. Consequently, the space in which oil generation occurs becomes continually smaller and filled to capacity. Insofar as it can now be determined, we may assume that the capacity to generate oil capable of movement from kerogen continues in general longer than migration channels are available.

Primary migration is thus closely linked with basin formation and is a process extending over spans of geologic time.

As to the kind and manner of oil migration in pore spaces of the source rock, opinions are divergent. Wassojewitsch (1960) envisages a type of gas differentiation through which the liquid hydrocarbons are dissolved in gas before movement can begin. The minimum depth for effective primary migration is believed by him to be 1,200 to 2,600 m. Sokolov *et al.* (1964) regard the two following processes as possible.

1. Liquid hydrocarbons are dissolved in gas and are transported in the gaseous phase to zones of lower pressure.

2. Gaseous and liquid hydrocarbons and other oil components are dissolved in water and flow in this form with the water.

Solution of oil components in formation water in large quantities is explicable only by the theory of Baker (1960, 1062), according to which salts of organic acids or other components can form micelles in formation water. In their turn the micelles themselves occur as colloidal

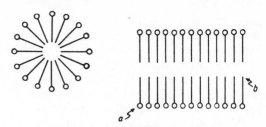

Fig. 6.—Types of micelles (after Baker, 1962). a— hydrophilic end. b—hydrophobic end.

particles in the water (Fig. 6). In or on the micelles the hydrocarbons can be enriched, and in this way their low solubility in formation water can be largely overcome. The hydrocarbons thus dissolved can now be set in motion toward the porous reservoir rocks by the expulsion of water as a result of compaction. The problem of freeing the dissolved oil remains to be explained. Perhaps this is possible through the increase in salt concentration or changes in pH of pore solutions from source to reservoir rock (Weller, 1959; Engelhardt and Gaida, 1963). Neumann (1964) distilled several crude oils and observed that the surface tension of the distillate decreased and the dielectric constants increased with increasing boiling point. This means that the polar surface-active molecules, which are necessary for the formation of micelles, are enriched in the high molecular-weight fraction. He found the surface-active components to be phenols. In his experiments the interfacial tension between oils and a buffered solution was strongly pH-dependent, probably because there is a pH-controlled equilibrium between phenols and phenolates. This suggests that pH differences in pore solutions can be very important for the stability of micelles and therefore for primary migration.

The distribution of hydrocarbon groups (paraffins, naphthenes, and aromatics) in the oils is related to their specific solubility in the types of micelles available for transport. Baker (1962) has fairly well confirmed this experimentally.

Hodgson *et al.* (1960) show, as already mentioned, another possibility of separation of dissolved hydrocarbons. On the basis of their porphyrin investigations they regard it as probable that decarboxylation also takes place in reservoir rocks. Through this, the micelle-forming components lose their hydrophilic groups and the micelles decompose. The hydrophobic residue of the molecules and the hydrocarbons transported in the micelles must then be separated as oil. Baker (1962) reports that, for example, the proportion of the higher n-paraffins in petroleums, which continually declines with increasing molecular weight, is exactly related to the increase in specific solubility through micelle growth.

The distribution curve of the n-paraffins in the related oil source rock can still, according to this theory, be entirely different from that in petroleum; for example, it may show a strong dominance of n-paraffins with odd C-numbers. As long as the supply of n-paraffins exceeds the absorption capacity of the formation waters through micelle building, the distribution curve in the corresponding oil decreases with increasing carbon numbers. Therefore, contrary to the opinion of Bray and Evans (1961), it may be possible for a sediment, extracts from which show an odd predominance, to be a true oil source rock.

The extraction experiments of Welte (1966) on bituminous rocks should be mentioned here. Two rock samples of Tertiary age were extracted, partly in the conventional manner with an organic solvent and partly stirred up over several hours with oil-field water at 60°C. The oil-field water was previously cleaned of traces of hydrocarbons by ether extraction for 50 hours. The normal extract of the samples showed a clear preference in the higher n-paraffins for odd C-numbers. From the oil-field water, on the other hand, after the hot-water treatment, n-paraffins with an approximately equal C-number distribution curve were isolated. Whether this effect is caused by the solubility relation in the sense of Baker, or by accessibility to different hydrocarbons in the sediments by the two extraction methods, can not yet be explained satisfactorily.

According to Baker, a similarity between the proportion of certain hydrocarbon groups in oil source rocks and in oil can be expected only if the supply in the sediment is less than the quantity which the formation water can remove by micelles. The distribution curve of oils with the maximum at the end of the n-paraffin series (Fig. 2, Uinta basin oil) can be explained in this way, because during the early stage in the history of a potential oil source rock the alteration of the organic materials has not yet proceeded so far. In this case, there is a greater supply of those molecules capable of micelle building, i.e., those whose

functional groups still are not split up. On the other hand, the supply of short-chain n-paraffins is relatively small, as the alteration through thermally induced reactions has hardly begun. The afore-mentioned oils might owe their origin to this initial stage of petroleum genesis and migration.

Besides the afore-described possibilities there is still the "globule migration" theory of Gussow (1954) and Hobson (1954), among others, to be considered.

According to Baker (1962), the size of the micelles ranges from 0.5 to 600 microns. Oil globules and gas bubbles are certainly of larger size. Though it is very difficult to give numerical values, it still is possible to give a maximum size for each particle of organic material from which an oil globule was formed. In potential oil source rock the size is in the range from 1 to 1,000 microns. There are no figures available about the size of the pores in the rock. A rough estimate, based on the size of the clay minerals, gives a value for buried formations of 0.1-10 microns.

These data naturally indicate only an order of magnitude, but we can conclude from them that micelle migration is more easily possible than globule migration, the geometry and wettability characteristics of the pores being of fundamental importance.

Considering the alteration which the oil source rock undergoes in the course of its development, and the migration mechanisms which have been summarized, the question arises whether the type and amount of migration are the same from the beginning to the end.

The three possibilities described, solution of oil in gas (Sokolov, Wassojewitsch), solution of gas and oil in water (Baker, Sokolov), and globule migration (Gussow, Hobson), are tied to different, variable processes. The solution of oil in gas is related chiefly to increase in pressure and temperature, the solution of oil in water to micelle-forming components, and the globule migration to the geometry and wettability of the pores. Temperature and pressure increase with increasing overburden thickness. As oil genesis progresses, relatively more low molecular-weight hydrocarbons are formed than at the beginning. Not only an increase in temperature and pressure, but also the presence of low molecular-weight components favor solution in the gaseous phase. This type of primary migration would then play a greater role with increasing sinking of the basin.

Those polar molecules with a hydrophobic and a hydrophilic end tend to form micelles. With advancing oil genesis, functional groups are detached in increasing quantities. The tendency toward micelle building must then diminish as the source rock is buried to greater depths.

The principal drawback to acceptance of globule migration is the restricted and highly variable pore diameters. As it is linked to increasing compaction, one would expect a decrease in this type of migration with increasing depth. On the other hand, compaction is the motivating force which presses the oil out of "oil nests." The question is therefore raised whether these two effects are not in opposition, and if globule migration in source rocks is of much significance. For micelle migration there are indirect evidences, such as the distribution of hydrocarbons in crude oils, the presence of surface-active components, and the experiments of Baker. The solution of oil in a gaseous phase is, on the basis of temperature and pressure relations at great depth, very probable, and this is supported by the natural occurrence of condensates. The same supporting evidence is lacking for globule migration. Besides, the fact that hardly any visible record of migration in porous carrier rocks has been found is not in favor of this kind of migration.

For age determinations and stratigraphic correlation work, the content of pollen, spores, *etc.* in crude oils has often been investigated (Tomor, 1964). It is said that these microfossils have been transported together with the migrating oil out of the corresponding source rock. However, it is highly unlikely that such large microfossils, with diameters of 100-500 microns, could ever find their way through a shale layer of even less than 10 cm. thickness, where the diameter of the largest pores probably will never reach 100 microns. Furthermore, it is known (Engelhardt and Tunn, 1954) that for migrating oil the whole pore volume of a sedimentary rock is not available because the polar water molecules interact with mineral components to form highly viscous water films on the pore walls which do not participate in migration.

The same probably holds true for the interaction between oil and mineral matter and the sta-

tionary organic matter along the migration path. It is to be expected that especially at these interfaces the strongly polar and mostly high molecular-weight hetero-compounds will be concentrated and adsorbed. Thus the oil will be depleted in polar molecules while moving through the narrow capillaries. This effect would be stronger with a longer migration path.

The type and quantity of migration should differ according to the development stage of the oil source rock and consequently change in the course of time. From this a certain selectivity of migration results.

2. SECONDARY MIGRATION

Secondary migration is a prerequisite of large oil accumulations. It begins where a potential oil source rock is in contact with a porous carrier or reservoir rock. Very thick and widespread complexes of potential oil source rocks can at first expel their oil and water content only from their marginal parts, unless fractures provide a channel for movement. The inner parts are generally only later drained of fluids, *i.e.,* with a certain delay or not at all. The ideal relationship for secondary migration then occurs when a large contact surface exists between the reservoir and source rocks. A strong interfingering of two facies, for instance in deltas, is an ideal model for this.

The transition from primary to secondary migration is linked to reduction of the surface of oil particles by coagulation of the finely dispersed or dissolved oil. During secondary migration the oil probably exists in an emulsion-like condition before it is united into larger accumulations. The numerous wells which encounter only oil emulsions in reservoirs testify to this.

For the formation of oil accumulations, the lithologic character and texture of the rock sequence below and above the source rock and the structural setting are of basic significance. It is beyond the scope of this paper to discuss these many-sided possibilities.

Strong alteration of crude oils during secondary migration is not to be expected, as the process requires a much shorter time than primary migration, and the oil particles increase significantly in size as a result of coagulation. Later alteration of oil in reservoir rocks can occur, but should generally, compared with the processes in the source rock, be of limited significance. Extreme conditions, such as the erosion of a filled reservoir or direct connection with the atmosphere, are not considered here.

There are occasional reports of finding hydrocarbon anomalies over oil reservoirs by geochemical prospecting. Leakage of at least the lighter hydrocarbon components must therefore have occurred. Philipp et al. (1964) have given examples from the Gifhorn trough of northwestern Germany, showing that there is a relation between the porosity of the cap rocks and the gas in the reservoir. They explain this phenomenon through a reduction of the outward diffusion of natural gas with reduction in porosity of the overlying shale.

Changes also can take place as a result of increase in depth of burial, besides alteration of the reservoir contents through selective loss. The oils are altered as a result of being depressed to a higher temperature level. In particular the heavier oils, with their asphaltic and other high molecular-weight components, are sensitive to increases in temperature. Just as with kerogen, we must expect the loss of functional groups, and splitting into smaller components, because the smaller molecules are thermodynamically more stable. We may regard this as an alteration process leading to the most thermodynamically stable products. The changes come about rapidly or slowly depending on the depth-related temperature. It is quite possible that the decrease of the C^{13}/C^{12} ratio with increasing age observed in crude oils by Silverman and Epstein (1958) likewise is the consequence of such an alteration process.

Through these processes the gas content is also increased. The formation of gas can not, however, be very important, as the limits of variation of the C^{13}/C^{12} ratio for oils are only about 10 per mil, and a high production of isotopically much lighter gaseous fractions would lead to a considerable enrichment of the C^{13} isotopes in the residual petroleum, and thus to a wider limit of variation within the oils (Silverman and Epstein, 1958).

Evidence for chemical changes of crude oils within reservoirs is also demonstrated by Prinzler and Pape (1964), especially with respect to the sulfur-containing hetero-compounds. Compiling available data on sulfur content in crude oils from all over the world, they show that there is a strong increase in sulfur (from 0.2-3.8 per cent by weight) from Tertiary to Cretaceous oils. After the Cretaceous maximum there is, with further increasing age, a continuous decrease in sulfur content of oils to the Cambrian. Prinzler and Pape explain this in the following way. The low sulfur content of the young oils corresponds with the low initial sulfur in the primary organic matter (sulfur in planktonic organisms amounts to as much as 0.5-1.5 per cent). With increasing age microbial reduction of $SO_4^=$ from the pore water produces sulfur and H_2S, which react with hydrocarbons and hetero-compounds abiotically. After the sulfate content of the pore water is exhausted, no further sulfur-containing oil components can be formed. But the aging process of the oil, which is basically a process of disproportionation, will also degrade sulfur-containing molecules by hydrogen transfer and generate mercaptans and H_2S. The H_2S can partly escape by diffusion out of the reservoir and thus with increasing age the total amount of sulfur in the oil is reduced. Reactions of the above type are highly probable, as it is a well-known fact that oils from limestone reservoirs (limestone has always a certain amount of $CaSO_4$) usually have a higher sulfur content.

RELATIONS BETWEEN PETROLEUM AND PETROLEUM SOURCE ROCKS

In the preceding sections of this paper, the most important data from the literature bearing on the relation between petroleum and petroleum source rocks have been considered. In the following sections, in order to develop a coherent picture, a synthesis is made of all these observations which have resulted from research on petroleum and sediments. For a better understanding of the text, Table I displays a very simplified summary of the historical course of oil origin and primary migration.

1. OCCURRENCE IN FRESH SEDIMENTS

The history of a potential oil source rock begins with the deposition of a compressible fine-grained sediment containing organic material. Deposition must take place in an aquatic environment in which the organic content can be preserved. This generally consists of aquatic (autochthonous) and terrestrial (allochthonous) components, whose proportions vary according to environment, facies, and type of sedimentation.

After death, the organic material undergoes a series of chemical transformations which so far are not fully understood. An important role is

TABLE I. SIMPLIFIED SCHEME OF SEQUENCE OF OIL GENESIS AND
PRIMARY MIGRATION IN COURSE OF TIME

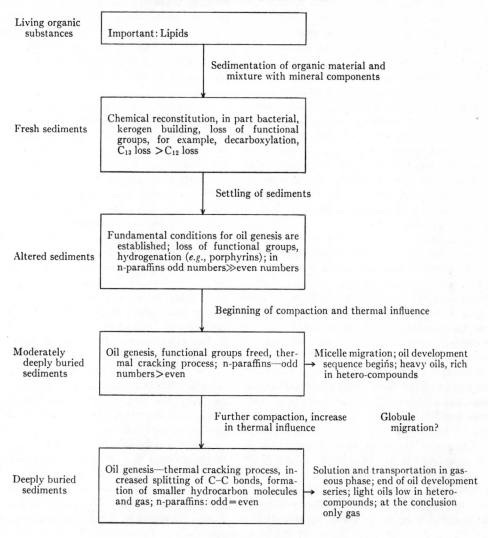

| Living organic substances | Important: Lipids |

Sedimentation of organic material and
mixture with mineral components

| Fresh sediments | Chemical reconstitution, in part bacterial, kerogen building, loss of functional groups, for example, decarboxylation, C_{13} loss $>C_{12}$ loss |

Settling of sediments

| Altered sediments | Fundamental conditions for oil genesis are established; loss of functional groups, hydrogenation (e.g., porphyrins); in n-paraffins odd numbers$\gg$even numbers |

Beginning of compaction and thermal influence

| Moderately deeply buried sediments | Oil genesis, functional groups freed, thermal cracking process; n-paraffins—odd numbers$>$even | $\rightarrow$ | Micelle migration; oil development sequence begins; heavy oils, rich in hetero-compounds |

Further compaction, increase   Globule
in thermal influence   migration?

| Deeply buried sediments | Oil genesis—thermal cracking process, increased splitting of C–C bonds, formation of smaller hydrocarbon molecules and gas; n-paraffins: odd = even | $\rightarrow$ | Solution and transportation in gaseous phase; end of oil development series; light oils low in hetero-compounds; at the conclusion only gas |

End of compaction; temperature
around 200°C.

played in this by the many unsaturated compounds of the lipid fraction, which, especially in primitive organisms, are extraordinarily numerous (Hilditch, 1956). Insoluble kerogen, which already is encountered in newly formed sediments, is the principal product of the reconstruction. The building blocks of the kerogen are the prefabricated, partly altered, molecular blocks and fragments of organisms. The cement consists of highly unsaturated compounds and other reactive compounds (e.g., aldehydes). Part of the organic matter is adsorbed to the clay minerals. The alteration of the organic material is most clearly shown by the strong decrease in soluble constituents from the living organisms compared with the organogenic detritus in the fresh sediment.

The hydrocarbons extracted during this phase of sediment formation are distinguished from those of petroleum by lack of lower molecular-weight components. As a result, the maximum of the n-paraffin distribution curve is not, as in most crude oils, in the low molecular starting phase, but at the end of the n-paraffin series (Fig. 2, Uinta basin oil). Besides there is a marked tendency in the higher n-paraffins of Recent sediments to show a distinct predominance of odd-numbered molecules, which is in general not to be seen in petroleum.

True petroleum porphyrins, even though only in small quantities, are found in addition to the products of chlorophyll decay, *i.e.,* the precursors of true petroleum porphyrins. The transformation of chlorophyll through different intermediate stages into a metal-bearing porphyrin complex can take place only in a reducing environment, that is, under exclusion of oxygen. The process requires decarboxylation and hydrogenation and should certainly be regarded as the first phase of incipient petroleum genesis.

Concurrently with these stages, which take place during deposition and shortly thereafter in the uppermost sedimentary layers, it appears that the organic carbon content becomes lighter through a preferred escape of the heavy C^{13} isotopes. The organic substance of fossil sediments and the petroleum formed from it are therefore seemingly more "terrestrial" than their source materials. As a result, the so-called "terrestrial" isotope ratios in most crude oils are not necessarily proof of a terrestrial origin of the primary organic detritus, although part of it can be of continental origin, depending on the facies and geographic location.

The extent to which bacteria participate in these processes in Recent sediments is largely unknown. In the upper sedimentary layers an extraordinarily strong bacterial activity certainly prevails, and in many fossil sediments that may be regarded as oil source rocks, evidence of bacterial feeding can be found by microscopic study. Still it appears that bacterial activity is limited to a few meters of burial depth. However, the remains of the dead bacteria certainly can be regarded as contributors to the oil-genesis process.

In summary it may be said that in the early stage of diagenesis no true formation of petroleum takes place, but that important steps are taken toward it during this phase.

2. OIL GENESIS AND MIGRATION IN MODERATELY BURIED SEDIMENTS

Further developments apparently are entirely under the influence of additional sinking of the sedimentary trough, with accumulation of new sedimentary material over the potential source rock; as depth of burial increases there is on the one hand an increase in overburden pressure leading to compaction of the sediments and thus to migration, and on the other an increase in temperature, causing reforming and disintegration of the organic material, resulting in petroleum formation.

Though the two processes of oil genesis and primary migration, proceeding according to the temperature gradient and compaction rate related to depth of burial, are basically unrelated, their synchronous course is of great significance in the development of petroleum deposits.

With increasing depth new hydrocarbons are formed in the fossilized sediments. Under the influence of increasing temperature, thermal disintegration of the dispersed organic material takes place in the sediments. The most loosely bonded groups, such as carboxyl and hydroxyl, are the first to be freed, at a low temperature level, from the kerogen and soluble substances. As the temperature increases and with the passage of time, still more bonds which require a greater expenditure of energy can be broken, for example, the splitting-off of carbon chains.

The geothermal gradient thus exerts a decisive influence. Whether sufficient oil that is capable of migrating has been formed at the beginning of the primary migration depends chiefly on this. At present we have still not reached the point where this stage can be identified. A possible approach to the problem is a systematic study of the extract/carbon ratio with increasing depth. A very rapid increase in this ratio could be indicative of the presence of oil capable of migrating. Bitterli (1963) suggested in this connection that an extract/organic C-ratio (10^3) below 5 could be attributed to coaly matter, whereas a ratio above 100 would indicate migrated oil.

Let us assume with Abelson (1964) that a temperature of 50°C. for a period of one million years is sufficient to form enough hydrocarbon capable of migration (*e.g.,* through decarboxylation) in a potential source rock. If the geothermal gradient in the corresponding sedimentary trough is 4°C. per 100 m.,[9] and the mean surface temperature is 11°C., this temperature would be

---

[9] W. Philipp, personal communication. The figure given is about that for the Gifhorn trough of northwest Germany.

reached at a depth of 1,000 m. If the average rate of sedimentation is 0.1 mm. per year, deposition of an additional 100 m. of sedimentary cover in one million years would result in an increase in the temperature to 54°C. Let us also consider a relationship such as that described in the Variscan border depression (Teichmüller, 1962), where there is a geothermal gradient of 7°C. per 100 m., and at a depth of not quite 600 m. a temperature of 50°C. is already reached. With a rate of sedimentation of 0.2 mm. per year, a thickness of 200 m. can be attained in one million years, and the temperature increases to 64°C.

Primary migration can begin at about the point when, through increasing compaction of the potential source rock, mechanical deformation takes place in the clay minerals. This can happen at a depth of 500-600 m.

In the first of the examples which have been given, true primary migration can hardly begin at this depth because of insufficient oil having been formed. On the other hand, in the second case, corresponding with the higher temperature, a larger quantity of oil would have formed, and under the influence of compaction migration could begin. In the latter case, the beginning of oil genesis and of primary migration are concurrent, whereas in the former the oil genesis can take place only later, *i.e.*, subsequent to a significant amount of migration.

The oil which begins to move at this stage, under relatively thin sedimentary cover, will correspond in its composition with the source material which is in the first stages of oil genesis. Thermal disintegration of the kerogen and of the soluble organic material has just begun, and because of the low temperature level carbon chains have split only in limited quantity. Consequently, only a very few low molecular-weight hydrocarbons have been formed, and the number of the branched iso-compounds is still relatively large. The decarboxylation and splitting-off of other functional groups began a considerable time before, but is by no means concluded. As a result, the supply of oxygenated molecules at the beginning of oil genesis is still relatively high. The first migrating oil is, in the absence of low molecular-weight constituents, therefore relatively heavy and extremely rich in iso-compounds and oxygen-bearing components (*e.g.*, naphthenic acids and fatty acids).

The n-paraffin distribution curve shows that the maximum lies in the higher molecular-weight range (Fig. 2, Uinta basin oil), because the new formation of short-chained n-paraffins, especially of C_4 to C_{11}, was still hardly possible, while long-chained molecules already were being supplied from animal and plant substances or could have been produced through decarboxylation of unbranched fatty acids. The immature oils described from Nigeria (Imo River) and the United States (State Line and Uinta basin) are of this type. A peculiarity of the Uinta basin oil is the large proportion of higher n-paraffins with odd C-atom numbers. This peculiarity is understandable if we consider that the oil was derived from a source rock in the initial stages of oil genesis, before the new formation of n-paraffins was of significance, and where the original biologically determined pattern of distribution of hydrocarbons or their immediate precursors was still clearly expressed.

In the search for a mechanism for primary migration of this earliest oil, the most understandable is the colloidally dispersed micelle movement in the sense of Baker. Polar molecules suitable for micelle building, with hydrophobic and hydrophilic terminal groups, are still present in large amounts at the beginning of oil genesis, whereas with the advance of thermal disintegration they become relatively rare. The newly formed hydrocarbons are concentrated in or on the micelles formed by these molecules and thus dissolved in formation water. The colloidally dispersed petroleum components then can be removed from the source rock through normal water-draining processes.

As long as the hydrocarbon-solubility capacity of the formation water caused by the micelles is larger than the supply, the distribution spectrum of the different hydrocarbons in the aqueous transport medium corresponds with that in the source substance. A still present imbalance in favor of the higher n-paraffins with odd-numbered C-atoms therefore can be carried over into the crude oil, as is apparently true in the case of the Uinta basin oil. However, as the supply of hydrocarbons exceeds the solubility capacity, the percentage of the transported components is related to their specific solubility in the micelles. The maximum of the n-paraffin curve in the lower molecular zone of the Darius oil (Fig. 2) is partly the result of this. The solubility of the n-paraffins declines in proportion to increasing molecular weight.

3. Basin Deepening and Oil Development Sequence

With a further deepening of the sedimentary trough, the source rock, especially in the center of the basin, is subjected to constantly increasing temperature and pressure. Under the influence of the temperature, the thermal disintegration of the source substances and accompanying oil genesis continue. With increase in pressure, porosity decreases and the motivating power for migration is maintained.

Depending on varying conditions, the oils which were produced during and after this new phase show a different chemical constitution from those formed earlier. The longer the source rock remains under cover of overburden and the more the temperature increases, the smaller becomes the ratio of iso- to n-paraffins, and in general the simpler becomes the molecular structure of the newly formed hydrocarbons. The content of hetero-compounds decreases and the average molecular weight of the different groups of materials becomes less. As a consequence of this process the insoluble organic residue remaining in the source rock has a continually smaller H/C ratio.

It is therefore evident that the source rock undergoes a methodical development, resulting inevitably in the formation of an evolutionary series of different crude oils. At the beginning of the series are the heavy, oxygen-rich oils with a small amount of low-molecular-weight n-paraffins, whereas at the end we find the light oils rich in low-molecular-weight n-paraffins and with a small amount of hetero-compounds. Between these lie the transitional stages.

An oil source rock can therefore produce different oils at different stages of basin development if the conditions favoring primary migration persist. The composition of this petroleum compared with the composition of the original source substance is a measure of the maturity stages conditioned by temperature and to a smaller degree by time.

4. Role of Migration

We have already dealt briefly with the effect of micelle migration on changes in the composition of primary oil. In the later stages of primary migration the solubility of the oil in a gaseous phase would be significant, and micelle migration would be of diminishing importance. The solubility in a gaseous phase depends mainly on the boiling point of the dissolved components. Heavy molecules have less chance of going into solution and being transported. This type of migration, too, is selective because of variable solubilities. It is questionable whether globule migration plays an important role. Whether or not this is true, a certain selectivity in the solubility factor must be considered here also. The different types of migration therefore permit selectively the preferred movement of certain molecules which contribute to the formation of petroleum accumulations.

Nevertheless, the characteristics of the oil in the source rock are clearly recognizable also in the crude oil in the reservoir; this resemblance is probably in no small degree a result of the fact that the dominant mode of migration is determined by the conditions in the source rock.

Primary migration becomes secondary migration when the hydrocarbons move from the fine-grained source rock into the coarse-grained reservoir rock. Usually this change is connected with an increase in salt concentration of pore waters, especially if the potential source rock is rich in base-exchanging clays (Weller, 1959; Engelhardt, 1961; Engelhardt and Gaida, 1963). It is additional to a gradual decrease in pressure and temperature, as migration generally takes place from lower to higher levels. In micelle migration the disintegration of the micelles and the coagulation of oil out of a colloidally dispersed condition can be caused by an increase in electrolyte content of the surrounding waters as well as by a progressive loss of functional groups in the micelle-forming blocks. Hydrocarbons dissolved in a gaseous phase tend to form droplets or bubbles as a result of progressive decrease in pressure and temperature. Gas bubbles and oil drops can move freely in the larger pore spaces of the reservoir rock, and upon contact with one another they become fused. During the process of primary and secondary migration to the beginning of accumulation in a reservoir, individual hydrocarbon bodies progressively increase in size. At the beginning they are colloidally dispersed; later they are changed to an emulsion-like form, and finally they develop into larger, coherent oil bodies and pools.

5. Geological Relationships Connected with Oil Genesis and Migration

We have seen that from the beginning of oil genesis and migration, until the end of both, several tens of millions of years may pass.[10] Conditioned by the tectonic events in the related sedimentary trough during this time interval, different structures may be alternately opened and closed to migrating fluids. The various kinds of oils given up during the development of the source rock may therefore be trapped in the individual reservoirs during different time periods. Different phases of the continually developing series thus can be separated into particular accumulations. It also can be concluded with certainty that by no means all of the hydrocarbons given up by the source rock came to rest in oil accumulations.

With the knowledge of the chemical composition of petroleums and the stratigraphy and tectonic evolution of the basins, even though gaps exist in our information, the oils can be classified into the natural series expressing their developmental history. Somewhat different relationships must be allowed for on the basin margins, because the evolution of the source rock lags behind compared with the conditions in the basin center where it is more deeply buried. Additional proof of this is given by the decrease in the specific gravity of oils toward the deepest parts of the basins.

Until now in this study we have implied that there is only one source rock in a basin. Actually there may be several such rock complexes. This complicates any evaluation because each separate source rock gives off an oil corresponding with its state of evolution.

Some important exceptions to these statements must be made with regard to the potential source rocks that are not, or are only slightly, compressible. The most important of these are the calcareous sediments (Weller, 1959). In these, primary migration may not necessarily occur. With depression to greater depths, oil genesis pursues its usual course, but the hydrocarbons remain in the already-consolidated rock after they are formed. They can escape only when oil genesis enters the final stage and many low molecular-weight liquid- and gas-forming hydrocarbons are produced. Migration in the gaseous phase should be possible

without compaction. As no migration has taken place previously, more material is available for gas formation than in compressible source rock. Calcareous sediments lithified at an earlier time, containing organic material, are thus destined to be suppliers of gas. It must be observed here that in the course of its development history coal also is capable of producing gas because its formation is attributable chiefly to thermal action.

Compaction may be delayed with argillaceous as well as with calcareous sediments. Engelhardt and Gaida (1963), in performing compaction experiments with clays, observed that compaction proceeds more rapidly if the pore solution shows a high content of electrolyte. Therefore, it may be possible that migration in certain brackish or marine-brackish, electrolyte-poor sediments is delayed. The oil finally migrating thus would be pushed toward the end of the chemical evolutionary series.

We have postulated as source rocks all fine-grained sediments with a certain quantity of disseminated organic material. According to our present knowledge, there must be both good and poor source rocks. For evaluation of source potential we must analyze the absolute content in organic carbon, establish the ratio of extractable carbon to total carbon, and determine the composition of the extract. At the present state of our knowledge it may be assumed that the minimum amount of organic carbon should be around 0.5 per cent and the ratio of extractable to total carbon (10^3) should lie between 20 and 100.

In conclusion, it must be pointed out that the step from the recognition of the existence of a potential source rock to that of the identification of an actual or specific source rock is beyond our currently available techniques. We can only set forth the opposite rule that: if oil is found, there must also be a related source rock.

CONCLUSIONS

The following conclusions can be drawn from the preceding discussion.

I. Oil genesis is primarily the result of a thermal disintegration of finely disseminated organic material in the source rock. The regional geothermal depth gradient is therefore of great significance in oil genesis.

II. The mechanism of primary migration changes with increasing depth of burial. At the beginning, at lesser depths, a type of micelle mi-

---

[10] This is only to give an order of magnitude. With young Tertiary oils, as in the Gulf Coast and in the Niger delta of western Africa, the oil generation and migration processes are far from being completed. This would be even more true of the Pleistocene oils of southern Louisiana.

gration is probably dominant, and at greater depths, solution in a gaseous phase.

III. Oil genesis and migration are temporally closely linked processes which are related to the formation of the basin. In the course of basin history the petroleum source rock can give up oils of different types. These oils express in their chemical composition the condition at that time of the source rock and form a natural development series. At the beginning of the series are heavy oils, and at the end are the light oils.

IV. Later thermal alteration of the petroleum is possible. The chief result of this is the conversion of heavy to light oil, but this is not reversible.

It frequently has been deduced that petroleum, as it rises to near-surface zones, is oxidized by circulating oxygen in aqueous solution, and as a result becomes heavier. This conclusion is not justified, because in the first place the distribution of hydrocarbons is different in heavy oils, and in the second place these heavier oils contain fairly large amounts of hetero-compounds containing sulfur and nitrogen besides oxygen-bearing hetero-compounds. Oxidation alone can not account for both these characteristics.

V. Potential petroleum source rocks from which primary migration has been prevented or retarded by lack of compactability have a good potential for supplying gas. This applies chiefly to limestones. It is especially significant in the search for gas and condensate deposits.

VI. A petroleum and the extract of the related source rock do not have to be similar in their chemical compositions. The petroleum corresponds with a certain phase in the development of the source rock, whereas the extract reflects its present condition but certainly not that existing at the time when migration occurred. Besides, primary migration is a selective process in which there is a separation of the available components capable of migration, unlike the products of extraction. On this basis it appears that for the correlation of a particular petroleum with its source rock the suitable indices to be selected are only such substances or groups of substances which on the one hand are the most specific ones possible and on the other are most surely able to maintain their character during oil genesis and migration. The ratio of Ni-V porphyrins, or that of the different isoprenoid components, should meet these specifications.

REFERENCES

Abelson, P. H., 1964, Organic geochemistry and the formation of petroleum: 6th World Petroleum Cong. Proc., 1963, Frankfurt/Main, Sec. 1, p. 397–407, Hamburg.

—— and Hoering, T. C, 1961, Carbon isotope fractionation in formation of amino acids by photosynthetic organisms: U.S. Natl. Acad. Sci. Proc., v. 47, p. 623–632.

—— and Parker, P. L., 1962, Fatty acids in sedimentary rocks: Carnegie Inst. Washington Year Book, v. 61, p. 181–184.

Athy, L. F., 1930, Density, porosity, and compaction of sedimentary rocks: Am. Assoc. Petroleum Geologists Bull., v. 14, p. 1–35.

Baker, E. G., 1960, A hypothesis concerning the accumulation of sediment hydrocarbons to form crude oil: Geochim. et Cosmochim. Acta, v. 19, p. 309–317.

—— 1962, Distribution of hydrocarbons in petroleum: Am. Assoc. Petroleum Geologists Bull., v. 46, p. 76–84.

Barton, D. C., 1934, Natural history of the Gulf Coast crude oil, in Problems of petroleum geology: Am. Assoc. Petroleum Geologists, p. 109–155.

Bendoraitis, J. G., Brown, B. L., and Hepner, L. S., 1962, Isoprenoid hydrocarbons in petroleum: Anal. Chemistry, v. 34, p. 49–53.

—— —— —— 1964, Isolation and identification of isoprenoids in petroleum: 6th World Petroleum Cong. Proc., 1963, Frankfurt/Main, Sec. V, p. 13–29, Hamburg.

Bitterli, P., 1963, Aspects of the genesis of bituminous rock sequences: Geol. en Mijnbouw, v. 42, p. 183–201.

Blumer, M., 1950, Porphyrinfarbstoffe und Porphyrin-Metallkomplexe in schweizerischen Bitumina: Helv. chim. Acta, Bd. 33, p. 1627–1637.

—— and Omenn, G. S., 1961, Fossil porphyrins: uncomplexed chlorins in a Triassic sediment: Geochim. et Cosmochim. Acta, v. 25, p. 81–90.

—— Mullin, M. M., and Thomas, D. W., 1963, Pristane in Zooplankton: Science, v. 140, p. 974.

Bray, E. E., and Evans, E. D., 1961, Distribution of n-paraffins as a clue to recognition of source beds: Geochim. et Cosmochim. Acta, v. 22, p. 2–15.

Breger, I. A., and Brown, A., 1962, Kerogen in the Chattanooga shale: Science, v. 137, p. 221–224.

Brieskorn, C. H., and Zimmermann, K., 1965, Über das Vorkommen von Pristan in der Anisfrucht: Experientia, Bd. 21, Basel (in press).

Colombo, U., and Sironi, G., 1961, Geochemical analysis of Italian oils and asphalts: Geochim. et Cosmochim. Acta, v. 25, p. 24–51.

Cooper, J. E., 1962, Fatty acids in recent and ancient sediments and petroleum reservoir waters: Nature, v. 193, p. 744–746.

—— and Bray, E. E., 1963, A postulated role of fatty acids in petroleum formation: Am. Chemical Soc., Los Angeles meeting, 31 p.

Dean, R. A., and Whitehead, E. V., 1964, The composition of high boiling petroleum distillates and residues: 6th World Petroleum Cong. Proc., 1963, Frankfurt/Main, Sec. V, p. 261–279, Hamburg.

Dunning, H. N., and Moore, J. W., 1957, Porphyrin research and origin of petroleum: Am. Assoc. Petroleum Geologists Bull., v. 41, p. 2403–2412.

—— —— and Myers, A. T., 1954, Properties of porphyrins in petroleum: Indust. Eng. Chemistry, v. 46, p. 2000–2007.

Dunton, M. L., and Hunt, J. M., 1962, Distribution of low molecular weight hydrocarbons in Recent and ancient sediments: Am. Assoc. Petroleum Geologists Bull., v. 46, p. 2246–2258.

Eckelmann, W. R., Broecker, W. S., Whitlock, D. W., and Allsup, J. R., 1962, Implications of carbon isotopic composition of total organic carbon of some Recent sediments and ancient oils: Am. Assoc. Petroleum Geologists Bull., v. 46, p. 699–704.

Eglington, G., Scott, P. M., Belsky, T., Burlingame, A. L., and Calvin, M., 1964, Hydrocarbons of biological origin from a one-billion-year-old sediment: Science, v. 145, p. 263–264.

Emery, K. O., 1960, The sea off southern California: John Wiley and Sons, 366 p.

Engelhardt, W. von, 1960, Der Porenraum der Sediments: Springer-Verlag, Berlin-Göttingen-Heidelberg, p. 33–50.

———— 1961, Zum Chemismus der Porenlösungen der Sedimente: Bull. Geol. Inst. Univ. Upsala, v. 40, p. 189–204.

———— and Gaida, K. H., 1963, Concentration changes of pore solution during the compaction of clay sediments: Jour. Sed. Petrology, v. 33, p. 919–930.

———— and Tunn, W., 1954, Über das Strömen von Flussigkeiten durch Sandsteine: Heidelberg Beitr. zur Mineralog. und Petrog., Bd. 4, p. 12–25.

Erdman, J. G., 1961, Some chemical aspects of petroleum genesis as related to the problem of source bed recognition: Geochim. et Cosmochim. Acta, v. 22, p. 16–36.

Gussow, W. C., 1954, Differential entrapment of oil and gas; a fundamental principle: Am. Assoc. Petroleum Geologists Bull., v. 38, p. 816–853.

———— 1955, Time of migration of oil and gas: Am. Assoc. Petroleum Geologists Bull., v. 39, p. 547–574.

Hedberg, H. D., 1926, The effect of gravitational compaction of the structure of sedimentary rocks: Am. Assoc. Petroleum Geologists Bull., v. 10, p. 1035–1072.

———— 1936, Gravitational compaction of clays and shales: Am. Jour. Sci., v. 31, p. 241–287.

Hilditch, T. P., 1956, The chemical constitution of natural fats: 3d ed., John Wiley and Sons, 664 p.

Hobson, G. D., 1954, Some fundamentals of petroleum geology: Oxford Univ. Press, 139 p.

Hodgson, G. W., Hitchon, B., Elofson, R. M., Baker, B. L., and Peake, E., 1960, Petroleum pigments from recent fresh-water sediments: Geochim. et Cosmochim. Acta, v. 19, p. 272–288.

———— and Peake, E., 1961, Metal chlorine complexes in Recent sediments as initial precursors to petroleum porphyrin pigments: Nature, v. 191, p. 766–769.

———— Ushijima, N., Taguchi, K., and Shimada, I., 1963, The origin of petroleum porphyrins—pigments in some crude oils, marine sediments and plant material of Japan: Sci. Rept. Tohuku Univ., 3d ser., v. 8, p. 483–513.

Hoering, T. C., 1962, The stable isotopes of carbon in the carbonate and reduced carbon of Precambrian sediments: Carnegie Inst. Washington Year Book, v. 61, p. 190–191.

Hunt, J. M., 1962, Geochemical data on organic matter in sediments: Internatl. Sci. Oil Conf. Proc., Budapest, preprint.

———— and Jamieson, 1956, Oil and organic matter in source rocks of petroleum: Am. Assoc. Petroleum Geologists Bull., v. 40, p. 477–488.

Illing, V. C., 1938, The migration of oil, in The Science of petroleum, v. 1, p. 209–215, Oxford Univ. Press.

Jurg, J. W., and Eisma, E., 1964, Petroleum hydrocarbons: generation from fatty acid: Science, v. 144, p. 1451–1452.

Krecji-Graf, K., and Wickmann, F. E., 1960, Ein geochemisches Profil durch den Lias alpha. (Zur Frage der Entstehung des Erdöls): Geochim. et Cosmochim. Acta, v. 18, p. 259–272.

Landergren, S., 1954, On the relative abundance of stable carbon isotopes in marine sediments: Deep-Sea Research, v. 1, p. 93–120.

Levorsen, A. I., 1954, Geology of petroleum: Freeman, San Francisco, 703 p.

Martin, R. L., Winters, J. C., and Williams, J. A., 1963, Distribution of n-paraffins in crude oils and their implications to origin of petroleum: Nature, v. 199, p. 110–113.

———— ———— and ———— 1964, Composition of crude oils by gas chromatography: geological significance of hydrocarbon distributions: 6th World Petroleum Cong. Proc., 1963, Frankfurt/Main, Sec. V, p. 231–260, Hamburg.

Meinschein, W. G., 1964, Biological remnants in a Precambrian sediment: Science, v. 145, p. 262–263.

Mold, J. D., Stevens, R. K., Means, R. E., and Ruth, J. M., 1963, 2, 6, 10, 14-tetramethylpentadecane (pristane) from wool wax: Nature, v. 199, p. 283–284.

Neumann, H. J., 1964, Grenzflächenspannung und Entölung von Lagerstätten: Erdöl und Kohle, Erdgas, Petrochemie, Jg. 17, p. 346–348.

Park, R., and Dunning, H. N., 1961, Stable carbon isotope studies of crude oils and their porphyrin aggregates: Geochim. et Cosmochim. Acta, v. 22, p. 99–105.

Philipp, W., Drong, H. J., Füchtbauer, H., Haddenhorst, H.-G., and Jankowsky, W., 1964, The history of migration in the Gifhorn trough (NW-Germany): 6th World Petroleum Cong. Proc., 1963, Frankfurt/Main, Sec. I, p. 457–481, Hamburg.

Philippi, G. I., 1957, Identification of oil-source beds by chemical means: 20th Int. Geol. Cong., Mexico (1956), Sec. 3, p. 25–38.

Prinzler, H. W., and Pape, D., 1964, Zur Entstehung und zum postgenetischen Verhalten der organischen Schwefelverbindungen des Erdöls: Erdöl und Kohle, Erdgas, Petrochemie, Jg. 17, p. 539–545.

Ronov, A. B., 1958, Organic carbon in sedimentary rocks (in relation to presence of petroleum): translation in Geochemistry, no. 5, p. 510–536.

Rossini, F. D., Pfitzer, K. S., Arnett, R. L., Braun, M. R., and Pimentel, G., 1953, Selected values of physical and thermodynamic properties of hydrocarbons and related compounds: Carnegie Press, Pittsburgh, 1050 p.

Sackett, W. M., 1964, The depositional history and isotopic carbon composition of marine sediments: Marine Geology, v. 2, no. 3, p. 173–185.

———— and Thompson, R. R., 1963, Isotopic organic composition of recent continental derived clastic sediments of eastern Gulf Coast, Gulf of Mexico: Am. Assoc. Petroleum Geologists Bull., v. 47, p. 525–528.

Silverman, S. R., and Epstein, S., 1958, Carbon isotopic compositions of petroleum and other sedimentary organic materials: Am. Assoc. Petroleum Geologists Bull., v. 42, p. 998–1012.

———— 1964, Investigation of petroleum origin and evolution mechanisms by carbon isotope studies, in Isotopic and cosmic chemistry, p. 92–102, Craig.

H., Miller, S. L., and Wasserburg, G. J., editors: North Holland Pub. Co., Amsterdam.

Smith, H. M., and Rall, H. T., 1953, Relationship of hydrocarbons with six to nine carbon atoms: Indust. Eng. Chemistry, v. 45, p. 1491–1497.

Smith, P. V., Jr., 1954, Studies on the origin of petroleum-occurrence of hydrocarbons in Recent sediments: Am. Assoc. Petroleum Geologists Bull., v. 38, p. 377–404.

Sokolov, V. A., 1959, Possibilities of formation and migration of oil in young sedimentary deposits: Proc. Lvow Conf., 1957, May 8–12, Moscow (in Russian).

—— Zhuse, T. P., Vassoyevich, N. B., Antonov, P. L., Grigoriyev, G. G., and Koslov, V. P., 1964, Migration processes of gas and oil, their intensity and directionality: 6th World Petroleum Cong. Proc., 1963, Frankfurt/Main, Sec. I, p. 493–505, Hamburg.

Teichmüller, R., 1962, Die Entwicklung der sub-varischen Saumsenke nach dem derzeitigen Stand unserer Kenntnis: Fortschr. Geol. Rheinland und Westfalen, Bd. 3, p. 1237–1256.

Tomor, J., 1964, Forschungsergebnisse über die Entstehung und Altersbestimmung ungarischer Erdöle: Zeitschr. f. angew. Geologie, Bd. 10, hft. 11, p. 579–585.

Treibs, A., 1936, Chlorophyll- und Häminderivate in organischen Mineralstoffen: Angew. Chem., Bd. 49, p. 682–686.

Wassojewitsch, N. B., translation and comments by Krejci-Graf, K., 1960, Mikronaphtha und die Entstehung der Erdöls: Mitt. Geol. Gesell. Wien, Bd. 53, p. 133–176.

Weller, J. M., 1959, Compaction of sediments: Am. Assoc. Petroleum Geologists Bull., v. 43, p. 273–310.

Welte, D. H., 1966, Extraktionsexperimente an bituminösen Gesteinen: Geochim. et Cosmochim. Acta (in press).

Reprinted from:
BULLETIN OF THE AMERICAN ASSOCIATION OF PETROLEUM GEOLOGISTS
VOL. 56, NO. 5 (MAY, 1972), PP. 925-940, 9 FIGS., 1 TABLE

Vegetation and Geochemical Prospecting for Petroleum[1]

LEO HORVITZ[2]
Houston, Texas 77042

Abstract   The value of geochemical prospecting for petroleum has been challenged by Smith and Ellis, who reported that soil hydrocarbons, saturated as well as unsaturated, result from vegetation in the form of grass and roots. These investigators claimed that 1 g of this type of vegetation is sufficient to produce the same quantity of saturated hydrocarbons found in 200 g of soil, and imply that it is the source of hydrocarbon anomalies. Unfortunately, no quantitative analytical data were supplied by them and, therefore, the claim is not substantiated.

Investigations by the present writer show that less than one part per billion by weight of saturated hydrocarbons, in the range from ethane through pentane, is contributed by 1 g of grass or roots to 200 g of soil. This concentration is much lower than that normally found in the soil of background or barren areas and, therefore, is of no importance in geochemical exploration. Actually, few significant anomalies contain ethane through pentane values lower than 25 parts per billion by weight.

Small amounts of saturated hydrocarbons may be generated by heating grass or roots in a partial vacuum, either alone or in the presence of phosphoric acid, but in the presence of concentrated nitric acid the production of saturated hydrocarbons is so inhibited that less than one part per billion of ethane through pentane is contributed to 100 g of soil by as much as 5 g of this vegetation. Therefore, the presence of any of this group of saturated hydrocarbons in grass or roots, in the natural state, is doubtful.

As additional evidence to condemn geochemical exploration, Smith and Ellis cited their unsuccessful attempts to find anomalous hydrocarbon values over oil fields. These failures are explained easily by the fact that their soil gas extraction technique is incapable of removing the major part of the saturated hydrocarbons that are adsorbed on the soil, making it difficult to discern an anomaly. Moreover, the near-surface anomalies, originally produced by the fields they sampled, may have become weakened or even have disappeared because of changes in reservoir conditions that take place as petroleum is being produced.

Introduction

Geochemical exploration is based on the premise that the lighter hydrocarbons contained in oil and gas deposits migrate vertically toward the surface of the earth where they manifest themselves in the form of recognizable distribution patterns in the soil air (Laubmeyer, 1933; Sokolov, 1935) or in the soil itself (Rosaire, 1938; Horvitz, 1939). An obvious requirement for successful application of the method of detection is that no significant amounts of saturated hydrocarbons, especially ethane through the pentanes, be present in near-surface soils from sources other than petroleum; their presence, in anomalous quantities, would obscure distribution patterns produced by the migrating gases.

Soil Organic Matter and Saturated Hydrocarbon Gases

In 1955, the writer investigated organic matter as an alternative source of soil hydrocarbons and reported (Horvitz, 1959) that it contributes negligible amounts of saturated hydrocarbons heavier than methane and does not interfere with the application of hydrocarbon geochemistry to exploration for oil and gas. In making the investigation, a series of boreholes was dug at 1,000-ft intervals along a profile crossing the Bonney oil field, Brazoria County, Texas, where a soil hydrocarbon anomaly had been observed in 1948, 2 years prior to discovery of the field. From each hole, samples of soil were collected from the surface (including the top inch) and at 2-ft intervals down to and including 12 ft. Glass jars, of the Mason type, were used to store the samples but plastic bags, now being used, are also suitable. The type of container is not critical, nor is it necessary to store samples at low temperatures. It is important, however, to keep to a minimum the amount of air in contact with the soil. When the container is completely filled, it can be stored for many months without significant loss of hydrocarbons. However, if it is only partly filled, some soils will lose most of their hydrocarbon content within a few days.

In the Bonney experiment, the organic matter was determined by the method of Walkley (1947) as modified by the U. S. Salinity Laboratory Staff (1954). Briefly, the method involves heating, to 150°C, a dried portion of soil with an excess of standard potassium dichromate solution to which concentrated sulfuric acid has been added, and then determining the unused dichromate by titration with standard ferrous sulfate solution. The amount of dichromate used up is a measure of the organic matter (organic carbon × 1.72) present in the soil.

[1]Manuscript received, October 27, 1970; revised, July 7, 1971; accepted, August 23, 1971. Read before the 55th Annual Meeting of the Association at Calgary, Alberta, June 22, 1970.

[2]Horvitz Research Laboratories, Inc.

Table 1. Analytic Data Used in Figure 1 and Some Significant Ratios

Sample Depth (Ft)	Methane (PPB by Weight)	Organic Matter* (% by Weight)	Ethane +** (PPB by Weight)	$\dfrac{Methane \times 10^6}{Organic\ Matter}$	$\dfrac{Ethane + \times 10^6}{Organic\ Matter}$	$\dfrac{Methane}{Ethane +}$
Surface	321	3.59	4	9.0	0.1	80.2
2	75	1.10	15	6.8	13.7	5.0
4	85	0.75	8	11.3	10.7	10.5
6	206	0.13	65	159.0	50.0	3.2
8	259	0.12	170	216.0	142.0	1.5
10	235	0.17	91	138.0	53.5	2.6
12	264	0.21	154	126.0	73.3	1.7

* Saturated hydrocarbon gases represent negligible part (less than 0.0001 percent) of organic matter.
** Ethane and heavier saturated hydrocarbons, through pentane.

The largest quantities of organic matter, ranging from 2.17 to 5.12 percent, invariably were found in the surface samples. With depth, the concentrations decreased very abruptly. At 6 ft, the organic matter ranged from 0.13 to 0.73 percent and at 12 ft the range was 0.05 to 0.21 percent. Within the upper 2-4 ft, high concentrations of methane were found which, undoubtedly, originated in organic matter; but the ethane and heavier hydrocarbon concentrations were low in this zone. In the background area, low ethane and heavier hydrocarbon values were found throughout the boreholes and, below 4 ft, the methane values were also low. Over the oil field, but more consistently in the boreholes near its edges, appreciable amounts of ethane and heavier hydrocarbons began to appear within the depth range of 6-10 ft and their values tended to be highest at 12 ft. Significantly, the high ethane and heavier hydrocarbon concentrations were found in soils containing very small amounts of organic matter. The methane was also high within the zones of high ethane and heavier hydrocarbons. The more common presence of anomalous hydrocarbon values in the soil near edges, rather than over the center, of petroleum accumulations is consistent with the distribution pattern referred to as the "halo."

The relation of the hydrocarbons to organic matter is illustrated in Figure 1. The highest values for organic matter (3.59 percent) and for methane (321 parts per billion by weight) were found at the surface where the ethane and heavier hydrocarbons (ethane +) totaled only 4 parts per billion and remained low to a depth of 6 ft. However, within the interval from 6 to 12 ft, where very small amounts (0.12-0.21 percent) of organic matter were found, high ethane and heavier hydrocarbon values were obtained. High methane values also obtained in the lower 6 ft suggested that the methane, as well as the ethane and heavier hydrocarbons within this section, may have originated in the petroleum accumulation and not in organic matter.

Table 1 contains the data used in preparing Figure 1, and the ratios of methane to organic matter, ethane + to organic matter, and methane to ethane +. These ratios may serve as indicators in petroleum exploration. For example, a high ratio of methane to organic matter suggests petroleum as the principal source of the methane; likewise, high ratios of ethane + to organic matter relate the ethane and heavier hydrocarbon gases to petroleum. Low ratios of methane to ethane + also suggest petroleum, and not organic matter, as the source of the soil hydrocarbons.

PURPOSE OF PRESENT REPORT

On the basis of chromatographic analyses, Smith and Ellis (1963) stated that vegetation, especially in the form of dried grass and live roots, contains saturated as well as unsaturated hydrocarbons and concluded that geochemical exploration is, therefore, of questionable value. Figure 2 shows the results obtained by these investigators. The top chromatogram is of natural gas and includes peaks for the saturated hydrocarbons ranging from ethane through normal pentane. The middle chromatogram is that of dried grass, the form of vegetation found to contain the largest amount of saturated hydrocarbons. According to this chromatogram, *propane, isobutane, normal butane, isopentane,* and *normal pentane* are present. The relative amounts of these constituents are small compared to the large peaks representing unsaturated constituents. Live roots showed the same hydrocarbons as those of dried grass but in smaller amounts. The bottom chromatogram is that of a soil sample; only a propane peak and a smaller one of normal butane are seen in addition to some larger, unsaturated hydrocarbon peaks. Because these chromatograms were not related to quantitative values, the significance could not be determined of those gases found in vegetation and considered to be saturated hydrocarbons. It is apparent, however, that the soil yielded the smallest amounts of hydrocarbons of any type.

Although the effect of organic matter had been shown previously to be unimportant insofar as

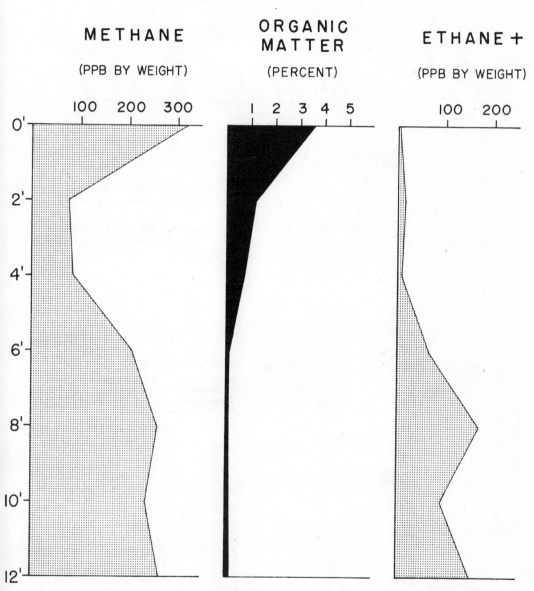

F𝗂G. 1—Relation of hydrocarbons to organic matter. Data from borehole over edge of Bonney oil field show near-surface methane to be related to organic matter. Low ethane and heavier hydrocarbon values in upper 4 ft, where organic content is high, and high ethane and heavier hydrocarbon values below this interval, where organic content is low, suggest petroleum as source of ethane and heavier hydrocarbons. Same source is suggested for methane below 4 ft. Prepared from data of Horvitz (1959).

geochemical exploration is concerned, dried grass or live roots were not studied specifically. Therefore, the main purpose of this report is to present results of new experiments which were conducted with these substances and which disagree with the results of Smith and Ellis. Other purposes are to point out the inadequacy of the procedure used by those investigators to extract hydrocarbons from soil, and to explain their unsuccessful attempts to obtain significant soil hydrocarbon data over oil fields.

Analytical Procedures

Smith and Ellis Technique

In their experiments, Smith and Ellis placed their samples in metal cans, sealed them by soldering with zinc chloride flux, and heated them at 85°C for at least 16 hours. Holes then were punched in the cans and samples of gas, withdrawn with a syringe, were injected into a Barber Coleman Model IDS 20 chromatograph. Any methane and ethane that may have been present could not be determined with this instrument.

Horvitz Technique

Figure 3 shows, diagrammatically, the soil gas extraction apparatus used, with minor variations, for more than 30 years. This technique was adopted after it was found that the major part of soil hydrocarbons is tightly held and not readily liberated by heat alone. Briefly, 50-100 g of undried sample are placed in flask A, which then is attached to the stationary extraction unit at B. After reducing the pressure in the system to about 50 mm of mercury, several ml of a 50-percent phosphoric acid solution are added; if carbonates are present, the carbon dioxide produced is removed by shaking flask F which contains a 50-percent solution of potassium hydroxide. Acid is added until the carbonates are decomposed and then an excess of 25 ml is added. If no carbonates are present, 25 ml of acid is added at the outset. Flask A is immersed in boiling water and heated for 30 minutes during which flask F is intermittently shaken. Because the pressure in A is always below atmospheric, the temperature of the soil remains below 100°C. At the end of the heating period, the gas mixture is transferred to flask F by filling flasks A and A' with distilled water. The various parts of the apparatus are calibrated so that the total volume of the hydrocarbon-air mixture, now in F, can be calculated and routinely is found to be of the order of 80 ml. The gas finally is displaced into a previously evacuated tube, T, which then is detached from the extraction unit.

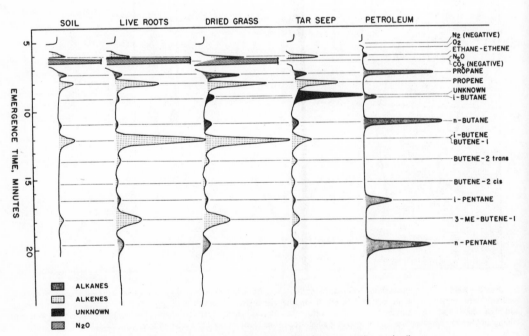

Fig. 2—Examples of chromatograms of gases from petroleum, vegetation, and soil sources. After Smith and Ellis (1963).

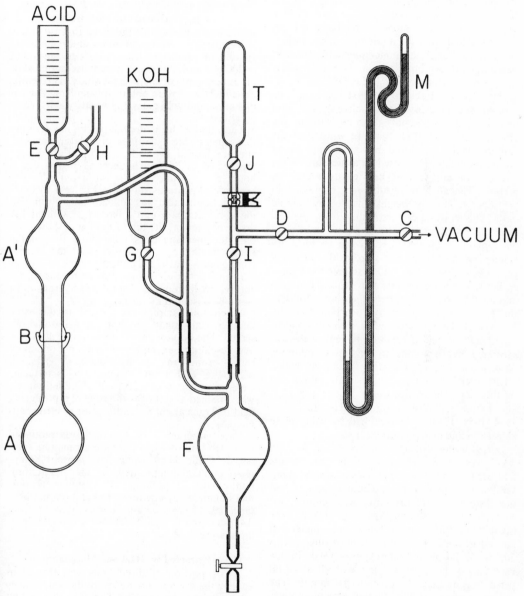

Fɪɢ. 3—Apparatus used in extracting gases from soils and vegetation. After Horvitz (1969).

Until 1963, vacuum techniques were employed to analyze the extracted soil gases. The gases were passed through potassium hydroxide solution, concentrated sulfuric acid and through phosphoric anhydride before they were separated into two fractions, one containing the methane and the second containing the ethane and heavier hydrocarbons. The separation was effected under a high vacuum at a temperature of $-196°C$. Methane is noncondensable whereas the ethane and heavier hydrocarbons are trapped at this temperature. The quantity of methane present was calculated from the amount of carbon dioxide produced after the noncondensable fraction was burned over a hot platinum wire in the presence of pure air or oxygen. The ethane and heavier hydrocarbon fraction was first volatilized, its volume measured with a sensitive gauge, and then burned over a hot platinum wire in the presence of oxygen. The gain in volume of the carbon dioxide produced by combustion, over the initial volume, was used as a measure of the quantity of ethane and heavier hydrocarbons present. The analytic results were expressed in parts per billion by weight on the dry sample basis. The moisture content required for the calculations was determined from a portion of sample separate from that used in the analysis. By employing the combustion step, oxides of nitrogen, when present, are not mistaken for hydrocarbons. Furthermore, this step is useful in estimating the average number of carbon atoms in the ethane and heavier hydrocarbon fraction. Passage of the extracted gases through concentrated sulfuric acid is required to reduce the unsaturated hydrocarbons to negligible amounts. Also, nonhydrocarbon organic substances which might interfere with the analysis are eliminated by this step. The hydrocarbon data of Figure 1, and of other published surveys (Horvitz, 1945, 1954, 1959, 1969), were obtained by using these extraction and analytic procedures.

In 1963, the hydrogen flame chromatograph was investigated and after comparable results had been obtained from analyses of several thousand soil samples by the vacuum technique and the chromatograph, the latter was adopted for routine work. A Varian Aerograph Model 600, fitted with a 10-ft column, 1/8 in. in diameter and packed with a 20 percent mixture of silicone (SE 30) on 60-80 mesh firebrick, was used. In the experiments the hydrogen flow was set at 20 ml per minute and the air at 300 ml per minute. The nitrogen carrier gas flow through the column was maintained at 24 ml per minute. A gas sample handling system, composed of a few stopcocks and a manometer, was constructed and con-nected to the sample valve of the chromatograph through a 1.0 ml sample loop. The combined volume of the sample loop and sample valve was 1.35 ml. The tube containing the extracted gases was attached to the sample handling apparatus through an interchangeable joint; the system was evacuated, the sample tube opened, and a measured part of the gas was admitted into the column which was maintained at 50°C.

EXPERIMENTS WITH DRIED GRASS AND SOIL

Gases Extracted by Heat Only

Smith and Ellis (1963) indicated that 1 g or less of vegetation is sufficient to produce the same amount of volatile hydrocarbons that are found in 200 g of soil. At least 5 g were used in most of the experiments of the present study. The results to be illustrated were obtained from Saint Augustine grass, but wild grass and weeds yielded comparable data.

The middle chromatogram of Figure 4 shows that appreciable quantities of volatile constituents are, indeed, evolved when 5 g of dried grass are heated at 85°C. The heating period was 8 hours, half the time used by Smith and Ellis. The extraction apparatus shown in Figure 3 was employed, but no acid was added. Comparison with the chromatogram of the known, saturated hydrocarbon mix at the top of the figure shows that only the methane peak falls under that of the known methane, and suggests that the heavier components (black) are either unsaturated hydrocarbons or nonhydrocarbon substances. Of course, small amounts of saturated hydrocarbons could be obscured by the large peaks.

When soil samples are treated in the same manner, only small amounts of gases are observed. The bottom chromatogram of Figure 4 is a representative example; two parts per billion by weight of methane, less than one part per billion of ethane, and a small amount of isobutylene were the only constituents found. The soil sample was obtained at 9 ft over the edge of a producing field in Texas about 8 months prior to the experiment.

Chromatogram A of Figure 4 was produced from 0.091 ml of a hydrocarbon-air mixture (prepared by Big Three Industrial Gas and Equipment Co.) containing: methane, 66 parts per million; ethane, 57; propane, 33; isobutane, 32; normal butane, 43; isopentane, 68; and normal pentane, 29. Chromatogram B was prepared from 0.448 ml, and C from 0.396 ml of extracted gas-air mixtures.

Gases Extracted by Heat and Phosphoric Acid

A 5-g portion of dried grass was treated in exactly the same manner as were the soil samples of the Bonney experiment. The grass was heated for 30 minutes at 100°C with 50 ml of a 50-percent solution of phosphoric acid, and the middle chromatogram of Figure 5 was produced. Some

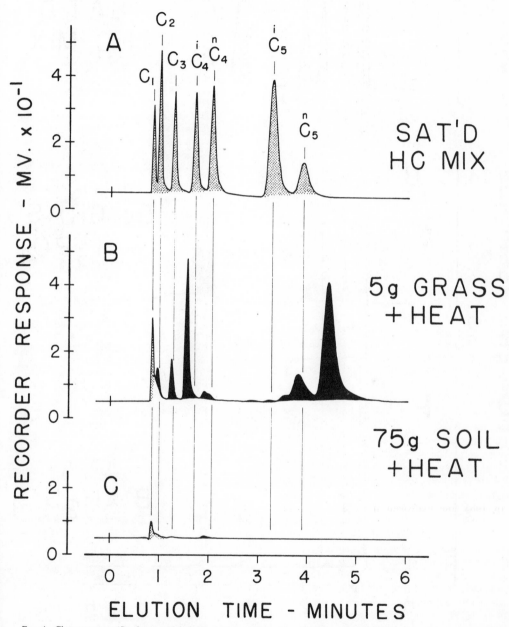

Fig. 4—Chromatogram (*B*) of gases produced by heating dried grass at 85°C for 8 hours reveals many peaks but only methane is clearly a saturated hydrocarbon. Chromatogram (*C*) of gases evolved from soil by same heat treatment shows negligible amounts of hydrocarbons.

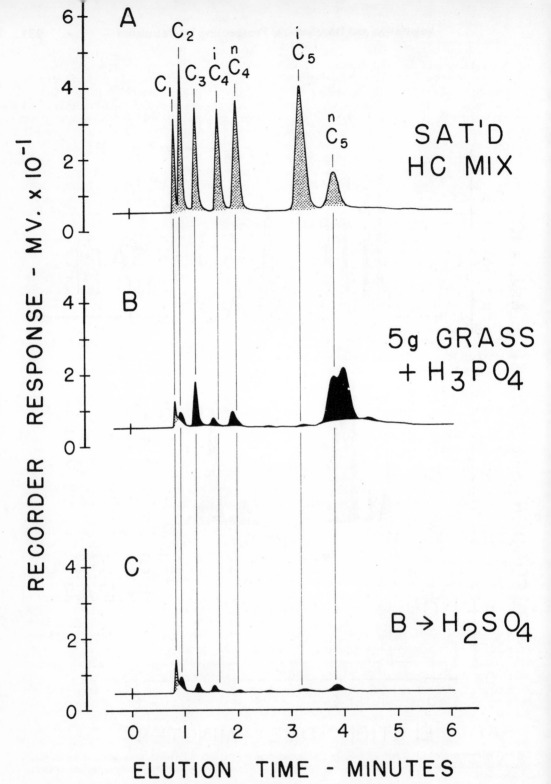

Fig. 5—Chromatogram of gases produced by heating dried grass for 30 minutes at 100°C with 50-percent phosphoric acid before (*B*) and after (*C*) contact with concentrated sulfuric acid.

of the peaks, except for relative sizes, are the same and some are different from those produced when no acid was used (Fig. 4B). To compare the relative sizes of peaks, it is necessary to know that the volume of gas used to produce the chromatogram of Figure 4B was approximately twice that used to produce the chromatogram of Figure 5B. When the gases produced by the phosphoric acid treatment were passed through concentrated sulfuric acid, the bottom chromatogram of Figure 5 resulted. The part of the sample that was treated with the sulfuric acid was 20 percent larger, by volume, than that used to produce the middle chromatogram. The concentration of the methane remained unchanged but most of the other gases were removed by the sulfuric acid indicating that they are composed principally of unsaturated compounds. A similar chromatogram was obtained when part of the gas evolved from the heat-treated sample (Fig. 4B) was passed through concentrated sulfuric acid.

A separate, undried portion (75.0 g) of the same soil sample which was heated without phosphoric acid and which yielded very low values (Fig. 4C) was put through the extraction process in the standard manner using 50 ml phosphoric acid. The middle chromatogram of Figure 6 was prepared from the extracted gases. The bottom chromatogram was obtained after the extracted gases were bubbled through concentrated sulfuric acid. The soil hydrocarbon peaks fall exactly under those of the saturated hydrocarbon mix. The values calculated from the part (0.270 ml) of the sample that was passed through sulfuric acid agreed closely with those of the part (0.214 ml) that bypassed it. In parts per million by volume, the composition of the extracted gas-air mixture, which produced chromatogram C, was methane, 925; ethane, 54; propane, 16; isobutane, 3.0; normal butane, 5.0; isopentane, 2.5; and normal pentane, 2.6. Expressed in parts per billion by weight in terms of dry soil (sample contained 24.3 percent moisture), chromatogram C represents the following values: methane, 956; ethane, 97; propane, 44; isobutane, 11; normal butane, 19; isopentane, 12; and normal pentane, 12. These values are almost the same as those obtained during the original survey for the 12-ft sample of the same location. During the actual survey, only the phosphoric acid extraction procedure was used; prior work had demonstrated that about 10 percent of the soil hydrocarbons is released when heat alone is employed. This loosely held fraction is lost within a short time after the soil is collected.

As most of the volatile gases produced by dried grass are removed by concentrated sulfuric acid,

they probably are composed principally of unsaturated compounds; but some of the substances that did resist the sulfuric acid may, in fact, be saturated hydrocarbons. If all of the peaks of the bottom chromatogram of Figure 5, that appear under or close to those of the known hydrocarbons, are assumed to be saturated hydrocarbons, then 5 g of dried grass would contribute only 5 parts per billion by weight of methane and 8 parts per billion of ethane through pentane to 100 g of soil. One gram of grass would contribute *less than one part per billion* of ethane through pentane to 200 g of soil.

QUANTITY OF VEGETATION IN SOIL

To determine the actual amounts of grass and roots in soil, the residues produced by acid extraction of 100-g portions of 890 different soil samples, collected from depths of 3 to 12 ft in several areas of Texas and Louisiana, were examined carefully. Only one sample contained an appreciable amount (0.687 g) of grass, roots, and other debris. The remaining 99.9 percent contained less than 0.200 g of vegetation and 93 percent of the samples contained less than 0.050 g for each 100 g of soil. The quantities of saturated hydrocarbon gases produced by these small amounts of vegetation would hardly be measurable.

EXPERIMENTS WITH LIVE ROOTS, DRIED ROOTS, AND LIVE GRASS

When 11 g of live roots were heated (without acid) for 16 hours at 85°C, only a few very small peaks appeared under those of known saturated hydrocarbons; even the peaks of unsaturated compounds were small. If the entire root sample was part of a 100-g soil sample, it would have contributed 1.8 parts per billion by weight of methane, 1.3 parts of ethane-propane-butane, and 1.3 parts of pentane. When a 3-g sample of dried roots was heated for 30 minutes at 100°C with 100 ml of 50-percent phosphoric acid, the quantity of hydrocarbons produced would contribute 2.0 parts per billion of methane, 2.7 parts ethane-propane-butane, and 1.9 parts of pentane to a 100-g soil sample.

In all experiments, the dried samples produced more "saturated" hydrocarbons, as well as unsaturated ones than did fresh. When 25 g of live, undried grass (containing 66 percent moisture) was heated for 30 minutes at 100°C with 150 ml of 50-percent phosphoric acid, the amount of hydrocarbons produced would contribute to 100 g of soil: 2.3 parts per billion by weight of methane, 1.0 part of ethane-propane-butane, and 2.0 parts of pentane. Apparently, when vegetation is

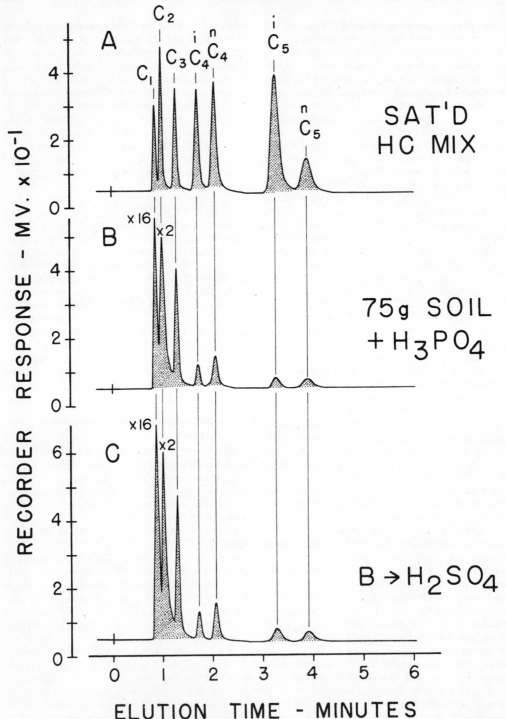

FIG. 6—Chromatographic peaks of gases extracted from soil, using 50-percent phosphoric acid solution and heat at 100°C for 30 minutes, coincide with those of known saturated hydrocarbon mix. Relative amounts of soil hydrocarbons are consistent with those in lightest end of petroleum; those of grass (Fig. 5C) are not.

dried, either in the open air or in the laboratory at room temperature, changes occur which cause relatively large quantities of unsaturated hydrocarbons and small amounts of saturated hydrocarbons to be produced as the dried vegetation is heated; the higher the temperature, the larger the quantity of hydrocarbons. Saturated hydrocarbons can be removed quantitatively from soil by heating it with 50-percent phosphoric acid solution at 100°C for only 10 minutes. If heated for 30 minutes, the hydrocarbons are removed at temperatures as low as 50°C. However, 5 g of dried grass, heated at 50°C under the same conditions, contribute only 2.0 parts per billion by weight of methane, and 2.0 parts of ethane through pentane to 100 g of soil, assuming all chromatographic peaks in the vicinity of those of known saturated hydrocarbons do, in fact, represent saturated hydrocarbons.

Experiments with Nitric Acid Prove Insignificant Amounts of Saturated Hydrocarbons are Generated by Vegetation

Use of elution times alone for the identification of substances is undependable. For example, the boiling points of trans-2-pentene, an unsaturated hydrocarbon, and normal pentane differ by only 0.2°C; their elution times are, therefore, nearly the same. Moreover, although sulfuric acid eliminates many of the unsaturated compounds, it does not remove them all; for instance, at room temperature, ethylene reacts only slightly with concentrated sulfuric acid. To obtain more information concerning the chromatographic peaks produced by grass and roots, use was made of the fact that unsaturated hydrocarbons are less resistant to strong oxidizing agents than are saturated ones. Thus, a series of experiments was conducted on samples of vegetation and soil in which concentrated nitric acid was employed in the extraction procedure instead of 50-percent phosphoric acid solution.

Interesting results were obtained when 5 g of dried grass was heated at 100°C for 30 minutes with 50 ml of concentrated nitric acid. When a part (0.376 ml) of the extracted gases was passed through concentrated sulfuric acid, the bottom chromatogram of Figure 7 resulted. The only saturated hydrocarbon recorded was methane, in an amount normally found in laboratory air. In addition, only ethylene and an unidentified substance (not a saturated hydrocarbon) whose peak appeared just before that of isobutane were apparent. The middle chromatogram is that of a much smaller part (0.132 ml) of the extracted gases that bypassed the sulfuric acid. In addition

to ethylene, some propylene was present and a large quantity of a substance appeared which has an elution time between that of propane and isobutane. An even larger peak appeared beyond normal pentane at 4.73 minutes. A sample of the known, saturated hydrocarbon mix was heated at 100°C with concentrated nitric acid for 30 minutes. The top chromatogram (Fig. 7), which shows that nitric acid has no effect on the saturated hydrocarbons, resulted.

The effect of heating 75 g of soil with 50 ml of concentrated nitric acid for 30 minutes at 100°C is shown by the middle chromatogram of Figure 8, produced by 0.198 ml of extracted gases. Apparently, a small amount of organic material was present in this sample for an extra peak was obtained between those of propane and isobutane. A second, even larger, extra peak appeared beyond normal pentane at 5.55 minutes. A rough calculation, based on the relative quantities of the extra peaks indicates that the amount of vegetation in the soil sample was less than 0.1 g. When part (0.256 ml) of the gas sample was passed through concentrated sulfuric acid, the bottom, clean chromatogram was produced. The calculated values of the saturated components of the two parts of the gas sample are in close agreement. When a separate, 75 g part of the same soil sample was heated with 50 ml of 50-percent phosphoric acid instead of with concentrated nitric acid, hydrocarbon values were obtained which agreed very closely with those of the nitric acid treated sample. However, the peaks that appeared between propane and isobutane, and at 5.55 minutes, when the soil sample was treated with concentrated nitric acid, did not appear on the chromatogram of the phosphoric acid treated sample. Obviously, concentrated nitric acid attacks organic matter more readily than does 50-percent phosphoric acid solution.

Because the peaks of ethylene and ethane appear close together when a 10-ft column is used, the ethylene peak may be obscuring a small amount of ethane. To assure an effective separation, the 10-ft column of the chromatograph was replaced by one 15 ft long, and the temperature was reduced to 39°C. The carrier gas pressure was increased from 16.5 to 20 pounds. At the top of Figure 9 is the chromatogram of the known hydrocarbon mix obtained with the longer column. Only the more widely separated peaks of methane through normal butane are included. Chromatogram B, prepared from dried grass that was treated with concentrated nitric acid, shows no saturated hydrocarbons other than a little methane. The presence of ethylene is confirmed by comparing this chromatogram with one of a

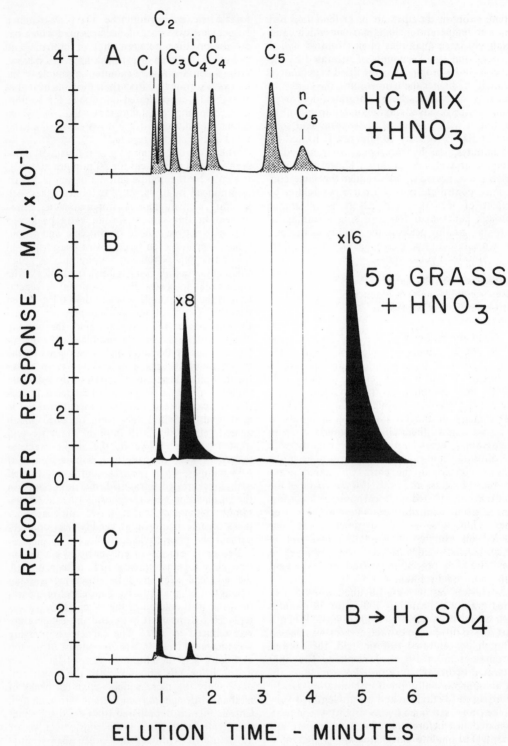

Fig. 7—Chromatograms showing that saturated hydrocarbons are not affected by concentrated nitric acid (*A*), whereas dried grass produces unsaturated constituents (*B*), most of which are removed upon contact with concentrated sulfuric acid (*C*).

mixture containing ethylene, propylene, and isobutylene, to which some methane was added and which is shown at C. The largest peak observed on the chromatogram of grass coincides with the known ethylene. The small peak before that of isobutane is the one that previously has been noted as being unidentified. To determine if the column is capable of separating ethane from ethylene, some of the saturated hydrocarbon mix was added to the one containing unsaturated hydrocarbons and a chromatogram of the composite mixture was made. It appears at the bottom of Figure 9 and shows clearly that ethane, if present in the gases that were extracted from grass, could have been detected.

The nitric acid treatment was repeated many times with dried grass, live roots, and dried roots, and the results were always the same. Only methane, ethylene, and the unidentified substance that appeared before isobutane were observed in measurable amounts. When the extracted gases were injected into a chromatograph having seven times the sensitivity of the one used in preparing the illustrations presented, the resulting chromatograms indicated that 5 g of dried grass in a 100-g soil sample would contribute *less than one part per billion by weight* of combined ethane and heavier hydrocarbons. The experiments with concentrated nitric acid suggest that vegetation, in its natural state, may contain no saturated hydrocarbons. Small amounts may be generated only after drying and heating the vegetation.

HYDROCARBONS HEAVIER THAN METHANE FROM FERMENTATION PROCESSES

Other experiments, relevant to the present discussion, are those of Davis and Squires (1954). They reported the production of gas containing, by volume, 49 percent methane, 7.0 parts per million of ethane, and 0.06 parts of propane after more than 2 months of microbial fermentation of paper, with cow dung as the inoculum. When municipal sewage sludge was used as the microbial inoculum, gas containing 80 percent methane, 0.3 parts per million ethane, and no propane resulted. Other experiments produced smaller amounts of saturated hydrocarbons heavier than methane. These small amounts would not interfere with geochemical prospecting, even if the same fermentation processes should occur in nature. Davis (1969), in discussing the application of microbiology in petroleum exploration, stated that ethane and propane have a high specificity with regard to petroleum, inasmuch as only a few parts per million of ethane and propane are detected in gas over the anaerobic decomposition of

organic matter. Furthermore, Davis stated that the actual mechanism of the biologic production of these gases is not known.

Stevens (1962) reported that ethane is a specific indicator for petroleum when found in soil air, and pointed out that the ratio of ethane to methane in the fermentation experiments of Davis and Squires (1954) is about 1 to 100,000 whereas in petroleum gas it is of the order of 1 to 20.

My experience is based on many thousands of analyses of samples, collected from many different petroleum provinces in several different countries, both onshore and offshore. These data consistently show ratios of methane to ethane, and to the other constituents, that are of the same order as those of the corresponding hydrocarbons in petroleum gas. Only in marshy areas do exceedingly high ratios of methane to ethane prevail; therefore, methane data obtained in such areas have no significance in geochemical exploration.

FAILURE TO OBTAIN HYDROCARBON ANOMALIES OVER OIL FIELDS

Smith and Ellis (1963) analyzed soil samples collected over oil fields and over barren areas and obtained similar results in both cases. Their extraction procedure, using heat alone, is incapable of removing the major part of saturated hydrocarbons from soil, and their analytic technique is incapable of determining methane and ethane, hence their results are not surprising. However, even if their extraction and analytic procedures were adequate, they still might have obtained negative data. Their results would be expected if a survey is made several years after a field has been developed. Experience shows that hydrocarbon anomalies tend to diminish and eventually disappear as the contents of the reservoir are withdrawn. This phenomenon was confirmed recently at the Hastings oil field, Brazoria County, Texas (Horvitz, 1969). Whereas a strong anomaly was obtained in 1946 from 8- to 12-ft samples, sampling below 20 ft is now necessary to obtain values of the order of those of the earlier survey.

CONCLUSION

Experiments described in this paper demonstrate that anomalous amounts of soil hydrocarbons, especially the saturated ones in the range from ethane through normal pentane, do not originate in soil organic matter, grass, or roots. Therefore, the presence of the latter substances in the soil does not interfere with the application of hydrocarbon geochemistry to petroleum exploration.

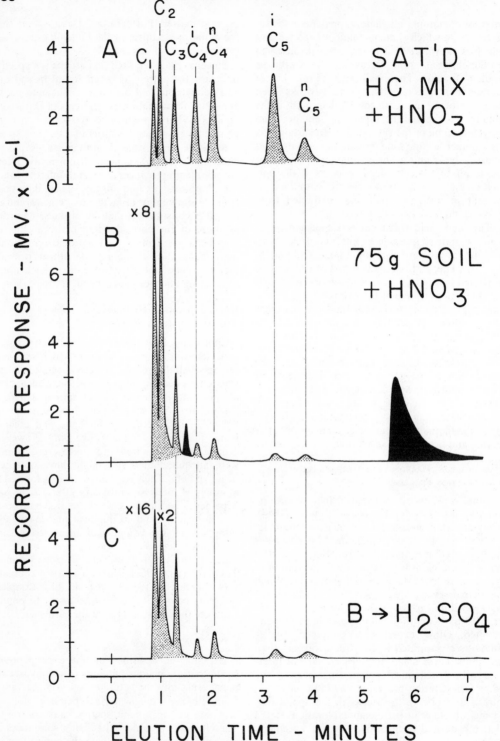

FIG. 8—Chromatogram of gases extracted by heating soil with concentrated nitric acid (*B*) confirms presence of saturated hydrocarbons. Organic matter present in soil produced two extra unsaturated hydrocarbon peaks (black) which were completely removed by concentrated sulfuric acid (*C*).

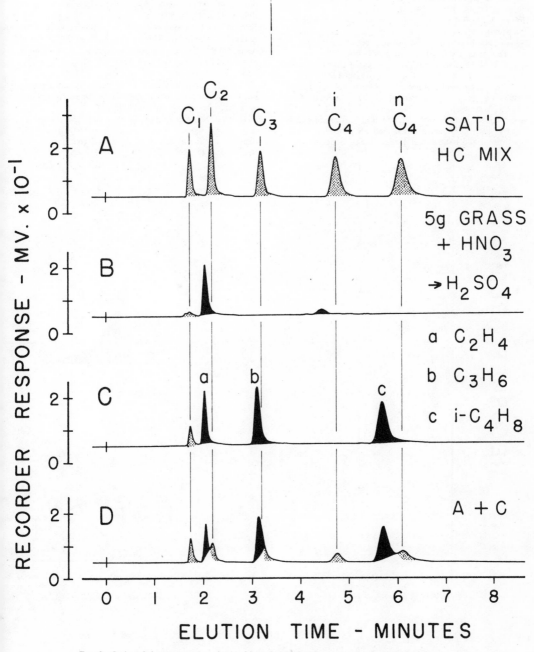

FIG. 9—Series of chromatograms, obtained by using 15-ft column, proves absence of ethane in grass.

REFERENCES CITED

Davis, J. B., 1969, Microbiology in petroleum exploration, *in* Unconventional methods in exploration for petroleum and natural gas: Dallas, Texas, Southern Methodist Univ., p. 139-157.

Davis, J. B., and R. M. Squires, 1954, Detection of microbially produced gaseous hydrocarbons other than methane: Science, v. 119, no. 3090, p. 381-382.

Horvitz, L., 1939, On geochemical prospecting: Geophysics, v. 4, p. 210-225.

────── 1945, Recent developments in geochemical prospecting for petroleum: Geophysics, v. 10, p. 487-493.

────── 1954, Near-surface hydrocarbons and petroleum accumulation at depth: Mining Eng., v. 6, p. 1205-1209.

────── 1959, Geochemical prospecting for petroleum: Symposium on geochemical exploration: 20th Internat. Geol. Cong., Mexico City, v. 2, p. 303-319.

────── 1969, Hydrocarbon geochemical prospecting after thirty years, *in* Unconventional methods in exploration for petroleum and natural gas: Dallas, Texas, Southern Methodist Univ., p. 205-218.

Laubmeyer, G., 1933, A new geophysical prospecting method, especially for deposits of hydrocarbons: Petroleum, v. 29, no. 18, p. 1-4.

Rosaire, E. E., 1938, Shallow stratigraphic variations over Gulf Coast structures: Geophysics, v. 3, p. 96-115.

Smith, G. H., and M. M. Ellis, 1963, Chromatographic analysis of gases from soils and vegetation, related to geochemical prospecting for petroleum: Am. Assoc. Petroleum Geologists Bull., v. 47, p. 1897-1903.

Sokolov, V. A., 1935, Summary of the experimental work of the gas survey: Neftyanoye Khozyaystvo, v. 27, no. 5, p. 28-34.

Stevens, N. P., 1962, Geochemical prospecting in the United States: Oil Producing Industry, Sci. Tech. Conf. Proc., Budapest, v. 2, p. 83-99.

U. S. Salinity Laboratory Staff, 1954, Diagnosis and improvement of saline and alkali soils: U. S. Dept. Agriculture, Agriculture Handbook No. 60, p. 105-106.

Walkley, A., 1947, A critical examination of a rapid method for determining organic carbon in soils—effect of variations in digestion conditions and of inorganic soil constituents: Soil Sci., v. 63, p. 251-264.

Reprinted from:
BULLETIN OF THE AMERICAN ASSOCIATION OF PETROLEUM GEOLOGISTS
VOL. 56, NO. 10 (OCTOBER, 1972), PP. 2029-2067, 5 FIGS., 6 TABLES

Depths of Oil Origin and Primary Migration: a Review and Critique[1]

ROBERT J. CORDELL[2]

Richardson, Texas 75080

Abstract Vast amounts of geologic and geochemical data bearing on depths of oil origin and primary migration have been published during the past 30 years. Although concepts and understanding have evolved substantially, contrasting views also have been reinforced.

Serious obstacles exist to theories of shallow "in-place" origin of reef oil and tar-sand oil. In reefs, environmental conditions would have dissipated and widely disseminated most of the organic matter. In sands, humic material could have accumulated abundantly, but its transformation to enormous quantities of heavy oil under the thin overburden characteristic of most tar-sand diagenesis is difficult to explain chemically.

Investigations of a variety of modern sediments demonstrate: (1) sparsity of liquid hydrocarbons and free hydrocarbon precursors; (2) absence or traces only of many hydrocarbon and other bitumen components which are common in ancient rocks and crude oil; (3) dilute occurrence of dissolved organic matter and only traces of liquid hydrocarbons in interstitial waters; and (4) major upward movement of water to the surface, representing a serious loss to proposed shallow primary migration mechanisms.

Experimental evidence indicates that effective barriers to mobility of liquid hydrocarbons and hydrocarbon precursors exist in source-type sediments under shallow burial. If hydrocarbons are in oil form, they apparently are immobilized by capillary attraction. In either oil or more finely divided occurrences, hydrocarbons are attracted to the bulk organic matter, which in turn is attracted to clays and other minerals. Hydrocarbon precursors in bitumen are subject to both organic and mineral barriers. At shallow depths, catalytic processes appear inadequate to explain any really significant hydrocarbon formation and primary migration.

It is concluded that most crude oil and its constituents resulted from thermocatalytic transformations and primary migration at depths ranging from a few thousand to about 10,000 ft. Chemical alteration of organic matter should be greatly accelerated in the relatively high temperature-pressure regime at these depths. Energy input from heat and pressure should greatly increase the mobility of bitumen constituents formed, including hydrocarbons. Sufficient water should be available as a migration medium where expandable clays are abundant. Clay mineral transformation should release water and organic matter, and improve source-bed drainage. Large amounts of colloid-forming substances would be produced, and gases would be abundant products of organic alterations. Primary migration of liquid hydrocarbons in colloidal, true-water, and/or gas solution would obviate the high-permeability requirements of migration in the form of oil.

Thermal investigations of oil shale and disseminated kerogen show that the heavy nonhydrocarbon part of the bitumen is produced initially, and at higher temperatures the bitumen is altered in part to hydrocarbons. The fact that relatively high temperatures are needed for these transformations, even in the presence of catalysts, argues for rather deep corresponding transformations in the sedimentary section. Bitumen contains colloid-forming fractions which would greatly increase the solubility of hydrocarbons in water. Carboxylated (colloid-forming) organic material is relatively resistant to temperature increase in a water-wet environment, even where catalysts (clays) are present. Hence colloids should remain active to considerable depth. Laboratory experiments indicate that thermocatalytic processes can best explain the origin of liquid normal paraffin hydrocarbons of oil from carboxylated organic matter.

Several studies of depth patterns in nonreservoir (source type) rocks of the geologic section reveal that both bitumen and the hydrocarbon component of bitumen increase markedly in the depth range of a few thousand to several thousand feet. The increase in hydrocarbon abundance is particularly impressive. In some examples, a zone of sharply decreasing bitumen and hydrocarbon occurrence has been observed directly below the zone of abundance, indicating that primary migration has evidently taken place. Much primary migration apparently is related to stages of "dehydration" of expanded clay minerals, the most important stage being temperature-dependent.

Coordinated studies of geology, geochemistry, and oil occurrence, particularly in the Soviet Union and northwestern Germany, support major origin–primary migration episodes at depths of a few thousand to several thousand feet or more. They also suggest continuous or recurrent primary migration through rather long time and broad depth intervals.

Ultra-deep origin of mobile oil, and upward migration through thousands of feet of shale are shown to be very unlikely. In examples such as the South Sumatra, Los Angeles, and Ventura basins, where the ultra-deep concept has been applied, the data can be explained better by primary migration at intermediate depths.

INTRODUCTION

About 30 years ago, the AAPG Research Committee published the results of a survey of prevailing views on time of origin and accumulation of petroleum (Van Tuyl and Parker, 1941). Contributions included statements by 117 well-recog-

[1] Manuscript received, May 13, 1971; revised and accepted, January 6, 1972.

[2] Sun Oil Company Production Research Laboratory.

The writer thanks Sun Oil Company management for permission to publish this paper. Editors and critics have furnished numerous helpful suggestions. The illustrations were drafted by George T. McCulloch.

nized geologists who were interested in research problems.

Opinions were divergent on the question of burial depth required for oil genesis and accumulation. Some geologists favored shallow origin and migration, on the basis of criteria such as paleogeography, structure, time of trap development, and overburden history. Others believed oil can form at almost any depth range in the sedimentary section, as indicated by the variable depths of present oil pools. A third group preferred relatively deep origin, emphasizing examples of extensive updip migration, late trap formation, and subsequent removal of overburden by erosion. The idea of ultra-deep sources received little support in light of the probable low-temperature history of oil genesis from sedimentary organic matter. The old distillation hypothesis, which postulated intense heat from metamorphic or igneous activity as the motivating energy for oil formation, had few adherents.

The depth problem should be analyzed in the context of three important stages: origin in source beds, primary migration (source beds to reservoir system), and secondary migration (within reservoir system to a trap). Long ago, Illing (1938) underscored these stages, yet even now migration commonly is expressed as a general process without clear distinction between source beds and reservoir systems. Interpretations of overburden depths during primary migration are mostly related to sites of oil accumulation, largely ignoring basinward source-bed facies. The term "oil formation" itself is ambiguous. Some views imply that an accumulation must occur before the organic substance is called oil, whereas others assume that oil develops in source beds, or during primary or secondary migration.

Only where oil reservoirs are extremely lenticular and lacking in downdip persistence can the interpreted overburden during oil accumulation also be reasonably applied to depths of migration in the associated source beds. Perhaps the best known example is the Raccoon Bend field in Austin County, Texas (Teas and Miller, 1933). The highly lenticular reservoir sandstone in this field apparently had only 400-500 ft of overburden when the oil accumulated, from interpretations of relative timing of oil emplacement, faulting, and incremental additions of overburden.

Difficulty arises, however, in proving that a reservoir was isolated in all directions during migration. The reservoir lithology may persist in the depositional or structural downdip direction in one or more tongues or channels. If so, the source beds supplying the fluids may have been considerably deeper than the reservoir accumulation. Moreover, the principal source-bed fluids could have migrated upward for appreciable distances through the compacting source-bed muds before entering the sand or other reservoir lens.

Recently, Meinhold (1969) stated that the problem of early diagenetic origin of oil versus late origin from kerogen (main insoluble bulk of the organic matter) has not been resolved. The objective of the present report is to shed further light on this question from many lines of evidence.

A better understanding of depths of origin, migration, and accumulation of hydrocarbons is fundamental to improved predictions of major oil occurrence. If these processes are typically shallow, there is little need for marked basin subsidence and thick overburden as essential concomitants to hydrocarbon origin and primary migration. Depositional and early diagenetic environmental patterns would, accordingly, assume primary importance. Conversely, if hydrocarbon origin and primary migration typically require considerable overburden, investigation of subsidence patterns must accompany knowledge of depositional and early diagenetic environments. In this event, one of the chief exploration aims should be to outline areas in which marked basin subsidence was accompanied or followed by relatively thick sediment accumulation.

Shallow Origin and Migration Views

During the middle and late 1940s, the idea of shallow origin and migration of oil began to gain further favor. Expansion of modern sediment studies reflected the importance attached to depositional processes and early diagenesis in understanding geologic processes. In part, this interest represented a view that oil origin and primary migration may be closely related to depositional and early diagenetic phenomena.

Investigation of bacterial activity in modern sediments, summarized by ZoBell (1945, 1946, 1947, 1951), revealed major transformations of organic matter, apparently representing important steps toward hydrocarbon formation. Certain bacteria were known to be instrumental in hydrogenation, decarboxylation, and other chemical reactions leading to petroleum-like products. In addition to methane, small amounts of heavier hydrocarbons had been obtained from bacterial cultures in the laboratory. These findings, reinforced by the commonly accepted view that most bacterial activity occurs at quite shallow depths, helped to popularize the shallow-origin concept.

Influential members of the geologic and geochemical communities offered their support to views of early origin and/or migration. Pratt (1947) stated that an impressive array of data supports the conclusion that petroleum accumulation is often completed soon after deposition of the source beds. He suggested the possibility of Pleistocene oil on continental shelves. W. Link (1949) thought processes of oil formation must begin as soon as sediments are deposited. Although he left the question of relatively deep origin open, he doubted whether compaction of organic shales is a typical cause of primary migration. T. Link (1950) espoused the idea of "in-place" origin of petroleum to explain reef oil. He believed that, because a reef environment teems with life, successive layers of organic-rich material should collect as the organisms die. He concluded that this indigenous organic matter could become petroleum after certain relatively minor modifications. Studies by Bergmann and Lester (1940) had shown that modern corals contain about 0.14 percent of wax material consisting largely of hydrocarbons and complex alcohols. T. Link (1950) questioned the source-bed role of compacting shales and believed that oil-like material in organic shales would be given up only under abnormal conditions. He viewed marsh gas of swamps as the first stage in petroleum hydrocarbon formation. From such considerations, oil origin and accumulation processes would evidently be regarded as relatively shallow, and rather early diagenetic phenomena.

One of the theories advanced for the origin of the Athabasca tar sand was "in-place" precipitation and accumulation of humic acid during sand deposition (Corbett, 1955). A corollary of this concept is that a thin overburden caused transformation of humic acid precipitate (humate) to an immature petroleum. Yurkevich (1962) described a modern-sediment example of downslope migration of organic matter from an algal ledge to an underlying sand. He believed this process obviates the necessity of compaction-migration theory relating to lithologically isolated source beds. Rainwater (1966) visualized a somewhat similar sand-algal contiguity as a possible explanation of petroleum origin in the Gulf Coast. Al'tovskii et al. (1958) and Cate (1960) postulated that the introduction of organic matter into reservoirs was a late process, but that it occurred at shallow depths after the sediments were uplifted and subjected to terrestrial groundwater activity. Their idea of source material was somewhat similar to Corbett's (1955), consisting of humic material carried into reservoirs by ground water.

Weeks (1952) stated that the key to a solution of the problem of petroleum origin probably lies in transformations that take place in organic matter during diagenesis in recent and near-recent sediments. He predicted that Pleistocene oil would be found on or beyond the continental shelves. Gussow (1954) thought that hydrocarbons may form any time after deposition and that they occur in marine clay and carbonate muds as minutely disseminated globules. Hobson (1954) considered bacteria important in oil origin, and visualized early migration before permeability becomes too low. He opposed the idea of thermal origin of hydrocarbons because of the high temperatures apparently required.

Geochemical support for the shallow origin came with the discovery by Smith (1952) of free liquid hydrocarbons in modern sediments of the Gulf Coast. Dott and Reynolds recently (1969) referred to this discovery as the beginning of a new era. Smith's later reports (1954, 1955) gave confirmation, based on carbon-14 analysis, that at least part of these hydrocarbons are comparable in age to the modern sediments in which they are found. His observation that organic matter is more oil-like in modern sands than in associated muds gave rise to speculation that primary migration already may have been initiated.

Subsequently, many other geochemists have reported liquid hydrocarbons in modern sediments, and further reports have been published of the more oil-like character of organic matter in modern sands, compared with that in associated muds. Karimov (1963) emphasized that organic matter indigenous to sands should be less, rather than more oil-like than that of clays, owing to the relatively oxidizing environment, water flushing, and lesser catalytic activity characteristic of sands, compared to clays. The fact that the opposite relation commonly exists in modern sediments suggests that early migration of oil-like material from muds to sands probably occurs. However, the very dilute concentrations of total organic matter found in most of these sands scarcely can be considered significant from a petroleum standpoint.

Smith's reports (1952, 1954, 1955) of free liquid hydrocarbons in modern sediments were noted and applied by several authors (in Weeks, 1958b). Cohee and Landes (1958) referred to Smith's interpretations as a syngenetic theory of oil origin. They believed that oil in the Dundee (Devonian) Limestone of Michigan may have migrated downward from the overlying Rogers City Limestone before the latter became consolidated. Dallmus (1958) stated that an increasing body of evidence indicates that hydrocarbons begin to concentrate in permeable beds very early in the compaction process, at depths of a few thousand feet. He implied that hydrocarbons in modern sediments constitute evidence of early hydrocarbon availability. Renz *et al.* (1958) remarked that there is no reason to doubt that petroleum in the Eastern Venezuela basin formed during or soon after deposition of source beds, in view of the hydrocarbons in modern sediments. Woodward (1958), though he did not specifically refer to modern sediments, concluded that early migration characterized the Appalachian basin, and that it began practically as soon as the sands were covered by other sediments. Dickey and Rohn (1958) believed that most migration probably occurs during the early stages of consolidation.

In the Pedernales area of Venezuela, Kidwell and Hunt (1958) described a concentration of hydrocarbons of near-recent generation in a Holocene sand lens at a depth of only 110 ft. Although this hydrocarbon occurrence amounted to only 160 ppm, it was quite high relative to concentrations in associated muds and sands.

Meinschein (1959, 1961) and Baker (1959, 1960, 1962) believed that adequate amounts of hydrocarbons are available in the original organic material, or as slight modifications therefrom, to account for known petroleum reserves. Meinschein thought primary migration was accomplished through solution of hydrocarbons in water with the aid of colloids. Baker presented evidence for colloidal migration of liquid hydrocarbons. Hodgson *et al.* (1964) stated that hydrocarbons need to be mobilized in concentrations of at least 1 ppm, to account for known petroleum deposits by migration of hydrocarbons in a water medium. With hydrocarbon concentrations of only a few parts per million, the volume of water required would be so large that most of it would necessarily originate at very shallow depths where extremely large percentages of water are present in sediments. According to Baker's (1967) calculations, if the waters expressed from muds carry only 1.8 ppm of hydrocarbons in colloidal solution, they could, if properly focused, explain the known oil accumulations. The vast amount of migrating water assumed by Baker would entail major shallow primary migration. Welte (1965) concluded that primary migration is functional to depths of many thousand feet, but he regarded organic colloidal solution as an important mode of primary migration, and he doubted that colloid-forming (carboxylated) organic material persists below relatively shallow depths.

Weeks (1958a, 1961) advocated the idea of rather shallow origin and primary migration, on the basis of some of the preceding evidence plus

additional points such as: (1) the vast amount of heavy oil and "tar" near basin margins, inferred to require a large drainage area, enormous amounts of water, and high permeability; (2) mechanical flattening of particles in marine deposits after very shallow burial, which would reduce upward loss of fluids; (3) greater evidence for lateral than vertical migration; and (4) the large amounts of oil and oil residues in late Tertiary sediments. The earlier predictions of commercial oil in the Pleistocene were verified by discovery of Pleistocene oil accumulations in offshore Louisiana, reported by Andrews (1961) and Andrews and Stipe (1961). Part of this oil was thought to be indigenous to the Pleistocene.

Landes (1960) concluded that the data on modern sediment hydrocarbons indicate that these hydrocarbons are present in sufficient amounts for eventual commercial accumulation. Rainwater (1963, 1966) believed that oil originated soon after deposition, as indicated by the numerous depositional traps in deltaic facies of the Gulf Coast that are sites of early oil accumulation. Other workers in the Gulf Coast recently have presented similar views (Martin, 1969; Rogers, 1969).

In modern sediment studies of offshore California basins, Emery (1963) noted rather erratic changes in petroleumlike compounds at shallow depths. He reasoned that, if petroleum is a product of diagenesis, petroleumlike substances should increase consistently with depth. In view of the observed irregularities, he attributed the presence of petroleumlike compounds to survival from deposition rather than diagenetic alteration. By implication, he apparently conceived of petroleum as originating at shallow depth from minor transformation of the surviving petroleumlike material.

Hedberg (1964) concluded that relatively shallow primary migration is required if the liquid hydrocarbons are already in oil form. He called attention to the abrupt downward decrease in permeability which would impede primary migration of oil at depth. Later (Hedberg, 1967), he stated his belief that a large part of primary migration is associated with moderately early stages of compaction and water expulsion.

Banks (1966) highlighted evidence for early origin and primary migration, including presence of hydrocarbons in modern sediments, rapid compaction and fluid loss in mud deposits, and early transformations of organic material. Hunt (1967) believed that catalytic action by clays may allow appreciable hydrocarbon formation and migration, even at depths less than 500 ft, and that deep migration would be impeded by low permeability and slow fluid movement. From modern sediment studies in the northeastern Gulf of Mexico, Swanson *et al.* (1968) concluded that processes of crude oil and gas accumulation in sands begin shortly after deposition. They noted concentrations of bitumen (the organic fraction soluble in organic solvents) in the sand, attributable to small hydrodynamic changes.

2,000-Ft Minimal Depth Theory

During the past half-century, many geologists have advocated the 2,000-ft minimal depth level for most oil accumulation. This concept is based on evidence such as: (1) present depths of oil accumulation; (2) amount of overburden apparently removed since time of oil emplacement; (3) apparent timing of trap development; and (4) relation of overburden pressure to gas saturation of the oil.

Gussow (1954) reemphasized the concept of appreciable depth during primary migration, stating that flush quantities of oil globules should migrate out of the source beds into reservoirs only after overburden pressures are sufficiently high to overcome capillary pressures. Up to that point the oil would be relatively immobile because of capillary attraction.

Gussow (1954) applied the term "filter-pressing" to the compaction effects which cause source-bed fluids to be expelled. Later (Gussow, 1955), he concluded that flush primary migration of oil should begin after the deposition of 2,000 to 3,000 ft of overburden. Referring to compaction-depth stages, established by Hedberg (1936) from shale porosity and other data, Gussow (1955) noted that oil should not migrate from source beds during Hedberg's stages 1 and 2 (mechanical rearrangement and dewatering, respectively). Only after Hedberg's stage 3 begins, with the sediment grains in close contact and being mechanically deformed, would capillary attraction be overcome by overburden pressure. The depth range of stage 3 is from 800 or 1,000 ft to about 6,000 ft (Hedberg, 1936). The actual overburden thickness was probably several hundred feet more than this range indicates. Weller (1959) concluded that about 500 ft of overburden has been removed from the Venezuela locality where Hedberg established his depth ranges. This postulated additional thickness is based on a comparison of Hedberg's data with data published by Terzaghi (1925) on artificially compacted clay samples.

Hedberg's (1936) porosity pattern in the depth span from 291 to 3,094 ft seems significant. In the interval 291–2,031 ft, the porosity of lithologically similar argillaceous sediments decreases moder-ately, and apparently rather regularly, from about 34 to about 27 percent. No really major compaction effect seems evident here. More surprising, however, is the very small porosity decrease of less than 2 percent between 2,031 and 3,094 ft. These data imply that the important compaction for primary migration probably does not take place until greater thicknesses of overburden accumulate. If 500 ft is added to the depth, as suggested by Weller (1959), the role of compaction in hydrocarbon mobilization would not appear very significant above a burial depth of 3,600 ft.

Studies of the montmorillonitic clays in the Gulf Coast support this conclusion. Parenthetically, although the mineral name "montmorillonite" is preoccupied by the name "smectite" (Grim, 1968), the former will be used in this report because of its familiarity. Powers (1967) reported that water in Gulf Coast clays remains abundant in the 1,500–3,000-ft depth range, but that most of the water is held in a viscous adsorbed condition which retards compaction. In a relatively pure montmorillonitic clay sediment, the adsorbed water together with the clay layers appears to fill nearly all the sediment volume, leaving very little effective pore space. Actually, intermixture of montmorillonite with other minerals is normal, and a montmorillonitic sediment with sizable proportions of other clays and silt might have more pore space and some permeability. Nevertheless, where the montmorillonitic content is high, conditions would appear unfavorable for compaction and fluid expression in the 1,500–3,000-ft interval.

As further substantiation of a 2,000-ft minimum depth for flush primary migration of oil, Gussow (1955) mentioned hydrocarbon studies by Smith (1954) in cores from the Pelican Island area of southern Louisiana (Table 1). Smith (1954, 1955) reported 11,700 ppm of hydrocarbons in a sandstone at the 2,233-ft depth, compared with 138 and 113 ppm in sands at 314- and 680-ft depths, respectively. Although the hydrocarbon concentration at 2,233 ft is suggestive of incipient petroleum accumulation, it does not necessarily confirm a minimal depth for flush primary migration and accumulation. Because there are no hydrocarbon data for sandstones between the depths of 680 and 2,233 ft, a wide range of minimum depth possibilities could be represented. Hydrocarbon data from clays in the section provide some evidence for the deeper end of the range. The clay core at 2,314 ft, below the high-hydrocarbon sand, contained a bitumen fraction with a much higher percentage of hydrocarbons than was found in the several shallower clays.

Actually, no age relation has been established in these Pelican Island cores between hydrocarbons in the clay and sand. Carbon-14 analyses, which indicated a Holocene age for hydrocarbons in cores from Grand Isle (Smith, 1954), apparently were not run on the Pelican Island core material. Accordingly, the marked concentration of hydrocarbons in the sandstone at 2,233 ft in the Pelican Island cores may be much older than the hydrocarbons subjected to age analysis. Thus, quite possibly this hydrocarbon concentration originated at a much deeper level, from which it could have migrated through a connected reservoir system.

Gussow (1955) utilized knowledge of pressure effects on gas saturation of oil to determine the depths at which various gas-unsaturated oil accumulations in Alberta were emplaced. In accordance with pressure-saturation principles, as an additional increment of overburden was deposited over the position of an oil accumulation, the secondary gas cap should have been dissolved by the oil because of pressure increase. Accordingly, the calculated depth (overburden pressure) level of gas cap disappearance was taken as the approximate depth of original oil accumulation. Most of the examples studied by Gussow showed overburden thicknesses of more than 2,000 ft at the time of oil emplacement, although a few depths were calculated to be much less.

It should be emphasized that application of any gas saturation-pressure method for determination of original depth of oil accumulation presupposes a closed and stable system. A recent detailed discussion by Hoshkiw (1970) stresses that this system is neither completely closed nor completely stable under most natural conditions.

Holmquest (1966) concluded from a field-depth study that minimal overburden for oil accumulation in the Gulf Coast and West Texas provinces was about 2,000 ft. He extended this minimal thickness figure to apply to primary migration as well. However, the corollary of a 2,000-ft or deeper level for most oil accumulation is that secondary migration, and therefore primary migration also, typically begins at still greater depths. Within a reservoir system, direction of migration should generally have an upward component. Hence the levels at which the hydrocarbons originally entered the reservoir system are likely to have been appreciably deeper than the sites of accumulation.

Perhaps the most thoroughly integrated and documented investigation of depths of accumulation was conducted by Philipp et al. (1963), in the producing Mesozoic section of the Gifhorn trough in northwestern Germany. They studied the subsidence history and time of trap development and accumulation, utilizing logs, isopach and paleostructural mapping, time of seal formation, sequential diagenetic changes in the sandstones, and shale depth-compaction-porosity data. They concluded that effective migration probably was initiated at a depth of approximately 500 m (about 1,650 ft), corresponding to a shale porosity of about 30 percent.

This implied relation to shale compaction, however, applies only to the vicinities of the oil accumulations, which generally are near the margins of the Gifhorn depositional trough. On the assumption that the source beds may be in the deeper part of the trough, the depth at which primary migration was initiated could be considerably greater. Discussion reveals (Fuechtbauer, in Philipp et al., 1963) that the trough could have subsided to an 800-m sediment depth or deeper before migration started.

In summary, there is considerable indication that the 2,000-ft minimal depth for primary migration may be too shallow. Depth figures of 3,000 or 3,500 ft, or even deeper levels, appear more likely to represent the beginning of signifi-

Table 1. Distribution of Hydrocarbons with Depth in Clays and Sands, Pelican Island Core, Southern Louisiana (Adapted from Smith, 1954)

DEPTH — FT.	20	72	95	118	168	233	314	354	456	680	1322	2196	2233	2314
SEDIMENT TYPE	SILTY CLAY	CLAY	CLAY	CLAY	CLAY	CLAY	SAND	CLAY	CLAY	SAND	CLAY	LIGNITE AND CLAY	SAND	CLAY
HYDROCARBONS P.P.M.	31	41	43		50	203	138	104	62	113	33	706	11,700	76
% PARAFFINIC-NAPHTHENIC HC'S IN EXTRACT	9.5	11.2	10.4	12.0	10.2	19.3	63.8	20.4	10.4	33.7	8.0	7.1	47.2	29.1
% AROMATIC HC'S IN EXTRACT	5.7	5.7	3.3		4.8	9.9	16.8	5.5	4.1	11.6	4.8	5.0	23.3	9.8
% EXTRACT LEFT ON ALUMINA	69.6	66.6	69.7		67.4	42.5	14.4	56.8	66.2	35.2	69.5	66.5	15.4	42.9

cant primary migration. Other evidence presented on subsequent pages suggests that most primary migration probably occurs at much greater depths.

PROBLEMS OF SHALLOW "IN-PLACE" ORIGIN

The idea that major reef oil can originate from abundant organic matter indigenous to the reef must be considered in light of several environmental factors. First, the reef environment is typically high-energy and oxidizing, whereas petroleum formation requires reducing conditions. Secondly, scavengers are an important part of the reef fauna, and these animals should prevent a thick concentrated accumulation of organic material.

Also, vast amounts of organic matter must accumulate within the reef to produce a relatively small amount of oil. Prior to burial, the bulk of organic constituents would be converted by bacteria to carbon dioxide, water, ammonia, and other products that would be largely dissipated. Only the fatty and waxy part of the organisms plus hard parts would resist aerobic bacterial attack. In the high-energy reef environment, organic accumulation would be disrupted by periodic or continuous flushing. Accordingly, it is difficult to envision a mechanism whereby reefs could ever accommodate enough indigenous organic matter to form more than a very small fraction of any major reef oil accumulation.

The theory that surface or near-surface deposits of humic acid (humate) may give rise to tar-sand oil is an interesting idea. Mainly, it is an attempt to account for the enormous quantity of bitumen in the Athabasca tar sands of Alberta and other similar deposits which have not been adequately explained. As Corbett (1955) pointed out, vast amounts of humic material could have accumulated in the Athabasca area as the Lower Cretaceous "tar"-containing sands originally were deposited. Possibly analogous humates are accumulating today in large amounts in modern sands along the northwest coast of Florida. Swanson and Palacas (1965) concluded that these humates form when fresh swamp-type water laden with humic material comes in contact with brackish or marine water.

However, certain chemical problems arise with the humic acid theory. As Welte (1965) and Fuloria (1967) pointed out, humic acid is an oxidized substance, scarcely harmonizing with the reducing conditions necessary to generate the ingredients for oil and primary "tar" material. Moreover, there apparently is no evidence of any chemical reaction for the transformation of hu-

mate to great abundances of petroleum substance under the shallow-depth conditions characteristic of most tar-sand diagenesis.

DIFFICULTIES WITH SHALLOW SOURCE-BED CONCEPTS

Many investigations have been conducted on the uppermost several feet of a large variety of modern fine-grained sediments. Much less information is available in the depth range from several feet to 1,500-2,000 ft. Therefore the dependability of conclusions as to extremely shallow processes is high, whereas the evidence for deeper processes is more limited.

From what is known, many characteristics and relations represented in sediments subjected to only shallow burial appear to militate against significant hydrocarbon origin and primary migration. Tables 2 and 3 summarize this information for terrigenous and carbonate source-type sediments. Utilizing the same format, Tables 4 and 5 outline the favorable aspects at intermediate and greater depths.

Dilute Hydrocarbons in Modern Sediments

Modern mud sediments average only about 1 percent organic matter (Gehman, 1962); and normally, less than 2 percent of the organic matter consists of hydrocarbons (Kidwell and Dickey, 1959). Concentration of hydrocarbons in most modern sediments is considerably less than 100 ppm. As examples, Gulf Coast sediments analyzed average 72 ppm of hydrocarbons (Smith, 1954), and samples from the Orinoco delta shallow subsurface average about 60 ppm (Kidwell and Hunt, 1958). Hydrocarbons in sediments of offshore southern California are unusually high for modern marine muds, averaging 168 ppm in three of the basins (Orr and Emery, 1956).

These hydrocarbons under shallow burial seem too dilute to constitute an adequate potential source for major hydrocarbon accumulations. The further problems of incorporating hydrocarbons into interstitial fluids at shallow depths and of preventing loss of fluids to the surface must be considered.

Some lacustrine muds have relatively high hydrocarbon concentrations (Smith, 1954; Judson and Murray, 1956). However, they tend to lack other source-bed attributes. Commonly the clay component is much less abundant than in terrigenous marine source beds; and this, in combination with high organic content, should reduce the water-retention capacity and ultimate catalytic activity. Accordingly, with moderate or deeper burial, the more highly organic lacustrine sediments should tend to become oil shales rather

Table 2. Summary Attributes of Terrigenous Source-Type Sediments under Shallow Burial

ORGANIC FRACTION		MINERAL FRACTION		SEDIMENT FLUIDS	
CHARACTERISTIC	CONSEQUENCE	CHARACTERISTIC	CONSEQUENCE	CHARACTERISTIC	CONSEQUENCE
1. LOW TEMPERATURE, LOW OVERBURDEN PRESSURE	a. THERMOCATALYTIC ACTIVITY FOR BREAKDOWN OF KEROGEN WEAK OR ABSENT	1. LOW TEMPERATURE, LOW OVERBURDEN PRESSURE	a. MINERAL CATALYTIC ACTIVITY FOR REDUCTION OF ORGANIC MATTER WEAK OR ABSENT b. COMPACTION WEAK	1. LOW TEMPERATURE, LOW OVERBURDEN PRESSURE	a. ORGANIC SOLUTION ACTIVITY SLIGHT
2. SPARSE BITUMEN	a. VERY LIMITED AND DILUTE SOURCE OF HYDROCARBONS b. SUSCEPTIBILITY TO ORGANIC ADSORPTION	2. ABUNDANT CLAYS	a. MUCH ADSORPTION OF ORGANIC MATTER b. TENDENCY TO RETAIN CONSIDERABLE PORE WATER c. PRESENCE OF THICK VISCOUS WATER LAYERS AROUND SEDIMENT PARTICLES AND BETWEEN CLAY MINERAL LAYERS	2. ABUNDANT PORE WATER	a. ABUNDANT HIGHLY MOBILE WATER b. DELAY OF CATALYTIC ACTIVITY FOR BREAKDOWN OF KEROGEN
3. VERY SPARSE LIQUID HYDROCARBONS	a. VERY DILUTE DISSEMINATION OF LIQUID HYDROCARBONS b. SUSCEPTIBILITY TO ADSORPTION c. LIMITED CONTACTS WITH INTERSTITIAL FLUIDS			3. ABUNDANT HIGHLY MOBILE WATER	a. RELATIVELY RAPID EXPRESSION OF WATER FROM MUDS DURING EARLY COMPACTION b. MUCH LOSS OF WATER BY ESCAPE TO THE SURFACE
4. VERY ABUNDANT KEROGEN	a. OLEOPHILIC TENDENCY b. IMMOBILIZATION OF BITUMEN AND HYDROCARBONS	3. MUCH ADSORPTION OF ORGANIC MATTER	a. IMMOBILIZATION OF BITUMENS AND HYDROCARBONS	4. DELAY OF CATALYTIC ACTIVITY FOR BREAKDOWN OF KEROGEN	a. LITTLE FORMATION OF BITUMEN AND HYDROCARBONS
		4. TENDENCY TO RETAIN CONSIDERABLE PORE WATER	a. DELAY OF CATALYTIC ACTION FOR BREAKDOWN OF KEROGEN	5. LOW ORGANIC CONTENT	a. LITTLE CONCENTRATION OF HYDROCARBONS
		5. PRESENCE OF THICK VISCOUS WATER LAYERS AROUND SEDIMENT PARTICLES AND BETWEEN CLAY MINERAL	a. DELAY OF CATALYTIC ACTION b. BARRIER TO MOBILITY OF BITUMEN AND HYDROCARBONS c. COMPACTION INSUFFICIENT TO OVERCOME SURFACE TENSION OF OIL		

Correction. 3d col., no. 5, last line should read: clay mineral layers

Table 3. Summary Attributes of Carbonate Source-Type Sediments under Shallow Burial

ORGANIC FRACTION		MINERAL FRACTION		SEDIMENT FLUIDS	
CHARACTERISTIC	CONSEQUENCE	CHARACTERISTIC	CONSEQUENCE	CHARACTERISTIC	CONSEQUENCE
1 LOW TEMPERATURE, LOW OVERBURDEN PRESSURE	a. THERMOCATALYTIC ACTIVITY FOR BREAKDOWN OF KEROGEN WEAK OR ABSENT	1. LOW TEMPERATURE, LOW OVERBURDEN PRESSURE	a. LITTLE OR NO MINERAL CATALYTIC ACTIVITY b. COMPACTION WEAK	1. LOW TEMPERATURE, LOW OVERBURDEN PRESSURE	a. PRESSURE SOLUTION ACTIVITY SLIGHT b. ORGANIC SOLUTION ACTIVITY SLIGHT
2. SPARSE BITUMEN	a. VERY LIMITED AND DILUTE SOURCE OF HYDROCARBONS b. SUSCEPTIBILITY TO ADSORPTION	2. LARGELY CARBONATES	a. WEAK ADSORPTION OF ORGANIC MATTER b. MOST INTERSTITIAL WATER LOST VERY EARLY c. SOME VERY EARLY CEMENTATION AND CRYSTALLIZATION d. CATALYTIC ACTIVITY WEAK OR ABSENT	2 LIMITED ORIGINAL AND SECONDARY PORE WATER	a. LIMITED WATER VOLUME FOR TRANSPORT OF BITUMEN AND HYDROCARBONS
3. VERY SPARSE LIQUID HYDROCARBONS	a. VERY DILUTE DISSEMINATION OF LIQUID HYDROCARBONS b. SUSCEPTIBILITY TO ADSORPTION c. LIMITED CONTACTS WITH INTERSTITIAL FLUIDS			3. HIGH ALKALINE EARTH CONTENT	a. COLLOIDAL MIGRATION UNLIKELY
4 FAIRLY ABUNDANT KEROGEN	a. OLEOPHILIC TENDENCY b. IMMOBILIZATION OF BITUMEN AND HYDROCARBONS	3. WEAK ADSORPTION OF ORGANIC MATTER	a. WEAK ADSORPTION OF BITUMEN AND HYDROCARBONS	4. LOW ORGANIC CONTENT	a. LITTLE CONCENTRATION OF HYDROCARBONS
		4. MOST INTERSTITIAL WATER LOST VERY EARLY	a. WATER FOR PRIMARY MIGRATION RATHER LIMITED		
		5. SOME VERY EARLY CEMENTATION AND CRYSTALLIZATION	a. COMPACTION RELATIVELY INEFFECTIVE b. ADDITIONAL LOSS OF WATER		
		6. CATALYTIC ACTIVITY WEAK OR ABSENT	a. LITTLE OR NO FORMATION OF BITUMEN AND HYDROCARBONS		

Table 4. Summary Attributes of Terrigenous Source-Type Sediments under Intermediate-Depth or Deeper Burial

ORGANIC FRACTION		MINERAL FRACTION		SEDIMENT FLUIDS	
CHARACTERISTIC	CONSEQUENCE	CHARACTERISTIC	CONSEQUENCE	CHARACTERISTIC	CONSEQUENCE
1. DOWNWARDLY INCREASING TEMPERATURE AND PRESSURE	a. RELATIVELY LARGE AMOUNTS OF BITUMEN, LIQUID HYDROCARBONS AND GASES, FORMED FROM KEROGEN b. KINETIC ENERGY INCREASED	1. DOWNWARDLY INCREASING TEMPERATURE AND PRESSURE	a. DECREASE OF VISCOUS WATER ASSOCIATED WITH CLAYS b. CHEMICAL ACTIVITY GREATLY INCREASED c. COMPACTION WITH GRAINS CONSIDERABLY DEFORMED	1. DOWNWARDLY INCREASING TEMPERATURE AND PRESSURE	a. INCREASED SOLUBILITY, LIQUID HYDROCARBONS, AND GASES IN WATER b. INCREASED SOLUBILITY OF LIQUID HYDROCARBONS IN GAS c. OVERBURDEN PRESSURE IMPARTED TO FLUIDS
2. ADEQUATE AMOUNTS OF BITUMEN	a. ADEQUATE AMOUNTS OF HYDROCARBON PRECURSORS b. ADEQUATELY CONCENTRATED SOURCE OF HYDROCARBONS c. LESS SUSCEPTIBILITY TO ADSORPTION d. COLLOID SOURCE FOR MIGRATION	2. DECREASE OF VISCOUS WATER ASSOCIATED WITH CLAYS	a. INCREASED CATALYTIC ACTIVITY b. ADDED SOURCE OF PORE WATER c. MAINTENANCE OF POROSITY d. IMPROVEMENT OF SOURCE BED DRAINAGE e. INCREASED FLUID PRESSURE	2. LOW ALKALINE EARTH CONTENT 3. MODERATELY HIGH ORGANIC CONTENT	a. COLLOIDAL SOLUTION ENHANCED a. ADEQUATE AMOUNT OF BITUMEN-HYDROCARBON MATERIAL TRANSPORTED BY WATER
3. ADEQUATE AMOUNTS OF LIQUID HYDROCARBONS	a. ADEQUATE AMOUNTS OF DISSEMINATED "MICRO-OIL" b. ADEQUATE CONTACT WITH INTERSTITIAL FLUIDS c. LESS SUSCEPTIBILITY TO ADSORPTION	3. CHEMICAL ACTIVITY GREATLY INCREASED	a. THERMOCATALYTIC TRANSFORMATION OF KEROGEN TO BITUMEN AND HYDROCARBONS	4. OVERBURDEN PRESSURE IMPARTED TO FLUIDS	a. OIL-IN-GAS MIGRATION ENHANCED b. FLUIDS FORCED TOWARD RESERVOIRS
4. LARGE AMOUNTS OF GAS	a. AVAILABILITY FOR OIL-IN-GAS MIGRATION	4. COMPACTION WITH GRAINS CONSIDERABLY DEFORMED	a. OVERBURDEN PRESSURE IMPARTED TO FLUIDS b. FLUIDS FORCED TOWARD RESERVOIRS		
5. INCREASED KINETIC ENERGY	a. LESS SUSCEPTIBILITY TO ADSORPTION b. AVAILABILITY TO MIGRATING FLUIDS				

Table 5. Summary Attributes of Carbonate Source-Type Sediments under Intermediate-Depth or Deeper Burial

ORGANIC FRACTION		MINERAL FRACTION		SEDIMENT FLUIDS	
CHARACTERISTIC	CONSEQUENCE	CHARACTERISTIC	CONSEQUENCE	CHARACTERISTIC	CONSEQUENCE
1. DOWNWARDLY INCREASING TEMPERATURE AND PRESSURE	a. RELATIVELY LARGE AMOUNTS OF BITUMEN, LIQUID HYDROCARBONS, AND GASES FORMED FROM KEROGEN b. KINETIC ENERGY INCREASED	1. DOWNWARDLY INCREASING TEMPERATURE AND PRESSURE	a. EXTENSIVE CEMENTATION AND CRYSTALLIZATION OF CARBONATE b. PRESSURE-SOLUTION ACTIVITY, STYLOLITE FORMATION c. FRACTURE FORMATION	1. DOWNWARDLY INCREASING TEMPERATURE AND PRESSURE	a. SOME INCREASE IN SOLUBILITY OF LIQUID HYDROCARBONS IN WATER b. MUCH INCREASED SOLUBILITY OF LIQUID HYDROCARBONS IN GAS AND GAS IN WATER c. INCREASED PRESSURE ON FLUIDS d. PRESSURE-SOLUTION ACTIVITY
2. ADEQUATE AMOUNTS OF BITUMEN	a. ADEQUATE AMOUNTS OF HYDROCARBON PRECURSORS b. ADEQUATELY CONCENTRATED SOURCE OF HYDROCARBONS c. LESS SUSCEPTIBILITY TO ADSORPTION	2. EXTENSIVE CEMENTATION AND CRYSTALLIZATION OF CARBONATE	a. INCREASED PRESSURE ON FLUIDS b. FLUIDS FORCED TOWARD LARGER PORE SYSTEMS	2. MODERATELY HIGH ORGANIC CONTENT ESPECIALLY RELATED TO GASES	a. BASIS FOR HYDROCARBON TRANSFER TO RESERVOIRS
3. ADEQUATE AMOUNTS OF LIQUID HYDROCARBONS	a. ADEQUATE AMOUNTS OF DISSEMINATED "MICRO-OIL" b. LESS SUSCEPTIBILITY TO ADSORPTION	3. STYLOLITE FORMATION	a. FLUIDS FORCED TOWARD CARBONATE RESERVOIRS	3. INCREASED PRESSURE ON FLUIDS	a. OIL-IN-GAS SOLUTION ENHANCED
4. LARGE AMOUNTS OF GAS	a. AVAILABILITY FOR OIL-IN-GAS MIGRATION	4. FRACTURE FORMATION	a. FLUIDS CHANNELED TOWARD CARBONATE RESERVOIRS	4. PRESSURE SOLUTION ACTIVITY	a. FLUIDS FORCED TOWARD CARBONATE RESERVOIRS
5. INCREASED KINETIC ENERGY	a. LESS SUSCEPTIBILITY TO ADSORPTION b. AVAILABILITY TO MIGRATING FLUIDS				

than important hydrocarbon source beds. An example is the Green River oil shale which, according to Weaver (1960), has a low expandable clay-mineral content, suggesting a weak catalytic role. Weaver (1967) further postulated that an early almost complete loss of water content in the Green River sediments may explain its low source-bed potential. Hill (1959) mentioned that nonmarine sediments are much less likely to be water-wet than are marine deposits; hence the former may tend to become oleophilic (oil-attracting). The oleophilic characteristic would restrict or prevent the source-bed function.

Among marine muds, even the offshore California hydrocarbons are much more dilute than those of many presumed ancient source rocks in petroleum areas (Hunt and Jamieson, 1956). Hunt (1961) reported an average (arithmetic mean) of 300 ppm of hydrocarbons in 360 shale samples and 340 ppm in nonreservoir limestone collections. Analysis by Gehman (1962) of a more extensive collection (1,066 shales and 346 limestones) gave comparable results, expressed as a geometric mean of 96 ppm for shales and 98 ppm for limestones. The more highly organic source-type rocks have far larger hydrocarbon concentrations. Hunt (1961) reported 2,000-4,000 ppm of hydrocarbons in several highly bituminous shales.

Thus, apparently the environment which accompanies increasing overburden during sediment consolidation causes considerable additional generation of hydrocarbons from the organic component. As a result of progressively higher temperatures and compaction, large amounts of hydrocarbons may continue to form and migrate to reservoirs. The ancient nonreservoir samples, therefore, may contain only a comparatively small remnant of the total hydrocarbons produced during diagenesis. This end product may have formed mostly *after* primary migration ceased. Hydrocarbons should continue to be generated in the active chemical environment at depth, long after further migration is precluded by diminished water content and nearly complete loss of permeability.

Problems with Bitumen Components of Modern Sediments

Important contrasts are found between bitumen (extractable organic matter) composition of modern sediments and that of ancient sediments and crude oil. These differences, together with the occurrence of dilute hydrocarbons in modern sediments, suggest that crude oil is not merely a concentration of hydrocarbons and related compounds from the original deposits. As a logical

alternative, new hydrocarbons apparently are formed from the main bulk of organic matter, and hydrocarbons already present may be altered to types similar to those of crude oil.

Stevens et al. (1956) concluded that the hydrocarbon-asphaltic (bitumen) mixtures in modern sediments possibly would never become crude oil. In ancient rocks and crude oils, these mixtures have a much higher hydrocarbon to asphaltic ratio, and a much more complex array of aromatic compounds. In modern sediments, the wax range of normal paraffin hydrocarbons has an odd-carbon preference, meaning that a majority of the molecules have an odd number of carbon atoms. In contrast, ancient rocks and crude oils tend to have smooth distributions of wax-range hydrocarbons.

Subsequently, Bray and Evans (1961, 1965) confirmed these findings with analyses of modern sediments from many environments and numerous ancient nonreservoir rocks. They found consistently high odd-carbon preference in modern sediments, and moderate, slight, or no preference in ancient rocks.

Dunton and Hunt (1962) observed that modern sediments do not contain light hydrocarbons in the C_4 to C_8 range, whereas in ancient rocks and oils, this range is well represented. These writers also summarized the following earlier work. Sokolov (1957) and Veber and Turkel'taub (1958) reported little or no representation of the C_2 to C_{14} hydrocarbons in modern sediments of the Black and Caspian Seas. Emery and Hoggan (1958) found less than 1 ppm of this range in modern sediments of offshore California basins. Erdman et al. (1958) did not detect the benzene-xylene suite in modern sediments, and found only methane and heptane in the C_1-C_8 range of paraffin hydrocarbons.

Kvenvolden (1962) was unable to find any of the C_8-C_{13} range hydrocarbons in modern sediments of San Francisco Bay. Bray and Evans (1965) emphasized the absence of gasoline hydrocarbons (C_5-C_{10}) in modern sediments. Kontorovich et al. (1965), as reported by Kartsev et al. (1971) could not detect any low-boiling (below 325°C) liquid hydrocarbons in clay samples from the upper 500 m of the depositional section.

These gaps in hydrocarbon composition are well filled in petroleum and for the most part in ancient rocks. Bray and Evans (1965) noted the apparently evolutionary nature of hydrocarbon compositions, beginning with modern muds and extending through older nonreservoir sediments to crude oil.

Theoretically, on the basis of relative solubilities, solution or colloidal migration processes could account for the light liquid hydrocarbon concentrations and the smooth odd-even carbon number distributions in wax-range normal paraffins of crude oil. This would leave an odd-carbon preference in the "immature" source bed. To explain the typical ancient source-bed pattern of smooth carbon number distributions, a post-migration maturation of source-bed hydrocarbons would have to be postulated. The absence or only trace occurrences of certain hydrocarbon ranges in modern sediments presents a serious concentration problem in accounting for the abundances of these same hydrocarbons in ancient rocks and crude oil, unless major alterations of the bulk organic matter take place.

An alternate suggestion is the shallow-depth migration of hydrocarbons and hydrocarbon precursors to reservoirs, where they could undergo the process of gradual maturation as the depositional overburden becomes thicker. However, where sediments have not yet been buried by considerable overburden, free or mobile occurrences of generally recognized hydrocarbon precursors are sparse. These occurrences are represented in the bitumen part of the organic matter. According to Bordenave (1967), bitumen concentrations, including hydrocarbons, average only about 5 percent of the organic matter in modern sediments. In contrast, Philippi (1957) reported 10-20 percent bitumens in the organic matter of many ancient source-type sediments.

The best explanation of bitumen and hydrocarbon differences between modern and deeper or older sediments seems to lie in the higher temperature-pressure environment at greater depth, where chemical changes are much more rapid and drastic. Breger (1960) and Hunt (1968) mentioned a rule of chemistry stating that reaction rate is doubled with every 10° rise in temperature. Philippi (1965) noted that for each rise of 10° C, the rate of most chemical reactions is increased by a factor of two to three.

Problems with Fatty Acid Role

It has long been recognized that fatty acids are abundant in organisms, relatively resistant to decay, and structurally similar to certain hydrocarbons. In many cases only a straightforward chemical transformation, largely decarboxylation, seems required to obtain hydrocarbons from fatty acids.

Cooper (1962) and Cooper and Bray (1963) proposed a model for decarboxylation of fatty acids to account for the odd-carbon preference in the relatively long-chain normal paraffin hydrocarbons in modern sediments. Kvenvolden (1965, 1966a, 1966b) and Kvenvolden and Weiser

(1967), through an analysis of normal paraffin and fatty acid data from modern and ancient sediments, evolved a model from the fatty-acid component of modern sediments which explains the origin of hydrocarbons and fatty acids in ancient sediment. Chemical transformations were believed to include decarboxylation, hydrogenation, cleavage of hydrocarbon chains, and possibly carbon addition.

Jurg and Eisma (1964) and Eisma and Jurg (1967) demonstrated that the carbon-number preference problem is resolved by thermocatalytic effects. They showed that smooth distributions of odd- and even-numbered carbon molecules of normal paraffins result when behenic acid (a fatty acid) is subjected to relatively high temperatures (200° C and higher) in the presence of clay minerals. Moreover, they demonstrated that both longer and shorter carbon chains develop during these treatments. The results help to explain the presence in crude oil of high-molecular-weight normal paraffins and their smooth carbon-number distributions. If only chain-splitting were involved, the long-chain paraffins would be residual, and they would retain the odd-carbon preference resulting from decarboxylation of preponderantly even-numbered-carbon fatty acid molecules.

It should be stressed that the foregoing chemical transformations take place in the laboratory only under relatively high temperatures. Hence they cannot be assumed to occur in the shallow part of the sedimentary section.

A serious objection to fatty acids as direct progenitors of petroleum hydrocarbons was emphasized by Abelson and coauthors (Abelson, 1963, 1967; Abelson and Parker, 1962; Abelson et al., 1964). They found that most of the free fatty acid of organisms disappears under reducing conditions, soon after deposition of the incorporating muds. Only about 0.1 percent of the organic matter in these sediments consists of fatty acids. It was concluded by inference that the disappearing fatty acids probably are transformed to kerogen, the main insoluble bulk of organic matter in sediments.

Several other investigators have noted the low incidence of free fatty acids in modern sediments. Trask (1932), in his classic studies of modern sediment organic matter, found only 20-60 ppm of fatty acids in marine muds. Emery (1960) and Cooper (1962) reported their scarcity in modern sediment samples. Kroepelin (1967) and Eisma (1967) reemphasized these findings in relation to oil origin problems. Peterson (1967) reported a maximum of only 90 ppm of fatty acids in modern shallow-water marine sediments, and most of his samples had a much lower fatty acid concentration.

The low free-fatty-acid content of modern sediments militates against any really significant fatty-acid contribution to oil source materials at shallow depths.

Barriers to Mobility of Hydrocarbons and Other Organic Components

Part of the dilute hydrocarbon and nonhydrocarbon bitumen content of sediments buried to shallow depths undoubtedly is encased within small or larger organic masses. The amount of organic matter through which these substances must move to be incorporated into migrating waters should constitute a significant barrier.

The oleophilic (oil-attracting) character of the carbonaceous part of the organic matter would tend to hold the hydrocarbons (Erdman, 1961). Much of the organic matter, because it is highly polar, is strongly attracted to clays and other minerals (Grim, 1953, 1968; Erdman, 1961; Van Olphen, 1963). From recent analysis of interstitial waters in Lake Maracaibo muds, it was inferred that the water-soluble organic matter of oil-progenitor type is strongly adsorbed on the sediment surface (Spencer and Harvey, 1971).

Experiments with a clay-organic (montmorillonite-pyridium) complex in the presence of hydrocarbons (benzene) show that the complexed material is oleophilic, and that hydrocarbons are adsorbed so strongly that desorption of the closest molecular layer of hydrocarbons probably would not occur within the normal sedimentary section (Van Olphen, 1963). Additional hydrocarbons are adsorbed less strongly outside this inner hydrocarbon layer. It follows that hydrocarbons in sediments under shallow burial should be attracted to clay-organic complexes where present.

It has been suggested (Erdman, 1965) that the ratio of hydrocarbons to kerogen may be an important determinant of hydrocarbon mobility. The smaller the ratio, the larger the attractive effect should be. Accordingly, under shallow burial where this ratio is extremely small, hydrocarbon mobility should be severely restricted.

Thus it seems likely that the organic and clay components of shallow sediments together constitute a barrier to hydrocarbon mobility which would not be overcome without considerable energy. At shallow depths, the energy input from overburden pressures and heat would be relatively small, and there appears to be no other mechanism to provide the energy required for making hydrocarbons and hydrocarbon precursors effectively accessible to migrating fluids.

Natural colloids should assist in the solution of hydrocarbons in pore waters, but according to Baker (1967), competition may be difficult for as little as 1 ppm to a few parts per million of hydrocarbons in water solution. Baker apparently visualized relatively shallow migration and did not discuss the probable acceleration of the solution process with higher pressures and temperatures at greater depths.

Hunt (1968) has suggested the possibility that the heavier hydrocarbons of crude oil may have originated from migrating ketones, acids, and esters. Migration of these nonhydrocarbon compounds is understandable because they are more water soluble than their hydrocarbon counterparts. However, these substances, which are rather dilute in mud sediments to begin with, are subject to organic barriers; and, because they are polar, they are also attracted to clay minerals. Therefore it is difficult to visualize really significant concentrations of these substances in shallow-depth pore solutions.

Many laboratory studies show that the catalytic action of clays on organic matter facilitates the formation of hydrocarbons together with other organic products which would increase the oil-like and oil-related fraction of organic matter soluble in water. However, the clays used in most of the experiments were relatively dry. In contrast, sediments are typically water-wet under marine conditions, and under shallow burial they have a high water content. Brooks (1952), a leading proponent of the catalytic role, admitted that presence of adsorbed water films on minerals would greatly reduce their catalytic effect. Landes (1959), Welte (1965), and Barbat (1967) also called attention to the probable barrier effects of these films. Prior to relatively deep burial, the thickness of water interlayers associated with expandable clay minerals, such as montmorillonite, should militate against intensive catalytic activity.

Califet-Debyser and Oudin (1966) reported a decrease in catalytic activity when water was added to dry montmorillonite in their experiments on evolution of an aromatic fraction of crude oil. Jurg and Eisma (1964) and Eisma and Jurg (1967) found that water greatly reduced the catalytic effectiveness of both montmorillonite and kaolinite. In their experiments, only about 180 ppm of behenic acid were transformed to low-molecular-weight hydrocarbons when a mixture of kaolinite, behenic acid, and water was heated at 200°C for 75 hours. This was only 2–3 percent as much light hydrocarbon product as was formed in the absence of water. Experiments by Erdman (1967) indicated some catalytic activity by water-wet clays in transformation of amino acids to hydrocarbons.

Relatively high temperatures were required to obtain even the small hydrocarbon product in the preceding investigations. Therefore it seems unwarranted to conclude that these laboratory results constitute evidence for a significant mineral catalytic role within the shallow part of the sedimentary section.

Hedberg (1964) postulated that mineral catalysts may have established direct contact with source material early in the history of a sediment. Although this appears to be a common occurrence, the importance of the catalytic role is doubtful. Typically, a marine sediment is derived from a nonmarine environment through the agency of stream transport. Some of the sediment may not be water-wet when it first encounters the marine influence, and possibly some marine organic matter is adsorbed before it becomes water-wet. Also, the clays may have retained marine organic matter from a previous depositional cycle or adsorbed nonmarine organic matter prior to entering the marine environment. In these cases catalysis possibly could begin quite early, but it would be limited because of low temperatures. Subsequently, if the sediments became more water-wet, catalysis would decrease. If the adsorbed organic component is sufficiently abundant, an effectively water-wet condition might not develop at all. In that event, catalytic activity could perhaps form some hydrocarbons from the organic matter, but the hydrocarbons would be so attracted to the organo-clay complex that they would not be accessible, at shallow depths, to migrating fluids.

In sediments with moderate or thicker overburden, the smaller amounts of water, higher temperatures and pressures, and more intimate overall association of constituents should accentuate the catalytic process. Much more oil-like material should be formed, and its concentration in pore solutions should be greatly increased. Laboratory studies, summarized by Sokolov et al. (1963), show that liquid hydrocarbons together with certain other organic materials (presumably including acids, esters, ketones, etc.) are much more water soluble as the temperature rises.

Water Problem

Pore water in muds is enormously abundant at very shallow depths. Some advocates of shallow migration and accumulation have apparently assumed that much of this shallow-sediment water is effective in hydrocarbon migration.

Baker (1967) concluded that a concentration of only 1.8 ppm of hydrocarbons in migrating waters theoretically could account for the known oil accumulations. This calculation was based on the availability of at least as much water, by weight, as the weight of the mineral fraction of the sediment. The assumption also was made that only about 3.6 percent of the hydrocarbons in a basin would ever reach reservoirs. This figure was taken from a study by Hunt (1961) of the nonreservoir and reservoir hydrocarbons in the Powder River basin.

Emery and Rittenberg (1952) estimated that only about 20 percent water by weight remains at the 150-ft depth in the Los Angeles basin. Because deeper sediments are reportedly finer grained on average, their water retention with similar overburden is probably somewhat greater. Kidwell and Hunt (1958) reported only about 32 percent water, by volume, in muds at the 190-ft depth in the Pedernales area of Venezuela. Although the sediments analyzed are probably coarser than average and the amount of retained clay-interlayer water is uncertain, these computations clearly demonstrate that a very large percentage of the water originally present in sediments is lost prior to the accumulation of 150-200 ft of overburden.

It is sometimes assumed that upward loss of fluids in shallow compacting fine-grained sediments is curtailed by flattening of clay or mud particles. This process would increase significantly the lateral to vertical permeability ratio. For example, Kidwell and Hunt (1958) found this ratio to be about two at a depth of 200 ft in the Pedernales, Venezuela, section. But regardless of the downward increase in this ratio, the destination of water movement in muds at such shallow depths is the surface. Fluids have a strong tendency to migrate in any direction of markedly reduced pressure, and in shallow compacting muds this direction typically would be upward. Temporary lateral detours would be expected beneath unusually impermeable beds and along aquifers, but if these strata are quite shallow, they normally extend to the surface within short distances.

Lateral fluid movement in shallow sands would be facilitated where fluid pressure in excess of hydrostatic develops in the associated muds. An example is the Pedernales, Venezuela, locality, where Kidwell and Hunt (1958) reported a maximum of 16 psi excess over hydrostatic pressure in muds within the 80-160-ft depth interval. According to Hedberg (1936) and Weller (1959), mud particles are separated more or less completely by water down to depths of several hundred feet. Fluid pressure in these shallow muds should be somewhat in excess of hydrostatic because overburden pressure is exerted not only by water but also by the denser sediment component. The excess pressure has been verified recently by laboratory experiments simulating comparable burial conditions (Smith et al., 1971). As fluid pressure in associated shallow sands is typically hydrostatic, the somewhat higher mud pressures may aid in transfer of fluids to the sands. However, these fluids normally would be delivered to the surface within limited distances, either through overlying muds or at outcrop.

At greater depths, most of the mud grains are in reasonably close contact. Under these conditions hydrostatic pressures would be present if the sediment were a persistent sand. In muds, however, the amount of pore water is much reduced and pore connections are narrower and more tortuous. An increasing percentage of the remaining water is bound to mineral surfaces in a more viscous form. Accordingly, any buoyant effects should be minimized. Moreover, clay mineral particles are relatively soft and pliable, and they are much less resistant to overburden compacting influences than are sand deposits. Hence, with increasing depth, mud fluids should be increasingly subject to the total weight of sediment overburden, of which the density progressively increases because of decreasing percentages of water. In contrast, pore water in sands maintains a relatively unobstructed continuity. Therefore sand fluid pressures should remain much closer to hydrostatic, provided the sand unit persists an appreciable distance updip. It follows that, as overburden thickness increases, fluid pressure in compacting terrigenous muds or shales commonly becomes much greater than in associated sands. Source-bed fluids would migrate into the sand bed, where the fluid pressure is lower, and would then move through the sand to areas with minimal overburden where pressures are also minimal. There the water would preferentially escape by upward leakage through the cap rock, leaving behind progressively increasing concentrations of hydrocarbons.

The above model recognizes that most of the original sediment water is lost before significant migration and accumulation of liquid hydrocarbons can take place. It also shows that burial to moderate depths is necessary to establish adequate pressure gradients and traps. However, though the role of water is specified, no guarantee is given that enough mobile water is present in the source beds for primary migration to be accomplished adequately.

Another weakness of the shallow migration concept is the trace occurrence of hydrocarbons and the dilute concentrations of total organic matter in shallow interstitial waters. Even in highly organic shallow muds, the interstitial waters are surprisingly low in total organic matter, much less hydrocarbons. Starikova (1959) reported a total organic content of only 14-32 ppm, and 32-44 ppm, respectively, in interstitial fluids squeezed from muds collected in the Black Sea and Sea of Azov. Later (Starikova, 1970), in a more comprehensive study of modern and "ancient" (Pleistocene?) sediments of the Black Sea and other areas, he found a maximum of 62.6 ppm of organic carbon in interstitial waters at a depth of about 18 ft below the water-sediment interface. However, down to the maximum depths sampled (25-30 ft), most of the interstitial fluid organic concentrations were less than 30 ppm, even in euxinic sediments. Presumably only a small part of these occurrences were hydrocarbons. Kidwell and Hunt (1958) analyzed waters from shallow sands of the Pedernales area in Venezuela, for hydrocarbons and found less than 1 ppm. They calculated that up to 16 ppm possibly are present in the interstitial waters because sand and clay produced with the formation water contained excess hydrocarbon concentrations which apparently had been absorbed from the waters. Nevertheless, the results suggest that the average concentration in interstitial waters is probably no more than a few parts per million, which appears inadequate for the accumulation of petroleum.

The sand-lens trap described by Kidwell and Hunt (1958) is quite shallow (110 ft deep), but the hydrocarbon concentration in the trap is only 160 ppm. Therefore, at best, this can be regarded as only a minimal tendency toward incipient accumulation. For a recognizable oil emplacement, enormously greater concentration would be required, and it is difficult to conceive of such a concentration by shallow fluid movements. One could calculate that this sand lens, which is about 5,000 years old, could in 1,000,000 years contain 32,000 ppm of hydrocarbons, by a continuation of the rate of concentration to date. However, progressively less water will move through the trap as compaction continues, and there is no evidence that water migrating toward the trap would become appreciably more enriched in hydrocarbons, even if as much as 1,500 ft of overburden were added. In other sands of the shallow Pedernales section, hydrocarbon concentrations are only about 40 ppm (Kidwell and Hunt, 1958), and the migrating waters in these sands apparently escape directly to the surface.

The most readily conceivable source of an adequate water supply at depth is expandable clay, such as montmorillonite and mixed-layer illite-montmorillonite. As already mentioned (M. C. Powers, 1967), montmorillonitic clays at depths of 1,500-3,000 ft have extremely large amounts of adsorbed water but only minor amounts of pore water. However, at depths exceeding several thousand feet, the amount of pore water in sediments containing abundant expandable clays apparently is sufficient for primary migration (Weaver, 1960).

Much of the depth-pattern investigation of water in montmorillonitic sediments has been done in the Gulf Coast. Most cores from a well in the Mississippi delta area had 15-20 percent pore water in the depth range of about 4,500-14,000 ft (Weaver and Beck, 1971). Porosity is about 20 percent at 8,000 ft depth in the Weeks Island area of southern Louisiana (Dickinson, 1953), and in general averages 10 percent at 10,000 ft in the Gulf Coast Miocene (Brown, 1969). In the Gulf Coast as a whole, about 10-15 percent of the sediment is inferred to consist of mobilized water at depths ranging from several thousand to 15,000 ft, depending on variations in the geothermal gradient (Burst, 1969). This water, plus much of the earlier migrating water, is believed to become available during the "dehydration" of predominantly montmorillonitic and mixed-layer clays, as overburden thickness increases (Burst, 1959, 1969; Weaver, 1960; Powers, 1967; Weaver and Beck, 1969, 1971; Perry and Hower, 1970a, b; Weaver and Wampler, 1970).

As to effectiveness for primary migration, Burst (1969) believed that a 10-15 percent volume of fluid in sediments is capable of redistributing hydrocarbons to the extent necessary for oil accumulations.

The contraction (dehydration) of layers of expandable clays should have a salutary effect on porosity and drainage of the source bed. The release of water is necessary before contraction takes place, and this water is added to the pores. The result is an increase of mobile water volume in the system, together with a decrease in mineral volume. Thus, this process should enhance porosity and set the stage for more complete, though admittedly slow, drainage. M. C. Powers (1967) suggested that release of the bound water from expandable clays also should raise fluid pressures. He concluded that, in the Gulf Coast, the depth zone in which the last few water interlayers are released coincides with the principal zone of abnormally high fluid pressures. Recently, in a well in Calcasieu Parish, Louisiana, Schmidt (1971) reported the coincidence of an accelerated conversion from expandable to nonexpandable clays with increases in fluid pressure and shale porosity.

With the formation of mixed-layer clays and illitic clays from montmorillonite, there may be some coalescence of the very fine-grained mineral component into larger grains. This process should accompany the greater tendency toward crystalline mineral character with depth, and it should provide larger passageways for migrating fluids. Heling (1970) has suggested a fusion of mineral particles with depth (below 1,000 m) in Tertiary shales of the Rhinegraben, as an explanation of decreased grain surface area in a given mineral volume. The clay mineral alteration accompanying this pattern is a change from smectite (montmorillonite) to illite.

With the simultaneous increase of porosity, drainage potential, and fluid pressure resulting from clay-mineral transformation, conditions should become favorable for expression of the fluids from source bed to reservoir. The increased fluid pressure accompanying release of bound water would be added to the pressure differential already existing between source bed and reservoir due to overburden effects.

Kartsev et al. (1971) concluded that the relatively pure character of water released by clay dehydration should favor an aggressive solution of organic material. Concurrent with the water release, locked-in organic matter should be mobilized. Mixed-layer (illite-montmorillonite) clay is formed from montmorillonite in the process, and illite layers are much less capable of retaining organic matter than is montmorillonite (Grim, 1953; many other authors). Thus, organic matter should be freed and become conveniently available to the newly mobilized water. As this organic component consists of polar material, parts of it may form soaps and other colloids for colloidal migration, and those parts which are more easily decarboxylated should become hydrocarbons.

Kartsev et al. (1970) found that highly compacted clayey rocks, upon being soaked in water and then squeezed, yield more organic matter of both gaseous and nongaseous types than do less compacted rocks. This evidence could support the conclusion that interstitial waters at considerable depth may encounter a greater accessibility of organic matter than does water in shallow sediments.

The water problem in carbonate source beds appears much more serious than in argillaceous sediments. In the muds of Florida Bay, which simulate some features of theoretically conceived carbonate source beds, most of the water is lost in the first foot of burial, according to Ginsburg (1957). Early consolidation of carbonates precludes both effective water retention and compaction mechanisms. Many students of modern carbonates have noted surface and near-surface consolidation where subaerial influences are present. Even under continuous water cover, the hardening process is rapid in some situations, as witness several reports of carbonate consolidation at or near the water-sediment interface (Milliman, 1966; Fischer and Garrison, 1967; Ginsburg et al., 1967; Thompson et al., 1968; Shinn, 1969; de Groot, 1969; Bartlett and Greggs, 1970; Mabesoone, 1971). Studies of ancient carbonates have revealed considerable evidence against compaction, hence an inference of early consolidation (Illing, 1959; Pray, 1960; Powers, 1962; Beales, 1965; Krebs, 1969; Purser, 1969; Zankl, 1969; Bathurst, 1970).

When a carbonate mud is thoroughly consolidated, the large amount of original intergranular pore space is almost entirely eradicated. Thus, water content is reduced essentially to the volumes of secondary voids which commonly are rather limited. Moreover, there is little favorable basis for further compaction to force this water and its organic content into reservoir-type rock.

On the other hand, Einsele (1966) found that calcareous oozes from the Red Sea have about the same depth profile of water content and plasticity in the top 2 m as do terrigenous clays from the Nile delta. Taft (1967) reported that carbonate environments with 8-10 ft of modern sediments and 6-10 ft of overlying water show no evidence of cementation. Shinn et al. (1968) observed an absence of cementation in subtidal sediments west of Andros Island in the Bahamas, where deposition is likewise moderately rapid. Apparently a continuous water cover, together with rather rapid deposition, retards the cementation process and facilitates some degree of compaction.

Where appreciable percentages of impurities are present in deeper water carbonate muds, consolidation may be postponed and water retention greatly increased. Newell et al. (1953) found that black fine-grained Permian limestones in the Guadalupe area of West Texas and New Mexico typically contain flattened fossils. The fact that the fossils also contain sediment matrix suggests that some compaction occurred. Sujkowski (1958) noted that even a small amount of foreign matter in carbonate mud markedly retards crystallization and permits compaction. Bausch (1968) reported that a clay mineral concentration of more than 2 percent inhibited recrystallization in the Malm limestones of southern Germany. Zankl (1969) concluded that presence of less than 2 percent of insoluble residue (especially clay minerals) would favor early cementation and

crystallization before compaction can take place. By implication, an insoluble residue of appreciably more than 2 percent may allow some prior water retention and compaction.

Nevertheless, even under the most ideal conditions, carbonate muds apparently are less compactable and less water-retentive than are predominantly argillaceous muds. Accordingly, significant primary migration of hydrocarbons at shallow depths in carbonates is difficult to visualize.

Carbonates at greater depths continue to be affected by crystallization and cementation, and these mechanisms probably assist in primary migration. The process whereby crystals form should force the fluids into the nearest larger pore spaces. Cementation decreases porosity and should therefore increase the pressure on confined fluids. These fluids should be forced into larger pore systems, particularly along developing stylolites and fractures.

Enough water must remain in carbonates to assist in the formation of stylolites. The pressure-solution mechanism of stylolite formation requires that rather large amounts of carbonate be dissolved and transported elsewhere, and the only readily conceivable agent is water. Stockdale (1926) estimated that the formation of stylolites may reduce the thickness of a limestone unit by as much as 40 percent. Obviously, appreciable amounts of water would be necessary to accomplish this solution. Ramsden (1952) suggested that source bed fluid from a sizable part of the Jurassic carbonates in the Middle East may have migrated along the abundant stylolites as they formed at the expense of the original carbonate volume. Dunnington (1967) questioned fluid migration along the tightly fitting stylolite surfaces in carbonate reservoirs, but suggested that fluids from carbonate source beds may have been squeezed into higher reservoirs by the stylolitic mechanism.

Thus cementation, crystallization, and pressure solution in source-bed carbonates may have performed the same function as compaction in argillaceous source beds. Perhaps small percentages of impurities forestalled water loss sufficiently to permit stylolite development and significant primary migration. Presumably an appreciable overburden would be necessary to provide enough pressure to activate the pressure-solution mechanism. Dunnington (1967) suggested an overburden of 2,000-3,000 ft for some stylolites in Middle East reservoirs. As consolidation in impure source bed carbonates should be more delayed, their stylolitization probably would require greater depths. Undoubtedly, the fractures characteristic of most reservoir-type carbonates helped channel the source-bed fluids upward to positions of hydrocarbon concentration, where overlying sealing beds are present.

Meteoric waters could migrate downdip in carbonate sections if the beds are exposed at the surface in places. In some places, this water could aid in the formation of stylolites and in primary migration. However, most carbonate oil reservoirs have briny rather than brackish formation waters, which suggests the absence of meteoric water influence. Moreover, there is no other apparent water source at depth. Expandable clays, if present at all, would occur only as minor percentages.

In view of these problems, it seems more rational to attribute primary migration in carbonates to a gas rather than water medium. At considerable depth, large amounts of gas would be produced during transformations of the organic component. Also, the carbonate might constitute an inorganic source of gases containing carbon and oxygen. The competence of a consolidated carbonate rock would permit the formation and retention of joints and fracture systems for upward migration of the gases with their load of dissolved liquid hydrocarbons.

PERMEABILITY AND PRIMARY MIGRATION

One of the chief geologic arguments for shallow primary migration is the need for considerable permeability for globules or small masses of oil to be moved progressively through the fine-grained source bed toward a reservoir. Compaction-depth investigations in the geologic section by Hedberg (1936), von Engelhardt (1960), and others, together with many laboratory studies of compaction effects, seem to indicate that permeability of fine-grained rocks to oil globules would be insufficient even at moderate depths.

Hobson (1954) proposed a model whereby oil globules might move progressively from pore to pore in compacting shales by being differentially deformed during the process. However, he admitted (Hobson, 1954, 1967) that the water being expressed probably would move faster than the globules, and he questioned whether sufficient amounts of hydrocarbons in globule form could migrate with the expressed source-bed fluid into reservoirs. Hunt (1967) stated that high interfacial tension would tend to counteract the globule distortion process. Von Engelhardt and Tunn (1954) and Landes (1959) emphasized the blocking effect caused by viscous water films surrounding mineral particles. In carbonate sediments, as Hunt (1967) pointed out, the rigid nature of the carbonate grains would not permit deformation

of globules to activate pore-to-pore movement. Accordingly, oil globule migration in carbonate source beds seems very doubtful.

Alternatives to the globule form have been suggested, such as oil threads (Hedberg, 1964), and the possibility of oil movement along thin strands of kerogen (Erdman, 1965). The formation of oil threads would require considerable permeability, and therefore would necessarily be a shallow phenomenon; and the low incidence of liquid hydrocarbons in shallow muds would not appear compatible with oil-thread formation. It has not been demonstrated how oil at shallow depths would move through and leave a kerogen-strand environment, in view of the oleophilic nature of kerogen and its lack of continuity in the sediments. At greater depths, low permeability also would impede movement of oil through kerogen.

Other theories for explaining primary migration of hydrocarbons or their precursors are much less subject to permeability problems. Although thorough consideration of these theories is beyond the scope of this paper, a short discussion follows.

Much support has been offered for hydrocarbon migration in soap or other colloid form. According to Baker (1959, 1960, 1962), the chief architect of the soap migration concept, patterns of proportional and quantitative distribution of certain homologous and isomeric hydrocarbons in crude oils are most precisely explained by assuming the soap migration theory. Hodgson et al. (1967) stated that evidence is accumulating in support of colloidal migration. Organic matter susceptible to colloid formation is deposited in quantity with many sediments. The bitumen fraction in sediments contains colloidal material, for example, the high-molecular-weight carboxylic acids in the asphaltene part which are soap forming (Rodionova et al., 1963).

Uspensky (1962) stated that a large amount of saponifiable material is present in sediment organic matter, and that it requires only a suitable alkali to form a soap solution. He also emphasized that small concentrations of soap can increase hydrocarbon solubility in water 10- to 100-fold.

Hedberg (1964) suggested that, if colloidal migration occurs, it probably is responsible for the deeper migration where porosity and permeability are much reduced. With porosity and permeability problems in mind, Welte (1965) presented a basis for favoring colloidal particles over oil globules. Apparently, the size range of organic particles is such as to give rise to globules with diameters measured in microns or appreciable

191

fractions of a micron. These sizes are of about the same magnitude as pore sizes in compacted clays. Thus it is difficult to imagine a mechanism whereby globules could be forced through the very much narrower pore throats. On the other hand, the micellar particles of colloids are measured in millimicrons (Baker, 1962), or even the smaller angstrom unit (Mysels, 1959). The usual range of colloid particle sizes is 1 millimicron to 1 micron (Fischer, 1950). Colloids could, therefore, pass through many compacted clays with facility if other factors are favorable.

Welte (1965) thought that colloidal migration might not be effective in the deeper section because carboxylated organic material probably would become decarboxylated and lose its colloidal properties before very great thicknesses of overburden could be deposited. However, Treibs (1936) and Hodgson et al. (1963, 1967) reported that decarboxylation requires a rather drastic chemical environment. Hodgson et al. (1967) could visualize no acceptable biologic or chemical pathways for early decarboxylation. Erdman (1967) noted that carboxylic acids are fairly stable under natural conditions because (1) their decarboxylation requires relatively high laboratory temperatures, and (2) they apparently persist, in part, through the migration period, for they are present in crude oil. Although nearly all crude oil porphyrins are decarboxylated, Morandi and Jensen (1966) stated that shale-oil porphyrins and porphyrins from Wilmington crude oil still are mostly carboxylated, despite overburden pressures and elevated temperatures to which they have been subjected. Surprisingly, Baker (1971) found that deep ocean sediments as old as Early Cretaceous still have an incomplete conversion of chlorine to porphyrin, apparently because of a burial depth of only a few hundred meters. A reasonable conclusion is that decarboxylation may not occur generally, or to a major extent, until considerable thicknesses of overburden are deposited. Thus, the colloidal theory should harmonize with migration at intermediate and greater depths. Moreover, on the basis of kinetic phenomena, the higher temperatures at depth should weaken the adsorptive power of the organic matter for colloidal and hydrocarbon components. Hence solution of these components should be facilitated.

Another migration theory is the solution of hydrocarbons directly in water. Meinschein (1959, 1961) elaborated this concept, stating also that presence of oxygen, nitrogen, and sulfur compounds in the water would enhance hydrocarbon solubility. Baker (1962) and McAulifffe (1963) showed that light gaseous hydrocarbons are appreciably water-soluble under low-temperature, low-pressure conditions. Their solubility should increase markedly with increasing pressure at depth. McAuliffe's (1963) analyses indicate that light aromatic hydrocarbons are highly soluble in water. At room temperature, benzene and toluene have water solubilities of 1,780±ppm and 538±ppm, respectively. According to Hunt (1967), water solubility of gas-gasoline hydrocarbons is 10-100 ppm. He believed that heavier fractions probably are carried to the reservoir as ketones, acids, and esters. As these compounds are more soluble than are their hydrocarbon counterparts, they could contribute more effectively to a solution migration process. The effectiveness of true solution of hydrocarbons in water should increase with depth, owing to increasing temperatures. Sokolov et al. (1963) and Zhuze et al. (1971) reported laboratory studies demonstrating that rising temperatures greatly increase the solubility in water of liquid hydrocarbons. Waters newly freed from dehydrating clays at depth should be good solvents, according to Kartsev et al. (1971). They stated that this water is relatively pure, and Zhuze et al. (1971) have shown experimentally that pure water is a better liquid hydrocarbon solvent than gas-saturated or mineralized water. Vyshemirskii et al. (1971) found that water squeezed from clays under pressures of 150-300 kg/sq cm, and at temperatures of 40-70° C, contained a maximum of 8,600 ppm of bitumen, of which 2,100 ppm were in true solution.

Another migration theory proposes solution of liquid hydrocarbons in gases. The natural occurrences of gas condensate prove the solubility of liquid hydrocarbons in gas at considerable depth. Sokolov et al. (1963) and Sokolov and Mironov (1962) described experiments indicating that large amounts of liquid hydrocarbons can be dissolved by carbon dioxide, methane, and other gases, under temperature-pressure conditions corresponding to depths of 6,000-10,000 ft. At still higher temperatures and pressure, even the high-molecular-weight bitumen is reportedly soluble in gas (Sokolov et al., 1963). Chemodanov (1967) discussed the thermodynamic environment and migration of oil in gas solution under natural conditions and in gas-bearing regions. A combination of gas solution and colloidal solution was advocated by Uspensky (1962), Sokolov et al. (1963), and Dvali (1967). Meinschein et al. (1968) evolved a gas-solution migration model based on chromatographic principles.

The problem of explaining primary migration in carbonates has been discussed by Welte (1965), who concluded that carbonates may be chiefly gas suppliers. Increased pressure of the gas at greater depths may provide the activating energy for migration. As previously stated, compaction pressure and water appear not to be sufficiently available in many carbonates, even at shallow depths, for effective primary migration. Because of the restricted water supply, neither the globule theory nor water solution alone or with colloids appears to explain adequately the primary migration in carbonates. Furthermore, colloidal migration appears unlikely, as colloids tend to precipitate in high-calcium fluids. These colloidal problems have been demonstrated by Kartsev et al. (1954) for soaps, and by Hunt (1967) from experimental evidence. Gehman (1962) reported that limestones have much less nonhydrocarbon organic matter than do shales; accordingly, hydrocarbons in carbonates would be less subject to organic adsorption. Moreover, attraction of polar organic matter to minerals would be relatively weak, because clays are not abundant. This combination of factors suggests that gas solution migration of liquid hydrocarbons in carbonates is reasonable.

One of the chief products of organic transformations is gas, particularly methane and carbon dioxide. According to Silverman (1964), large quantities of methane should be formed at depth. He showed that the carbon-13 to carbon-12 ratios of the different cuts of crude oil can be explained by continued loss of methane during maturation of the original complex organic compounds. Abelson (1967) observed that large amounts of carbon dioxide and hydrocarbon gases are given off during thermal treatments of Green River kerogen. McIver (1967) concluded that carbon dioxide and hydrocarbon gases should be abundant maturation products of organic matter at depth, on the basis of chemical changes observed during kerogen breakdown. Furthermore, it long has been postulated that oil may be transformed to gas in the deeper parts of subsiding basins. This gas should be instrumental in facilitating oil migration, as it commonly contains large quantities of liquid hydrocarbons.

Once liquid hydrocarbons migrate in solution from source bed to reservoir, some mechanism must activate their release if oil concentration and accumulation are to follow. When argillaceous source bed fluids enter a sand reservoir, for example, they encounter an abrupt change in physical and chemical conditions. Likewise, reverse fluid movement, that is, leakage of formation water from reservoir into a capping fine-grained sediment, involves widely different environments. Baker (1959) suggested that hydrocarbons may be released from colloids by increase in fluid salinity, dispersal and dilution of colloid content of the fluid, or decrease in temperature.

A strong case can be made for salinity effects. Under long-continued compaction and increasing burial, reservoir fluids tend to become highly saline, if they are relatively isolated from meteoric water influx (White, 1965; Dickey, 1967, 1968). Conversely, fluids of muds tend to decrease in salinity as they are expressed during compaction (von Engelhardt and Gaida, 1963; Chilingar et al., 1966, 1969; Chilingar and Rieke, 1968a, b). During later compaction, a salinity reversal should develop, owing to semipermeable shale membrane effects. These effects have been reported from laboratory compaction (von Engelhardt and Gaida, 1963), and at depth in the subsurface (Weaver and Beck, 1971). However, on balance, compaction fluids probably are considerably less saline before and during primary migration than are typical high-salinity reservoir waters. Simulated seawater accommodates progressively increasing amounts of soap colloid in micellar solution, in the range from brackish to normal-marine salinities (Gordon and Thorne, 1967). At high salinities, however, such colloids are "salted out" (Jirgensons and Straumanis, 1962). The high salinity characteristic of many reservoir rock waters commonly is accompanied by high calcium content. Thus, further impetus should be provided to the salting-out process with its accompanying release of hydrocarbons. Accordingly, primary migration of hydrocarbons in colloidal form seems reasonable, as does hydrocarbon release from the colloid in reservoirs.

For colloid-breaking at the reservoir boundary, Hill (1959) suggested the Donnan equilibrium mechanism, and Pirson (1964) emphasized electrical streaming potential and cap-rock sieving effects. Welte (1965) listed changes in salinity and pH as possible mechanisms, in addition to progressive loss of functional groups from the colloid.

Mechanisms for release of hydrocarbons from true water solution are less tenable. Hunt (1968) pointed out that no definite acceptable explanation is yet available. Reduction of pressure appears to be an adequate cause for liquid hydrocarbon release from gas solution. As already discussed, a reduction in pressure commonly is encountered as source bed fluids enter a reservoir. Where upward migration of wet gas occurs, the accompanying pressure decrease should ensure a significant release of liquid hydrocarbons.

In summary, evidence supporting colloidal solution, true solution in water, and gas solution theories, under given regimes and for various oil-forming constituents, appears sufficiently acceptable to obviate the necessity of explaining globule or other forms of primary oil migration in terms of a shallow, high-permeability environment.

PETROLEUM ORIGIN RELATED TO KEROGEN AND BITUMEN

The likelihood that the great bulk of abundant fatty acids of organisms becomes incorporated into kerogen has already been discussed. Further evidence for such reactions is seen in numerous studies of oil shale kerogen. This kerogen has been characterized as a polymer in which structures bearing some resemblance to fatty acids occur (Kogerman, 1931; Cane, 1946, 1948; Schnackenberg and Prien, 1953; Kasatochkin and Zilberbrand, 1956). The binding of fatty-acid type structures in oil-shale kerogen may explain the low concentrations of free fatty acids in oil shales.

The kerogen commonly dispersed in sedimentary rocks appears to be formed by condensation or polymerization of microbially altered products originating from carbohydrates, proteins, lipids, and perhaps other organic material (McIver, 1967). Two general types or end members of kerogen are recognized, based on preponderant composition. Different name-pairs have been used, such as coal-like and oil-shale-like (Forsman and Hunt, 1958), humic and sapropelic (Breger and Brown, 1962), and carbohydrate-lignin-rich and lipid-rich (McIver, 1967). Apparently, the oil-shale-like, sapropelic, or lipid types contain fatty-acid type structures, in view of the statement by Eisma and Jurg (1967) that fatty acids are an important constituent of kerogen. The fact that Hoering and Mitterer (1967) found fatty acids to be a product of thermal decomposition of modern sediment kerogen further suggests the presence of fatty-acid structure in kerogen. Van der Weide (1967) reported that thermal treatments of modern sediment kerogen produce normal paraffin hydrocarbons with nearly smooth carbon-number distributions. Thus the odd-number preference problem in modern sediment hydrocarbons is resolved by attributing major hydrocarbon formation to origin from kerogen at elevated temperatures.

Because hydrocarbons, free fatty acids, and the bitumen component as a whole are so sparse in modern sediments, it seems unlikely that they can account for major oil genesis. A reasonable alternative is an origin of petroleum from kerogen. Obviously, for the insoluble and relatively inert kerogen to give rise to bitumen, including liquid hydrocarbons, a rather drastic chemical environment is required. The most easily conceivable environment would involve a thick overburden, which would be accompanied by elevated temperatures and pressures, and a much accelerated catalytic activity.

Thermal experiments on kerogen of the Green River and other oil shales have been conducted for many years. Several investigators have stated that the initial thermal decomposition product is bitumen, and that bitumen subsequently decomposes to oil or tar (McKee and Lyder, 1921; Franks and Goodier, 1922; McKee and Manning, 1927; Cane, 1948; Hubbard and Robinson, 1950; Kollerov, 1956). In this context, the term bitumen was intended to include only constituents other than oil and tar among the components soluble in organic solvents.

Later, Forsman and Hunt (1958) and Forsman (1963) investigated the thermal decomposition products of disseminated kerogen from marine rocks. Analysis of the products after hydrogenolysis at high temperature and pressure revealed bitumen components consisting of nitrogen-sulfur-oxygen compounds, asphaltics, and hydrocarbons.

From analyses and thermal studies at 180°C, of the Toarcian shales (Jurassic) of the Paris basin, Louis and Tissot (1967) reported that bitumens (soluble in methanol-acetone-benzene mixture) seem to represent important intermediate products between kerogen and hydrocarbons. Similar conclusions were drawn by Bajor et al. (1969) from thermal studies of organic matter from "immature" sediments. They found that the initial products are heavy heteroatomic compounds which at higher temperatures are changed progressively to hydrocarbons.

Because the sequence, kerogen to nonhydrocarbon bitumen to hydrocarbons, is formed in laboratory thermal experiments, similar reactions may be postulated over a long period under the lower temperature conditions in the sedimentary section. At depths of a few thousand feet, the original pore water of compacting shales, together with the more loosely bound water of the constituent clay minerals, largely has been squeezed out (Burst, 1966, 1969; Powers, 1967). As previously discussed, this decrease in water content, together with increasing temperatures, should intensify the catalytic effects of the clay minerals on organic matter. Therefore, transformations of kerogen to bitumens and production of the hydrocarbon component could be accelerated.

CONCEPT OF REPEATED OR LONG-CONTINUED PRIMARY MIGRATION

A widely accepted adage is the so-called "depth rule," which states that crude oil accumulations in a basin typically become lighter and more paraffinic with increasing depth. Barton (1934, 1937) established this relation in his classic study of the Gulf Coast crude oil distributions. McIver *et al.* (1963) compiled data on 3,000 domestic crude oils and concluded that the depth rule has consistent application within given reservoir systems. Numerous examples have been noted elsewhere, as well as some exceptions.

Many investigators have attributed the occurrence of progressively lighter and more paraffinic oils with depth to a maturation of reservoir-emplaced oil which was originally heavy and asphaltic. A question arises, however, in explaining how an asphaltic oil, which has already accumulated, can become a paraffinic oil. Hlauscheck (1950) and others have emphasized the thermodynamic and chemical problems in any proposed transformations of this type. Conversely, Kartsev (1964) reported laboratory experiments indicating that ring structures of naphthenic and naphthenic-aromatic hydrocarbons can open to form paraffinic structures. This laboratory transformation was stated to develop under relatively low-temperature conditions in the presence of clays. However, more evidence seems necessary before accepting this explanation for general application to natural rock-fluid-temperature conditions, and for large-scale bulk transformations. Many reservoirs have only very small amounts of clay minerals; and commonly these clays are not the most actively catalytic types. As evidence against radical maturation, an extensive study of U. S. Bureau of Mines data by Barbat (1967) in the Rocky Mountain region showed that upper Paleozoic crude oils retain their cyclic character regardless of depth. Barbat also summarized evidence from various areas indicating that nature of the source material, rather than depth, is the controlling factor.

Baker (1960) thought that the ordering of hydrocarbon distributions in crude oil is too consistent to be the result of in-place thermodynamic maturation in reservoirs. He also suggested that a strictly thermodynamic explanation apparently would require much higher temperatures than those to which most crude oils have been subjected. Later (Baker, 1967), he noted that individual hydrocarbons are surprisingly stable, and that little compelling evidence exists of maturation of oil after it has accumulated. He believed that the narrow range of carbon-12 to carbon-13 ratios among crude oils suggests minimal post-accumulation modification of their hydrocarbons. Welte (1965) referred to earlier work which appeared to question the significance of catalytic cracking in crude oil maturation. The main objection was

that this process should lead to thermodynamic equilibrium, which is not represented in isomeric distributions in crude oil.

On the other hand, little is definitely known about factors affecting crude oil maturation. It should not be entirely thermodynamic, because mineral catalysis, however weak, would be involved. The relatively high temperatures at depth should accentuate the catalytic activity, and important transformations may be possible over long spans of geologic time. Perhaps the time required to reach equilibrium of crude oil components is so long that few oils have attained it.

Other writers (Brooks, 1948, 1949, 1952; Bornhauser, 1950; Haeberle, 1951) have attributed the oil-type changes with depth to greater catalytic activity in the more shaly downdip sections. Another theory advocated by many workers is that, downdip, more marine environments generate a lighter, more paraffinic oil because of the high percentage of planktonic progenitors. This hypothesis is supported by a recent analysis of North African clay-bearing samples by Giraud (1970), showing that the marine clays, upon being heated, yield a paraffin-rich product, whereas continental samples yield an aromatic-rich hydrocarbon suite. The facies of downdip marine environments commonly become deeply buried by subsequent regressive deposition, seaward gravity slumping, or turbidite deposition. Accordingly, thick sedimentary sequences generally accumulate above deposits with paraffin-oil-forming potential.

Regardless of differing views on reservoir maturation and depositional effects, alteration with depth in crude oil hydrocarbons can be explained by chemical transformations of organic material in the associated source beds, and by changes in selective migration with increasing overburden or age. It is widely recognized that naphthenic material is abundant in asphaltic oils. Naphthenic acids are much better colloid formers than are paraffin-related acids (Kartsev *et al.*, 1954). Accordingly, naphthenic soaps (acid salts), because of their solubility in aqueous solutions and their chemical similarity to naphthenic hydrocarbons, probably could incorporate naphthenic hydrocarbons effectively during the rather early, lower temperature stages of primary migration. For this reason, naphthenic (asphaltic) material could contribute heavily to oil accumulations at relatively shallow or intermediate depths. The high solubility of benzene and related aromatic hydrocarbons at room temperature (McAuliffe, 1963) suggests that a significant aromatic hydrocarbon component may migrate rather early if it is present in sufficient amounts in the source beds prior

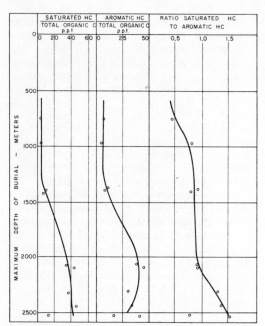

FIG. 1—Relation of saturated and aromatic hydrocarbon concentrations to maximum depths of burial in "schistes cartons" formation (Toarcian) of Paris basin (10 localities). Modified from Louis and Tissot (1967).

result of a cyclic to paraffinic transformation, it also seems possible that much of the aromatic hydrocarbon fraction migrated toward reservoirs during the increase of burial depth from 2,000 to 2,500 m. Conversely, the continued increase in saturated hydrocarbons through this interval may indicate a progressive concentration of these hydrocarbons prior to their primary migration.

This evidence does not necessarily conflict with the view, mentioned by Welte (1965), that many or most aromatic oils are late evolving. Large amounts of aromatic material apparently become available late in the process of kerogen breakdown. Silverman (1964) concluded that because of progressive loss of light hydrocarbons cyclic compounds may develop from straight-chain compounds. Barbat (1967) illustrated earlier work showing a basis for laboratory conversion of paraffins to naphthenes and aromatics.

In the Jurassic of the West Siberian lowlands, Kontorovich et al. (1965) found that hydrocarbon depth patterns in shales are similar to those in reservoirs. In both shales and reservoirs, a downward decrease in aromatics is accompanied by increases in lighter fractions and paraffin hydrocarbons. This change possibly represents a maturation process within the source beds which is reflected in the hydrocarbons recurrently expressed from source beds into reservoirs. Conversely, Buchta et al. (1963) reported that Vienna basin oils show an increase with depth in abundance of long-chain paraffins which does not occur in the Neogene marls at corresponding depths. Perhaps in the latter case, there is a downwardly increasing contribution from older Tertiary source beds deposited in nearshore environments and subject to nonmarine environmental influence. Hedberg (1968) demonstrated the relation of high-wax oils (with abundant long-chain paraffins) to nonmarine influence, using many examples.

Several detailed studies of geologic evidence bearing on depth and time of migration have been conducted in northwest Germany. Bentz (1958) suggested that migration in the Northwest German basin began shortly after formation of the source rocks and was long-continued or revived during tectonic movements. Brand and Hoffman (1963) suggested that oil migration in the Jurassic of northwest Germany occurred during all stages from mid-Lias to Tertiary, at least insofar as subsidence of the basin was continuous. Their conclusion was based on interpretation of the times of trap development and oil

to deep burial. In contrast, adequate solution migration of liquid paraffin hydrocarbons might not be possible without the higher temperatures associated with greater depths. The fact that deep samples generally have a large percentage of the light, more highly soluble hydrocarbon component could be explained, for the most part, by maturation in the source beds. Where source beds are argillaceous, large clay percentages should be instrumental in the thermocatalytic process leading to production of increasingly larger amounts of light hydrocarbons with increasing depth. Migration of these light hydrocarbons could have led to progressively lighter hydrocarbon accumulations.

A possible example of aromatic hydrocarbon primary migration preceding primary migration of other hydrocarbons is represented in the "schistes cartons" formation (Toarcian) in the Paris basin (Fig. 1). Analyses by Louis and Tissot (1967) show that the ratio of aromatic hydrocarbons to total organic carbon decreases downward in the 2,000-2,500-m maximum burial depth interval, even though the lithology and original organic makeup seem similar throughout. On the other hand, saturated hydrocarbons continue to increase in this interval. Although this may be the mobilization in the Lower Saxony basin and Gifhorn-East Holstein trough part of the Northwest German basin. Time of trap development, together with bubble point and gas-oil ratio patterns, led Brand and Hoffman (1963) and Boigk et al. (1963) to recognition of a considerable range of migration and accumulation depths. In a study of the Gifhorn trough, Philipp et al. (1963) utilized a combination of sonic logs and sand diagenesis data to determine subsidence history and timing of oil accumulation. They concluded that migration was confined mainly to a subsidence (depth) interval of 500-2,000 m, corresponding to a shale porosity range of 30-10 percent. Welte (1966, 1967) also investigated the Gifhorn trough and concluded that the evolutionary state of crude oil is determined mainly by the subsidence history of the source rock. He conducted laboratory experiments, analogous to subsidence effects, on sediments containing hydrocarbons. The results provided corroboration of the findings in the geological section.

Many additional investigations have produced evidence or views favoring long-continued and/or repeated primary migration. Gussow (1955) believed primary migration may have spanned a subsidence interval of a few thousand or several thousand feet on the basis of compaction studies of Hedberg (1936), surface tension considerations, and relative gas saturation of various reservoirs. Weeks (1958a) thought migration of heavy oils probably occurred very early while permeability of sediments was still high, whereas lighter oils could have migrated somewhat later. Pow et al. (1963) stated that the vast quantities of Athabasca "tar" in Alberta must represent focused drainage of large volumes of sediments over long periods of time. Hedberg (1964) suggested that primary migration of hydrocarbons in oil form was relatively shallow, and that in the deeper low-permeability section, colloidal or solution migration would be more logical. Although he believed most migration was early, he cited 10 examples of apparent late migration. Welte (1965) thought hydrocarbons can be expelled from "oil nests" in source beds during an extended part of the subsidence-compaction history, beginning with the first mobilization of the oil nests and extending to the time when water no longer can escape. He viewed primary migration as a selective process which varies with the developmental stage of source beds.

Thorough study of many areas of the USSR led Sokolov (1967) to conclude that oil and hydrocarbon gases are formed in the thermocatalytic zone at depths of from 1 or 2 km to 6 or 7 km. An implicit corollary is that primary migration probably occurs over a long time and broad depth span. Vassoyevich et al. (1967) proposed that hydrocarbon migration proceeds at all stages of lithogenesis. They viewed migration of micro-oil (disseminated oil-like material in source beds) as beginning in the brown coal stage of diagenesis, before the micro-oil becomes similar to the hydrocarbon composition of crude oil. Deeper burial of the source rock, according to Vassoyevich et al., causes migration of micro-oil of an increasingly lighter more paraffinic character. McIver's (1967) chemical analyses of organic material showed progressively decreasing percentages of hydrogen with depth (4,730-7,520 ft) in the Devonian shale of western Canada. From percentage patterns of hydrogen, carbon, and oxygen, McIver concluded that hydrogen decrease with depth probably represents loss of hydrocarbons by primary migration. The depth interval for this postulated migration would be about 2,800 ft or more. Bordenave (1967) believed that the same source rock can yield different oils at different times in its history. Based on investigations of changes in carbonaceous material with depth and time, Lopatin (1969) inferred that primary migration from Devonian source beds began in Late Devonian time and continued through the Paleocene. Kartsev et al. (1971) designated the depth interval of 1,500-3,000 m as the interval of formation and primary migration of hydrocarbons.

The weight of evidence and interpretation in the reports cited above provides strong indication that primary migration normally spans considerable intervals of time and burial depth.

BITUMEN-HYDROCARBON DEPTH PATTERNS IN GEOLOGIC SECTION

Long before the advent of appreciable organic geochemical depth data, Cox (1946) advocated a critical depth of 5,000 ft for oil formation, on the basis of geologic observation. He stated that the temperature for initiation of oil formation would, accordingly, be about 65°C, as determined from the geothermal gradient. That these conclusions have stood the test of time is shown by the large amount of corroborative evidence from organic geochemical investigations of the past few years.

A study by Albrecht and Ourisson (1969) provides key data on the depth patterns of bitumen

Table 6. Depth Patterns of Bitumen and Alkane Hydrocarbon Concentrations, Turonian to Paleocene Section, Douala, Cameroun (Adapted from Albrecht and Ourisson, 1969)

DEPTH IN METERS	TOTAL ORGANIC CARBON (NON-CARBONATE) %	TOTAL EXTRACT (BITUMEN) %	TOTAL ALKANE HYDROCARBONS P.P.M.	TOTAL ALKANES / TOTAL ORG. C.
775	0.36	0.082	10	0.0028
910	0.76	0.093	10	0.0013
1200	1.46	0.149	18	0.0012
1500	1.06	0.085	79	0.0075
1875	1.87	0.123	235	0.0126
2170	1.38	0.184	550	0.0400
2500	1.41	0.119	410	0.0290
2740	1.21	0.020	20	0.0017
2850	1.25	0.004	5	0.0004
3150	2.02	0.003	3	0.0002
3420	1.21	0.005	10	0.0008
4025	0.50	0.005	5	0.0010

and hydrocarbon constituents (Table 6). The sampled stratigraphic section, located in Douala, Cameroun (West Africa), is relatively thick, ranging in depth from 775 to 4,025 m. The time interval is comparatively short, from Turonian to Paleocene. The depositional environment is reportedly similar throughout, and the original nature of the organic matter apparently was roughly comparable in all parts of the section. Hence, patterns of change with depth should be essentially depth-dependent, meaning temperature- and pressure-dependent.

Although only 12 samples were analyzed within the 3,250-m interval, they are rather evenly spaced and show very definite pattern changes with depth. The total bitumen averaged about 1,180 ppm in the seven samples from depth range 775-2,500 m, and it showed a slight tendency to increase with depth. However, at 2,740 m there was an abrupt decrease to 200 ppm, and a further drop to 40 ppm at 2,850 m. The concentration was reduced further in the three deeper samples, at 3,150, 3,420, and 4,025 m. A comparative examination of the alkane (paraffin hydrocarbon) figures shows very low concentrations at 775, 910, and 1,200 m, but an abrupt increase at 1,500 m (to 79 ppm), and also at 1,875 m (to 235 ppm). Even higher readings were recorded at 2,170 and 2,500 m (550 ppm and 410 ppm, re-

spectively). Thus in the range of 1,500-2,500 m, where the bitumen percentage rises only moderately, the alkane hydrocarbon concentrations increase enormously. It appears that the large amounts of alkanes in this interval formed at the expense of the nonhydrocarbon part of the bitumen. At levels below 2,500 m, the alkane concentrations decrease markedly, in keeping with the sharp drop in bitumen percentages.

Albrecht and Ourisson (1969) concluded that these apparent losses are a function of either (1) removal through primary migration, or (2) cracking to form lighter hydrocarbons than those included in the analysis (lighter than C_{10}).

From the data, primary migration appears to be the more likely explanation. At 2,740-m depth, where the largest downward decrease in the C_{10} and heavier alkane hydrocarbon abundance occurs, the carbon-number distribution graph shows a maximum intensity at C_{13} and C_{14}. If cracking were the reason for the sharp decrease in this hydrocarbon abundance, one would expect the maximum representation among the C_{10} and heavier alkane fraction to be at the C_{10} position or near it. This would be in keeping with a progressive increase in hydrocarbon abundance toward some lighter than C_{10} maximum. The fact that C_{13} and C_{14} hydrocarbons are the most abundant at 2,740-m depth suggests that a hydrocarbon suite with a large percentage of greater than C_{10} molecules migrated out of the sediments to account, at least in part, for the marked decrease in hydrocarbon abundance. Also, the fact that total bitumen concentration decreases manyfold in the same depth interval with the sharp drop in alkanes suggests contemporaneous migration of nonhydrocarbon bitumen and alkanes. Otherwise it might be difficult to explain the fate of the lost bitumen.

In other areas, investigators have suggested depths of origin and primary migration similar to those inferred in the Cameroun locality. Vassoyevich et al. (1967) found that an abundance of hydrocarbons in source beds begins at a depth of 1.5-2.0 km. From comparative studies of coal and hydrocarbon evolution, they proposed that temperature is the principal causative factor, and that the minimal temperature for petroleum-significant hydrocarbon formation is 60-65°C.

In clay-type rocks of the Volga-Ural region, Kovacheva (1968) noted that bitumen increases downward in quantity and chemical reduction, together with hydrocarbon increase. This pattern was found at depths of 0.8-3.2 km.

Louis (1966), Louis and Tissot (1967), and Tissot et al. (1971) reported that the "schistes cartons" formation of the Toarcian (Jurassic) in the Paris basin shows important generation of hydrocarbons beginning at 1,400-1,500 m, equivalent to temperatures of about 60-65°C (Fig. 2). The "schistes cartons," like the Turonian to Paleocene section in Cameroun (Albrecht and Ourisson, 1969), shows little variation in rock composition and depositional environment. Hence, with increasing depth, this shale should reflect mainly temperature and pressure effects. At a burial depth of about 2,000 m, the hydrocarbon-to-bitumen ratio begins to decrease, after reaching a maximum value in the 1,700-2,000-m range. The decrease in ratio becomes progressively more pronounced downward to the deepest samples, at slightly below 2,500 m. Probably the pattern beginning at 2,000 m reflects primary migration of hydrocarbons toward reservoirs.

Meinhold (1969) concluded that a highly intensified transformation of bitumen to hydrocarbons rather definitely occurs in a depth interval of 1,500-2,500 m, apparently resulting from thermocatalytic processes. In a study of the Jurassic terrigenous platform rocks of the Fore-Caucasus, Konyukhov and Teodorovich (1969) found that the 1,500-1,800-m depth interval is optimal for oil generation, but that the generation process could continue downward to 4,000 m. The depth zone for major primary migration was inferred to be 2,000-3,500 m. The basis for these interpretations was a combined organic and mineralogical analysis. From bitumen analyses of the Middle Jurassic section of the Ural-Volga region, Timofeev (1970) concluded that primary migration began at 1,000-1,200 m and increased with increasing depth. Vasseyevich *et al.* (1971) identified the critical zone of oil formation in western Kuban as about 70-90°C, with depth partly dependent on geothermal gradient. S. G. Neruchev (in Kartsev *et al.*, 1971) observed a mass appearance of what he interpreted to be transported bitumens in source beds, beginning at depths of 1,500-2,000 m. Also, at 2,000-2,200 m, Neruchev (1970) found an abrupt decrease in volatile hydrocarbons, as well as in carbon and hydrogen content of the bitumen. He concluded that these decreases reflect major primary migration of hydrocarbons.

Another depth aspect is related to the initiation of tectonic activity. Dallmus (1958), from studies of basin development and concomitant stress patterns, calculated that a basin floor can subside nearly 2,000 m before any deformation takes place. This conclusion suggests that, beginning at about 2,000-m depth, tectonic pressures may become effective. Because these pressures are added to the overburden effects, the squeezing which facilitates fluid expression from source beds should be enhanced. It seems more than coinci-

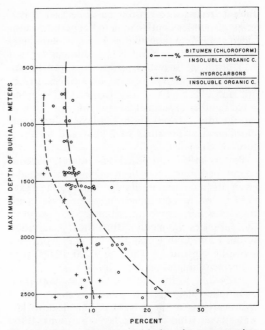

Fig. 2—Relation of bitumen and hydrocarbon concentrations to maximum depth of burial in "schistes cartons" formation (Toarcian) of Paris basin (19 localities). Adapted from Louis and Tissot (1967).

dental that this 2,000-m depth is close to the top of the depth range of flush primary migration of hydrocarbons postulated from other depth-pattern evidence.

IMPORTANCE OF CLAY-MINERAL DEPTH PATTERNS

Significantly, the primary migration depth intervals discussed previously show broad correspondence to depths at which expanded clay minerals, particularly montmorillonite, lose much of their interlayer water in the Gulf Coast area. In the Wilcox (Eocene) sediments, Burst (1959) observed downward decreases in montmorillonite, accompanied by increases in illite, within the 3,000-10,000-ft depth range. In the downdip Gulf Coast, Powers (1959) found that montmorillonite, the chief mineral in the shallow interval, begins to decrease in abundance below 5,000 ft. Conversely, below 5,000 ft, a pronounced and progressive downward increase was noted in mixed-layer illite-montmorillonite.

Weaver (1960) suggested that a statistical relation may exist between hydrocarbon production and expandable clays, on the basis of clay studies in most of the major basins of the United States. He also reported that montmorillonite mineral layers in Gulf Coast clays contract systematically below a depth of 8,000 ft. Laboratory studies by Khitarov and Pugin (1966) showed that montmorillonite changes to hydromica (illite and related minerals) under depth-simulated conditions, provided potassium is available. In an experimental diagenetic study of Gulf Coast sediments mixed with artificial seawater, Hiltabrand (1970) observed the formation of illite at temperatures of 100 and 200°C. Eberl and Hower (1971) monitored another experimental study in which montmorillonite changed to mixed-layer illite-montmorillonite at pressures of 2 kbar and temperatures of 150-490°C.

Powers (1967) proposed that an argillaceous source bed must have been deposited originally as an organic montmorillonitic mud. In the Texas Gulf Coast, he concluded that the release of interlayer water, as montmorillonite changes to illite, furnishes the medium for primary migration of oil. From Gulf Coast clay mineral depth patterns, he suggested 6,000 ft to 9,000-10,000 ft as the primary migration depth range.

According to Burst (1966, 1969), the important contraction stage related to primary migration of hydrocarbons is the release of the next to last water layer from clay interlayer positions. Burst showed that this water-loss stage is a function of the geothermal gradient. Where the geothermal gradient is highest, this stage begins at a depth of only about 4,000 ft, and where lowest, it extends to about 16,000 ft. Burst found that the preferred levels of major Gulf Coast oil accumulations consistently average about 1,500 ft higher than the position of this stage. He attributed the depth offset to upward flush migration of oil concurrent with mobilization of the next to last water interlayer. Conceivably, the water loss associated with oil migration could have begun somewhat earlier.

An investigation by Weaver and Beck (1969, 1971) of a Mississippi delta well showed that montmorillonite alters with depth to mixed-layer clay with downwardly decreasing percentages of montmorillonitic layers. As a product of this dehydration and of mineral change, the pore water percentage in the shales is relatively high in the 4,500-14,000 ft range. At about 10,000 ft, an abrupt decrease in pore water was observed. Quite possibly this decrease represents an important stage of primary migration. In this same well, Weaver and Wampler (1970) reported progressively increasing illite to montmorillonite ratios in mixed-layer clays with increasing depth, over a depth span from 4,233 to 16,450 ft. In a study of five Gulf Coast wells, Perry and Hower (1970a, b) found that mixed layer illite-montmorillonite is the predominant clay mineral. They observed a progressive downward increase in the illite to montmorillonite ratio beginning at 6,000-8,000 ft, depending on the geothermal gradient.

Perry and Hower concluded that the pattern represents progressive diagenetic loss of water from the mineral with increasing depth.

The Gulf Coast clay mineral contraction patterns span the same general depth intervals as are inferred for flush primary migration in several geologic provinces on the basis of changes in organic component data. The beneficial effects of clay contraction on pore water volume, porosity, fluid pressure, source-bed drainage, and organic availability already have been discussed. The logical conclusion is that organic transformations and clay mineral contractions are mutually associated with primary migration.

An interesting relation of clay-mineral depth patterns also is found in the Turonian to Paleocene beds of Douala, Cameroun. According to Dunoyer de Segonzac (1964), montmorillonite is the predominant mineral in this section downward to a depth of 1,500 m. The interval 1,300-3,800 m is characterized by mixed-layer illite-montmorillonite, and below 3,500 m illite is representative. The work of Albrecht and Ourisson (1969) in the same area and section (see Table 6) showed that the first major increase in alkane hydrocarbons occurs between 1,300 m (18 ppm) and 1,500 m (79 ppm). Thus, there apparently is a general depth coincidence between the beginning of mixed-layer clay development and the beginning of important hydrocarbon formation.

Recently, three other foreign examples have been reported of apparent clay-mineral dehydration with depth. In the Rheingraben, Heling (1970) found a downward change at about 1,000 m from smectite (montmorillonite) to illite. Van Moort (1971) analyzed samples from Mesozoic shales of Papua, New Guinea, and found a 40:60 illite-montmorillonite ratio at 3,500 ft and an 80:20 ratio at 10,200 ft. In the early Paleozoic rocks of the Algerian Sahara, Abu Amr (1971) noted that mixed-layer illite-montmorillonite is present only above depths corresponding to the 95° C-isotherm. At greater depths, illite is present but not the mixed-layer clay.

PROBLEMS WITH ULTRA-DEEP MIGRATION

It has been proposed that oil may originate quite deep within a sedimentary section and migrate upward through thousands of feet of shale. In the South Sumatra basin, Dufour (1957) observed that most of the oil accumulations in the older transgressive sands of the Tertiary section are located at intermediate depths of the basin. These hydrocarbon concentrations have a much heavier paraffin base than the crude oil in the younger regressive sands, which are relatively shallow.

Dufour concluded that the heavier paraffin-base oil may have migrated laterally from source-bed shale into the transgressive sands, and that the depth of origin and migration were insufficient for extensive thermocatalytic cracking. He believed the lighter oil may have formed in the deep part of the basin, perhaps by catalytic cracking at the higher temperatures, and that they later migrated upward through a very thick shale section to the sites of accumulation, with faults as the major conduits. Dufour admittedly drew on unpublished work of G. T. Philippi for source bed and migration ideas. Philippi (1957) published his conclusions in a paper on source bed recognition.

Although the thermocatalytic aspect of this model seems reasonable, difficulties arise in explaining the great distance of vertical migration through shales. First, at the depths of origin specified, the supply of water as a migration medium in shale would be severely limited. Migration of oil as a separate phase, without benefit of water, appears implausible because of the disseminated nature of the hydrocarbons in the shale and the lack of a suitable mechanism for hydrocarbon collection and concentration. Primary migration in a limited water medium could explain movement of hydrocarbons to nearby sands, but vertical migration of enormous amounts of hydrocarbons through great thicknesses of shales is a different matter. Weeks (1958a), following a worldwide study, doubted the efficiency of faults as migration avenues, stating that faults more typically are barriers to migration. The permeability along a fault zone is a questionable factor, particularly in shale sections. At considerable depths the Tertiary shales in the South Sumatra basin may be relatively well consolidated, and probably could support fault openings if the overburden were relatively thin. However, the great confining pressures at the depths involved should reduce severely any tendency toward maintenance of extensive fault zone permeability.

Snarsky (1961) believed that the high fluid pressure in source rocks at depth is sufficient to rupture the rocks and cause extension of fractures. These fractures would then serve as conduits for migration. In one example, he calculated that total pressure on fluids at a depth of 3,000 m is slightly higher than total overburden pressure. Recently, Snarsky (1970) emphasized upward movements of strata following subsidence which, according to his calculations, might increase pore pressures to as much as five times the hydrostatic pressure. He believed pressures of this magnitude would fracture source rocks sufficiently for oil and gas migration. However, it seems probable that gradual upward or lateral migration of fluids would prevent this degree of pressure in most situations. Where very high fluid pressure has been recognized in source-type rocks, they are typically very argillaceous and water-wet. Although such rocks may fail by fracturing, their development of extensive connected fracture systems for effective long-distance vertical migration seems unlikely.

Another possible oil migration mechanism is solution of oil in gas. Upward migration of oil in gas would require much less permeability than in liquid form. However, in examples such as the South Sumatra basin, reduction of pressure during postulated upward migration of oil in gas solution should cause all liquid hydrocarbons except the very lightest to be released far below the relatively shallow levels of oil accumulation. In the South Sumatra basin, the oil, though of light paraffin base, contains significant fractions of heavier constituents which cannot be explained adequately by a gas transport hypothesis.

In the Los Angeles and Ventura basins of California, Philippi (1965, 1969) applied the same general origin and migration model as discussed for the South Sumatra basin. After intensive investigation and analysis, he concluded that most crude oil in these California basins originated in shales several thousand feet below oil accumulation zones. Representative oil-origin depths were interpreted to be approximately 11,000 ft in the Los Angeles basin, and 15,000 ft or more in the Ventura basin. Evidence was based on the increasing degree of "maturity," with increasing depth, of the shale hydrocarbon assemblages. Maturity here refers to development of a hydrocarbon composition comparable to that of crude oil. Only at depths of several thousand feet below most oil accumulations have the increasing temperatures and pressures altered shale hydrocarbon constituents to an approximation of crude oil hydrocarbon composition.

Philippi (1965, 1969) assumed that primary migration consists essentially of a bulk transferral, without major selectivity, of the shale hydrocarbon components to reservoirs. On this assumption and the almost identical composition of crude oil from zones at widely varying depths in two of the oil fields, a deep origin was required to fit the data. He believed these similarities, together with the shale hydrocarbon depth patterns, indicated a common origin at great depth. He did not consider the possibility that a more "mature" hydrocarbon assemblage may have developed in the shales *after* primary migration ceased.

Philippi postulated that hydrocarbons are re-

Fig. 3—Depth patterns of hydrocarbon concentrations and hydrocarbon to non-carbonate carbon ratios, Los Angeles basin, California. Hydrocarbons plotted have boiling points > 325°C. Adapted from Philippi (1965).

tained by the associated organic material until heat and pressure at increasing depth create an abundance of hydrocarbons sufficient to exceed the hydrocarbon adsorption capacity of the organic matter. This idea is somewhat similar to Erdman's (1965) suggestion that hydrocarbon concentrations in kerogen may become sufficiently high to cause a break-out from the kerogen complex. According to Philippi's concept, the excess hydrocarbons above the adsorption capacity would be free to migrate. Moreover, because of the very weak attraction of inorganic components of the shale for hydrocarbons, upward movement of hydrocarbons could occur along the decreasing pressure gradient, with faults as the main conduits.

On the other hand, the question of migration medium presents serious problems in the deeper parts of the Los Angeles and Ventura basins. Water availability in shales at depths of 11,000-15,000 ft is rather restricted, even in areas such as the Gulf Coast, where clays with very abundant interlayer water constitute the chief original mineral components. Also these expandable clays with their high water content are probably not sufficiently competent for extensive upward conduits to develop through them.

Difficulty with a gas solution migration would

be even greater in the California basins than in the South Sumatra basin. Crude oil from the Los Angeles and Ventura basins contains much higher percentages of high-molecular-weight constituents than does the light paraffin crude of the shallow regressive facies of the South Sumatra basin. Because the heavier constituents are soluble in gas only under high temperature-pressure conditions, the relatively large amount of these constituents in the California crudes scarcely could have moved far upward in gas solution to the known oil accumulation levels, most of which are only about 2,500-7,500 ft deep.

ALTERNATIVE ORIGIN-MIGRATION MODEL

An alternative model can be suggested for the Los Angeles and Ventura basins. It is proposed that most of the primary migration probably occurred at depths beginning between 4,000 and 6,000 ft in the Los Angeles basin, and between 7,000 and 8,000 ft in the Ventura basin. The shale sediments overlain by overburden of these thicknesses still should have been actively compacting, and a sufficient supply of water for migration still should have been available.

During the post-migration stage, hydrocarbons would have continued to form in the shales under the higher-temperature conditions accompanying increased overburden. These hydrocarbons would have remained in the shales because of the much reduced quantities of mobilized water and decrease in permeability. Thermocatalysis associated with continued high temperatures would have caused this hydrocarbon assemblage to approach crude oil composition.

The migration depths suggested above can be inferred from Figures 3 and 4, which show Philippi's (1965) data on hydrocarbon abundance and hydrocarbon to noncarbonate (organic) carbon ratios plotted against depth. Philippi stressed that the data points shown have an exclusively indigenous hydrocarbon component; that is, essentially no migrated hydrocarbons appear to be present to mask the hydrocarbon-forming effects of increasing temperature and pressure. The pattern of hydrocarbon to noncarbonate carbon ratios seems particularly significant. In general, these ratios should increase with increasing depth if sediment and original organic component remain similar, and if the hydrocarbons remain in the source beds. The reason, as Philippi pointed out, is that hydrocarbons should continue to form from organic matter with increasing temperatures and pressures arising from additional increments of overburden. However, the graph for the Los Angeles basin (Fig. 3) shows smaller values for the hydrocarbon to noncarbonate carbon ratio at

6,000 and 6,700 ft than at 3,900 ft. Moreover, the ratio values at 8,000, 9,000, and 9,300 ft are almost identical.

Obviously, the relatively small number of data points does not provide a detailed pattern. Also, irregular variation in types and constituents of the organic matter may be a factor, as they would have a bearing on the amounts of hydrocarbons formed. However, with increasing depth in the Los Angeles basin, the paleoenvironment apparently had progressively less nonmarine and shallow-marine influence. Hence the potential for hydrocarbon formation conceivably should improve with depth for paleoenvironmental as well as diagenetic reasons. The observed pattern would result, however, if the hydrocarbons migrated out of the shales in the depth intervals in which the hydrocarbon to noncarbonate carbon ratio either decreases or remains relatively constant.

A somewhat similar pattern is found in the Ventura basin (Fig. 4), but in a deeper interval than in the Los Angeles basin. In the Ventura basin, the hydrocarbon to noncarbonate carbon ratio increases to a depth of about 7,400 ft in the upper Pliocene section. Conversely, in the lower Pliocene section, at depths of about 7,100, 8,000, and 10,800 ft, a lower ratio is present than in the deepest upper Pliocene sample. Again, this anomalous reduction in ratio may reflect primary migration of hydrocarbons. The fact that the sample at 7,400 ft has a higher ratio than the sample at 7,100 ft is probably a matter of age. These samples come from two different wells, and the data from the shallower lower Pliocene point should be expected to form a pattern with other lower Pliocene samples rather than with upper Pliocene data.

Total depth range over which primary migration occurred can also be postulated (Figs. 3, 4). In the Los Angeles basin, the main reversal or inflection intervals of the hydrocarbon to noncarbonate carbon ratio profile extend downward from about 3,900 ft to about 7,000 ft, and from about 8,100 ft to about 9,600 ft. A sharply progressive increase in ratio does not begin again until a depth of about 9,500 ft. In the Ventura basin, the inflection begins around 7,000-7,500 ft and extends downward for several thousand feet. The reason for the major irregularities between 15,000 and 16,000 ft in samples from Ventura field is unknown.

The actual depth intervals proposed for primary migration need not span the entire several thousand feet of reversals in the hydrocarbon to noncarbonate carbon ratio. From Figure 3, it can be postulated that primary migration in the Los

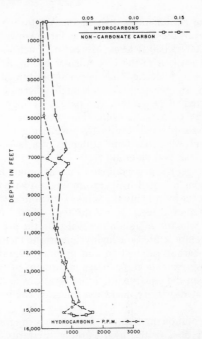

FIG. 4—Depth patterns of hydrocarbon concentrations and hydrocarbon to non-carbonate carbon ratios, Ventura basin, California. Hydrocarbons plotted have boiling points >325°C. Adapted from Philippi (1965).

Angeles basin may have been initiated with an overburden of between 3,900 and 6,000 ft. This is the depth interval within which the reversal in the profile begins. As additional increments of overburden were deposited, the position of the original primary migration was deepened progressively, and consecutive levels of primary migration may have developed in the overlying section. At some point in the subsidence of the initial migration zone, this zone probably ceased to contribute appreciably. In the Los Angeles basin, the depth below which primary migration became less significant would be about 9,500 ft. Below this depth, the ratio increases rapidly, as does the hydrocarbon concentration in the shale. Both increases could represent hydrocarbon retention.

In the Ventura basin (Fig. 4), the reversal in hydrocarbon to noncarbonate carbon ratio begins at about 6,500-7,500 ft. The data points at about 7,100, 7,900, and 10,700 ft constitute a reversal interval, possibly caused by primary migration.

Geothermal gradients in the Los Angeles and Ventura basins are 1° C per 84 ft, and 1° C per 123 ft, respectively (Philippi, 1965). As the geothermal gradient is much higher in the Los Angeles basin, a more abrupt downward increase might be expected in the hydrocarbon to noncarbonate carbon ratio than in the Ventura basin. However, the opposite is true as far downward as the 8,000-9,000 ft level. Moreover, much of this depth interval in the Los Angeles basin consists of lower Pliocene and upper Miocene shales which are bituminous and might therefore be expected to generate a comparatively high percentage of hydrocarbons. This puzzling aspect can be explained readily by postulating the migration of hydrocarbons at depths of a few thousand to several thousand feet.

The geothermal gradients indicate that depth positions of equal temperature in the two basins are offset by roughly the same vertical interval (a few thousand feet), as are the tops of the postulated primary migration intervals. On the premise that temperature is a highly important factor in hydrocarbon production and primary migration, the approximate coincidence in amount of offset between the equal-temperature positions and the suggested primary migration intervals is comprehensible.

A solution mechanism for primary migration would obviate the hydrocarbon "maturity" requirement stipulated by Philippi (1965) as a premigration condition. The effect of solution is to incorporate selectively more of the lighter than heavier liquid hydrocarbons in the migration fluid. As a general rule, an inverse relation exists in hydrocarbons between solubility and molecular weight (Baker, 1959, 1960, 1962). Thus, the liquid hydrocarbon content of the migrating fluid would consist preponderantly of the lighter rather than heavier molecules, and it would, therefore, be somewhat comparable to the hydrocarbon distribution in most crude oils. The large percentages of intermediate-length hydrocarbon chains characteristic of most crude oils could originate from the structure of colloidal micelles. The hydrocarbon-chain (internal) parts of a micelle must be reasonably long to satisfy the requirement of a very large hydrophobic part necessary for micelle formation (Jirgensons and Straumanis, 1962). As a result of true solution and/or colloidal solution migration, the hydrocarbons remaining in the shale would be preponderantly heavy and therefore "immature" from a molecular weight standpoint.

A solution mechanism of migration also can explain the odd-carbon preference of the heavier normal-paraffin hydrocarbons in shallow- and intermediate-depth shales, which Philippi (1965) mentioned as a mark of "immaturity." A solution process would extract a smooth distribution of molecules with comparable amounts of odd- and even-numbered carbons (Baker, 1959, 1960, 1962), as is found in most crude oils. However, an odd preference would remain in the shales, because they had an odd preference prior to the solution migration process.

After migration and with progressively deeper burial, the normal paraffin hydrocarbon assemblage remaining in the shale gradually would approach similarity to that of crude oil, owing to heat-pressure maturation. Laboratory studies (Jurg and Eisma, 1964; Eisma and Jurg, 1967) show that fatty acids (behenic acid), when heated in the presence of clays, are transformed to paraffin hydrocarbons with little or no odd-carbon preference among the heavier molecules, the condition found in the deep shales of the California basins. A similar result was observed in hydrocarbons produced by thermal treatment of modern sediment kerogen (Van der Weide, 1967). A gradation from high to very low odd preference with increasing depth has been noted in the Eagle Ford (Cretaceous) shale of North and East Texas (Bray and Evans, 1965), the Green River oil shales of the Rocky Mountains area (Robinson et al., 1963, 1965), and the Toarcian shales of the Paris basin (Califet-Debyser and Oudin, 1969). Similar trends toward hydrocarbon maturation with increasing depth could be expected in other shales, including those of the California basins.

This migration model could apply also to the South Sumatra basin. The relatively heavy paraffin-base oil in the transgressive sand facies of that basin could reflect the availability of wax-supplying plant material from adjacent nonmarine sources. Hedberg (1968) has documented the relation between high-wax (heavy paraffin) oils and nonmarine environments. The light paraffin base oil of the regressive facies in the South Sumatra basin could have originated from the marine shales directly underlying the regressive sands. These shales probably contained much less of the plant wax component than did the transgressive section. Therefore, if these shales were the source beds for the regressive sand oil, they could have provided the lighter paraffin-base oil. In both cases, primary migration could have begun at sediment depths of a few thousand feet or so, and continued through a burial interval of several thousand feet.

Conclusion

Although advocates of shallow oil origin and primary migration can marshall logical support for their views, much geologic, geochemical, and experimental evidence leads to interpretations of deeper burial for these processes. Figure 5 demonstrates, in general, the depth pattern indicated by inferences from the available data.

From subsurface investigations, there is no evidence that important amounts of liquid hydrocarbons originate or migrate through source beds within the top 1,500 ft of burial depth. It is admitted that only a few investigations have analyzed the organic constituents between the 10-20-ft depth and 1,500 ft. The dilute occurrence of free liquid hydrocarbons and suggested precursors in numerous very shallow fine-grained sediments is well documented. In interstitial waters of these sediments, the total organic component is sparse, even in euxinic samples. With one exception, Holocene and near-recent reservoir-type sediments are not known to contain concentrations sufficient to qualify as incipient oil accumulations. The exception is a core sample from 2,233 ft at Pelican Island, Louisiana, and that sample strongly indicates possibility of upward secondary migration.

Also significant are the important differences between modern sediment hydrocarbons and their petroleum counterparts. To concentrate an approximation of petroleum-type hydrocarbon assemblages from modern sediments, a solution mechanism must be invoked. However, as the interstitial waters are manifestly very low in hydrocarbons and hydrocarbon precursors, significant concentration by these waters seems unlikely. Also, the extremely dilute presence or complete absence of petroleum-type hydrocarbons in modern muds seems to defy meaningful concentration. Moreover, most of the original pore water is lost to the surface during the initial 150-200 ft of burial.

A strong deterrent to liquid hydrocarbon origin in shallow source-type sediments is the lack of a rigorous chemical environment. Bacteria and other organisms apparently effect some important transformations of hydrocarbon precursors, but there is no evidence that very significant amounts of these products are preserved in a free state. The low temperatures and pressures characteristic of shallow depths retard chemical breakdown of organic matter; and these conditions together with water abundance apparently preclude important catalytic transformation to hydrocarbons. In the laboratory, chemical mechanisms for hydrocarbon formation, such as decarboxylation, require relatively high temperatures, even in the presence of catalysts.

Among the barriers to shallow primary migration are the oleophilic character of the carbonaceous part of the organic matter, and the adsorption capacity of source bed minerals, especially

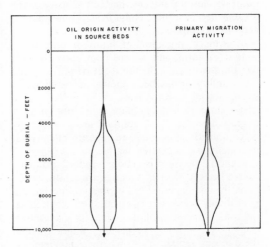

Fig. 5—Comparative activity of oil origin and primary migration as related to burial depth. Width of graph represents postulated amount of activity and worldwide basis.

clays. Because these attraction phenomena are powerful, they probably help to explain why interstitial fluids of modern sediments are so low in dissolved oil-related substances. Under these conditions primary migration in oil form should be minimal, even if oil globules somehow could develop from the sparse hydrocarbon content of the organic matter. Capillary attraction would immobilize any oil globules in shallow source-type muds.

The conclusion that appreciable primary migration may begin at 1,500-2,500 ft seems reasonable. However, *appreciable* migration should be carefully distinguished from *significant* and *flush* migration. Besides, two of the important lines of evidence supporting the 1,500-2,500-ft interval are subject to some qualification. First, the hydrocarbon enrichment of sand at 2,233 ft in the Pelican Island, Louisiana, core may not have migrated from shallow muds. Quite possibly this concentration resulted from upward migration from much greater depths. Secondly, the conclusion of a 500-m (1,650 ft) minimal depth in the Gifhorn trough of Germany assumes that early expression of source-bed fluids occurred in marginal parts of the basin, where most of the producing reservoirs are located. If the initial source-bed role were restricted to the trough, significant primary migration may not have begun until the deposition of 800 m (2,625 ft) or more of overburden. In provinces where the source-bed section is montmorillonitic, such as the Gulf Coast, a much greater depth seems likely from knowledge of clay-water relations.

If primary migration of liquid hydrocarbons were in oil form, only source beds under shallow overburden would have the requisite porosity and permeability. It has been shown that solution of hydrocarbons directly in water, in colloids, or in gases more reasonably can constitute primary migration media. These solution processes, singly or combined, should be much more effective at intermediate and greater depths.

A special problem is encountered in carbonate source beds. Unless carbonates are relatively impure, the compaction-fluid expression mechanism is probably ineffective. Cementation, crystallization, and pressure-solution activity (stylolitization) may contribute to primary migration, but the question remains as to whether sufficient water was available, and if so, in what form the migration took place. Colloids, which greatly enhance solubility of hydrocarbons in water, are demonstrably ineffective in carbonates. Most hydrocarbons of crude oils are not sufficiently water-soluble without colloidal enhancement, particularly in carbonates, where water apparently is quite limited. Pore-to-pore movement of globules seems unlikely because of weakness or absence of compaction or other mechanism progressively to move the globules. Solution of carbonate hydrocarbons in gases appears much more promising. However, this process requires considerable depth where elevated temperatures would transform parts of the organic component into large amounts of gas and liquid hydrocarbons, and would increase the kinetic energy of the hydrocarbons to facilitate their desorption from the bulk organic matter.

Because origin and primary migration processes seem inadequate at shallow depths, an ample source of bitumen-hydrocarbon substances apparently becomes available at greater depths. The most conceivable source is kerogen, the relatively insoluble bulk of the organic material. Laboratory studies indicate that sapropelic or lipid types of kerogen contain structures which are similar to fatty acids and therefore hydrocarbons. Thermal experiments show that parts of the kerogen can be transformed to bitumen, and that parts of the bitumen become more oil-like at higher temperatures. Thus hydrocarbon origin and primary migration mechanisms involving both nonhydrocarbon bitumen and hydrocarbons can be postulated. However, in the sedimentary section, these transformations and processes would be expected only at considerable depth, whose relatively high temperatures would permit adequate chemical activity.

The idea of in-place reservoir maturation of crude oil from a heavy asphaltic (naphthenic) substance to lighter paraffinic character is open to question. Maturation could, instead, take place in source beds. Since naphthenic acids are much more soluble than corresponding paraffinic compounds, the former could migrate earlier from source beds, carrying a preferred load of naphthenic hydrocarbons. These solubility differences could explain the downward increase in liquid paraffinic components which would require a more drastic chemical environment for their formation and migration. Although most aromatic hydrocarbons are probably late forming, evidence has been presented supporting some earlier formation and migration.

Investigations in several petroleum provinces show that primary migration commonly occurred in stages, or at least was long term during thousands of feet of basin subsidence. Important evidence for relatively deep origin and primary migration comes from analyses of source-type rocks through large depth intervals with little or no lithologic change. The depth patterns of bitumen and liquid paraffinic hydrocarbon concentrations at Douala, Cameroun, suggest that flush paraffinic hydrocarbon formation developed in the 1,500-2,500-m (about 4,900-8,200 ft) depth range, and that large quantities of paraffinic hydrocarbons migrated out of the source sediments at approximately 2,500-2,850 m (8,200-9,300 ft), together with some of the nonhydrocarbon bitumen present at these depths. In the Paris basin, a similar study showed that important generation of hydrocarbons apparently began at about 1,400 m (4,600 ft) depth, and that flush primary migration began at about 2,000 m (6,560 ft). These shallower depth figures in the Paris basin probably can be attributed to its greater age (Jurassic), and to the fact that all hydrocarbons, instead of paraffins only, were involved in the analysis. In the Cameroun section, only the paraffin hydrocarbons, which are probably late migrating, were investigated. The Paris basin data suggest that the main hydrocarbons migrating at 2,000-2,500 m may be aromatic rather than saturated. Thus it is possible that a deeper migration of saturated hydrocarbons in the Paris basin could be comparable in depth to the 2,500-2,850-m range in the Cameroun section. Unfortunately, in the Paris basin, data are not available from below 2,500 meters. Reports of interpreted depths of origin and primary migration in several other regions are roughly comparable to those of the Cameroun and Paris basin examples.

Progressive downward dehydration of expandable clay minerals in the Gulf Coast and elsewhere appears to be related to primary migration. Two interpretations of the dehydration process have been reported: (1) loss of consecutive layers of interlayer water from expandable mineral structure; and (2) progressive decrease of the montmorillonite-to-illite ratio in mixed-layer clays. In the former situation, a consistent relation of dehydration depth to depth of oil accumulation was noted. In general, it appears that flush primary migration of liquid hydrocarbons begins in about the same depth range (mostly 5,000-8,000 ft) as does the release of much of the more tightly held clay interlayer water. It seems evident that integrated studies of the organic component and clay-mineral patterns should be undertaken to gain a better understanding of primary migration.

Proponents of an ultra-deep sedimentary origin of oil (depth on the order of 11,000-15,000 ft) suggest organic shales as source rocks, and postulate upward migration for several thousand feet through the shale section. This theory has been applied particularly to the South Sumatra, Los Angeles, and Ventura basins. Although this model can fit much of the data, it does not explain several important points. First, although hydrocarbons certainly should be formed in large quantities at the depths postulated, it is difficult to understand how the hydrocarbons would be mobilized and concentrated. Water in shales at these depths is typically sparse, and permeability is quite low. It is not clear how faults and fractures could remain open as conduits for liquid hydrocarbons at these depths. Gas with dissolved hydrocarbons might find its way upward, but only the lightest liquid hydrocarbons would remain in gas solution to a sufficiently shallow depth to feed oil reservoir rocks, many of which are only a few thousand feet deep.

The Los Angeles and Ventura basin data can be better explained by postulating a flush primary migration beginning at depths of 4,000-6,000 ft in the Los Angeles basin, and 7,000-8,000 ft in the Ventura basin, and extending downward for a few thousand feet. Most of the depth patterns of hydrocarbon to organic carbon ratios, relative hydrocarbon maturation, and geothermal gradient data are accounted for more satisfactorily by proposing these intermediate migration depths.

REFERENCES CITED

Abelson, P. H., 1963, Organic geochemistry and the formation of petroleum: 6th World Petroleum Cong. Proc., Frankfurt, Germany, sec. 1, p. 397-407.
—— 1967, Conversion of biochemicals to kerogen and n-paraffins, in Researches in geochemistry, v. 2: New York, John Wiley and Sons, p. 63-86.
—— and P. L. Parker, 1962, Fatty acids in sedimentary rocks: Carnegie Inst. Washington Year Book, no. 61, p. 181-184.
—— T. C. Hoering, and P. L. Parker, 1964, Fatty acids in sedimentary rocks, in Advances in organic geochemistry: New York, Macmillan Co., p. 169-174.
Abu Amr, A. R., 1971, Spatial and temporal anchimetamorphism, and petroleum occurrence in the Algerian Sahara: Neues Jahrb. Geologie u. Paläontology Abh., Bd. 138, Hft. 3, p. 259-268.
Albrecht, P., and G. Ourisson, 1969, Diagénèse des hydrocarbures saturés dans une série sédimentaire épaisse (Douala, Cameroun): Geochim. et Cosmochim. Acta, v. 33, no. 1, p. 138-142.
Al'tovskii, M. E., Z. I. Kuznetsova, and V. M. Shvets, 1958, Origin of oil and oil deposits: Vses. Nauchno-Issled. Inst. Gidorgeol. Inzh. Geol., Moscow, 168 p. (in Russian); 1961, Engl. trans. by Consultants Bureau, 107 p.
Andrews, D. I., 1961, Indigenous Pleistocene production in offshore Louisiana: Gulf Coast Assoc. Geol. Socs. Trans., v. 11, p. 109-119.
—— and J. C. Stripe, 1961, In offshore Louisiana, some commercial oil, gas indigenous to Pleistocene: World Oil, v. 152, no. 7, p. 122-123, 126-127, 130.
Bajor, M., M. H. Roquebert, and B. M. van der Weide, 1969, Transformation de la matière organique sédimentaire sous l'influence de la température (Transformation of sedimentary organic matter under the influence of temperature): Centre Recherches Pau Bull., v. 3, no. 1, p. 113-124.
Baker, E. G., 1959, Origin and migration of oil: Science, v. 129, no. 3353, p. 871-874.
—— 1960, A hypothesis concerning the accumulation of sediment hydrocarbons to form crude oil: Geochim. et Cosmochim. Acta, v. 19, p. 309-317.
—— 1962, Distribution of hydrocarbons in petroleum: Am. Assoc. Petroleum Geologists Bull., v. 46, no. 1, p. 76-84.
—— 1967, A geochemical evaluation of petroleum migration and accumulation, in Fundamental aspects of petroleum geochemistry: New York, Elsevier, p. 299-329.
Baker, E. W., 1971, Fossil porphyrins in deep ocean sediments: Am. Chem. Soc., Petroleum Research Fund 15th Ann. Rept., abs. no. 4177-AC2, p. 81.
Banks, L. M., 1966, Geologic aspects of origin of petroleum: Am. Assoc. Petroleum Geologists Bull., v. 50, no. 2, p. 397-400.
Barbat, W. N., 1967, Crude-oil correlations and their role in exploration: Am. Assoc. Petroleum Geologists Bull., v. 51, no. 7, p. 1255-1292.
Bartlett, G. A., and R. G. Greggs, 1970, The Mid-Atlantic Ridge near 45°00′ north, VIII. Carbonate lithification on oceanic ridges and seamounts: Canadian Jour. Earth Sci., v. 7, no. 2, pt. 1, p. 257-267.
Barton, D. C., 1934, Natural history of the Gulf Coast crude oil, in Problems of petroleum geology: Am. Assoc. Petroleum Geologists, p. 109-155.
—— 1937, Evolution of Gulf Coast crude oil: Am. Assoc. Petroleum Geologists Bull., v. 21, p. 914-946.
Bathurst, R. G. C., 1970, Problems of lithification in carbonate muds: Geol. Assoc. London Proc., v. 81, pt. 3, p. 429-440.
Bausch, W. M., 1968, Clay content and calcite crystal size of limestones: Sedimentology, v. 10, no. 1, p. 71-75.
Beales, F. W., 1965, Diagenesis in pelleted limestones, in Dolomitization and limestone diagenesis: Soc. Econ. Paleontologists and Mineralogists Spec. Pub. 13, p. 49-70.
Bentz, A., 1958, Relations between oil fields and sedimentary

troughs in northwest German basin, *in* Habitat of oil: Am. Assoc. Petroleum Geologists, p. 1054-1066.

Bergmann, W., and D. Lester, 1940, Coral-reefs and the formation of petroleum: Science, v. 92, no. 2394, p. 452-453.

Boigk, H., H. U. Hark, and W. Schott, 1963, Oil migration and accumulation at the northern border of the Lower Saxony basin: 6th World Petroleum Cong. Proc., Frankfurt, Germany, sec. 1, p. 435-455.

Bordenave, M., 1967, La genèse organique des pétroles; 2, De l'enfouissement de sédiments aux migrations des hydrocarbures: Sci. Progrès-La Nature, no. 3392, p. 451-458.

Bornhauser, M., 1950, Oil and gas accumulation controlled by sedimentary facies in Eocene Wilcox to Cockfield Formations, Louisiana Gulf Coast: Am. Assoc. Petroleum Geologists Bull., v. 34, no. 9, p. 1887-1896.

Brand, E., and K. Hoffman, 1963, Stratigraphy and facies of the northwest German Jurassic and genesis of its oil deposits: 6th World Petroleum Cong. Proc., Frankfurt, Germany, sec. 1, p. 223-246.

Bray, E. E., and E. D. Evans, 1961, Distribution of n-paraffins as a clue to recognition of source beds: Geochim. et Cosmochim. Acta, v. 22, no. 1, p. 2-15.

—— and —— 1965, Hydrocarbons in non-reservoir-rock source beds: Am. Assoc. Petroleum Geologists Bull., v. 49, no. 3, p. 248-257.

Breger, I. A., 1960, Diagenesis of metabolites and a discussion of the origin of petroleum hydrocarbons: Geochim. et Cosmochim. Acta, v. 19, no. 4, p. 297-308.

—— and A. Brown, 1962, Kerogen in the Chattanooga Shale: Science, v. 137, p. 221-224.

Brooks, B. T., 1948, Active-surface catalysts in formation of petroleum: Am. Assoc. Petroleum Geologists Bull., v. 32, no. 12, p. 2269-2286.

—— 1949, Active-surface catalysts in formation of petroleum—II: Am. Assoc. Petroleum Geologists Bull., v. 33, no. 9, p. 1600-1612.

—— 1952, Evidence of catalytic action in petroleum formation: Ind. Eng. Chem., v. 44, no. 11, p. 2570-2577.

Brown, P. R., 1969, Compaction of fine-grained terrigenous and carbonate sediments—a review: Bull. Canadian Petroleum Geology, v. 17, no. 4, p. 486-495.

Buchta, H., R. Leutner, and H. Wieseneder, 1963, The extractable organic matter of pelite and carbonate sediments of the Vienna basin: 6th World Petroleum Cong. Proc., Frankfurt, Germany, sec. 1, p. 335-354.

Burst, J. F., 1959, Postdiagenetic clay mineral environmental relationships in the Gulf Coast Eocene, *in* Clays and clay minerals: New York, Pergamon Press, Internat. Ser. Mons. Earth Sci., v. 2, p. 327-341.

—— 1966, Diagenesis of Gulf Coast clayey sediments and its possible relation to petroleum migration (abs.): Am. Assoc. Petroleum Geologists Bull., v. 50, no. 3, p. 607.

—— 1969, Diagenesis of Gulf Coast clayey sediments and its possible relation to petroleum migration: Am. Assoc. Petroleum Geologists Bull., v. 53, no. 1, p. 73-93.

Califet, J., and J. L. Oudin, 1966, Influence of temperature, pressure, and a clay mineral on the evolution of the chemical structure of an aromatic fraction of a crude oil (abs.): 3d Internat. Geochem. Soc., European Branch Div. Mtg., London; Petroleum Abs., v. 6, no. 45, abs. 74,176, p. 2659.

Califet-Debyser, Y. D., and J. L. G. Oudin, 1969. Influence of depth and temperature on the structure and distribution of alkanes and aromatic hydrocarbons from rock extracts (abs.): Am. Chem. Soc. Petroleum Chemistry Div. Preprints. v. 14, no. 4, p. E16-21; also 1969, *in* Petroleum Abs., v. 9. no. 41 abs. 118,683, p. 2802.

Cane, R. F., 1946, A note on the chemical constitution of torbanite kerogen: Soc. Chem. Ind. Jour., v. 65, p. 412-414.

—— 1948, The chemistry of the pyrolysis of torbanite: Australia Chem. Inst. Jour. and Proc., v. 15, p. 62-68.

Cate, R. B., Jr., 1960, Can petroleum be of pedogenic origin?:

Am. Assoc. Petroleum Geologists Bull., v. 44, no. 4, p. 423-432.

Chemodanov, V. S., 1967, Difficult questions on the migration of petroleum in gas solution and its application to actual conditions in petroleum- and gas-bearing areas: USSR All-Union Oil and Gas Genesis Symposium Proc., Moscow, p. 436-440; 1968, Abs., *in* Petroleum Abs., v. 8, no. 9, abs. 94,146, p. 458.

Chilingar, G. V., and H. H. Rieke, III, 1968a, Chemical alterations of subsurface waters during diagenesis (abs.): Kansas Univ. Symposium on Geochemistry of Subsurface Brines; also *in* Petroleum Abs., v. 8, no. 20, abs. 97,083, p. 1147.

—— —— 1968b, Data on consolidation of fine-grained sediments: Jour. Sed. Petrology, v. 38, no. 3, p. 811-816.

—— —— and E. K. Olson, 1966, High-pressure compaction studies and chemistry of solutions squeezed out of muds at different stages of compaction (abs.): Am. Assoc. Petroleum Geologists Bull., v. 50, no. 3, p. 608.

—— —— S. T. Sawabini, and I. Ershagi, 1969, Chemistry of interstitial solutions in shales versus that in associated sandstones: 44th Ann. Petroleum Eng. Soc. AIME Mtg. Preprints, 8 p.

Cohee, G. V., and K. K. Landes, 1958, Oil in the Michigan basin, *in* Habitat of oil: Am. Assoc. Petroleum Geologists, p. 473-493.

Cooper, J. E., 1962, Fatty acids in recent and ancient sediments and petroleum reservoir waters: Nature, v. 193, no. 4817, p. 744-746.

—— and E. E. Bray, 1963, A postulated role of fatty acids in petroleum formations: Geochim. et Cosmochim. Acta, v. 27, no. 11, p. 1113-1127.

Corbett, C. S., 1955, *In situ* origin of McMurray oil of northeastern Alberta and its relevance to general problem of origin of oil: Am. Assoc. Petroleum Geologists Bull., v. 39, no. 8, p. 1601-1649.

Cox, B. B., 1946, Transformation of organic material into petroleum under geological conditions ("the geological fence"): Am. Assoc. Petroleum Geologists Bull., v. 30, no. 5, p. 645-659.

Dallmus, K. F., 1958, Mechanics of basin evolution and its relation to the habitat of oil in the basin, *in* Habitat of oil: Am. Assoc. Petroleum Geologists, p. 883-931.

DeGroot, K., 1969, The chemistry of submarine cement formation at Dohat Hussain in the Persian Gulf, *in* Lithification of carbonate sediments, 1: Sedimentology, v. 12, nos. 1-2, p. 63-68.

Dickey, P. A., 1967, Petroleum hydrogeology (abs.): Am. Assoc. Petroleum Geologists Bull., v. 51, no. 8, p. 1687-1688.

—— 1968, Increasing concentration of subsurface brines with depth: Kansas Univ. Symposium on Geochemistry of Subsurface Brines, 29 p.

—— and R. E. Rohn, 1958, Facies control of oil occurrence, *in* Habitat of oil: Am. Assoc. Petroleum Geologists, p. 721-734.

Dickinson, G., 1953, Geological aspects of abnormal reservoir pressures in Gulf Coast Louisiana: Am. Assoc. Petroleum Geologists Bull., v. 37, no. 2, p. 410-432.

Dott, R. H., Sr., and M. J. Reynolds, 1969, Sourcebook for petroleum geology: Am. Assoc. Petroleum Geologists Mem. 5, 471 p.

Dufour, J., 1957, On regional migration and alteration of petroleum in south Sumarta: Geologie en Mijnbouw (n. ser.), v. 19, p. 172-181.

Dunnington, H. V., 1967, Aspects of diagenesis and shape change in stylolitic limestone reservoirs: 7th World Petroleum Cong. Proc., Mexico, v. 2, p. 339-352.

Dunoyer de Segonzac, G., 1964, Les argiles du crétacé superior dans le bassin de Douala (Cameroun): Problemes de diagenèse: Alsace-Lorraine Service Carte Géol. Bull., v. 17, no. 4, p. 287-310.

Dunton, M. L., and J. M. Hunt, 1962, Distribution of low-molecular-weight hydrocarbons in recent and ancient sediments: Am. Assoc. Petroleum Geologists Bull., v. 46, no. 12, p. 2246-2248.

Dvali, M. F., 1967, Possible processes, geological conditions and time of primary migration: USSR All-Union Oil and Gas Genesis Symposium Proc., Moscow, p. 364-381; 1968, Abs., in Petroleum Abs., v. 8, abs. 909,906, p. 3242.

Eberl, D. D., and H. Hower, 1971, The diagenetic conversion of smectite to illite/smectite: experimental determination of reaction rates (abs.): Geol. Soc. America Abs. with Programs, v. 3, no. 7, p. 552.

Einsele, G., 1966, Sedimentary processes and physical properties of cores from the Red Sea, Gulf of Aden, and off the Nile delta, in Marine geotechnique: Illinois Univ. Press, Internat. Resources Conf. Marine Geotech. Proc., p. 154-162.

Eisma, E., 1967, Panel discussion, origin of oil and gas: 7th World Petroleum Cong. Proc., Mexico, v. 2, p. 73-77.

——— and J. W. Jurg, 1967, Fundamental aspects of the diagenesis of organic matter and the formation of petroleum: 7th World Petroleum Cong. Proc., Mexico, v. 2, p. 61-72.

Emery, K. O., 1960, The sea off southern California: a modern habitat of petroleum: New York, John Wiley and Sons, 366 p.

——— 1963, Oceanic factors in accumulation of petroleum: 6th World Petroleum Cong. Proc., Frankfurt, Germany, sec. I, p. 483-491.

——— and D. Hoggan, 1958, Gases in marine sediments: Am. Assoc. Petroleum Geologists Bull., v. 42, no. 9, p. 2174-2188.

——— and S. C. Rittenberg, 1952, Early diagenesis of California basin sediments in relation to origin of oil: Am. Assoc. Petroleum Geologists Bull., v. 36, no. 5, p. 735-806.

Erdman, J. G., 1961, Some chemical aspects of petroleum genesis as related to the problem of source bed recognition: Geochim. et Cosmochim. Acta, v. 22, no. 1, p. 16-36.

——— 1965, Petroleum—its origin in the earth, in Fluids in subsurface environments: Am. Assoc. Petroleum Geologists Mem. 4, p. 20-52.

——— 1967, Geochemical origins of the low molecular weight hydrocarbon constituents of petroleum and natural gases: 7th World Petroleum Cong. Proc., Mexico, v. 2, p. 13-24.

——— E. M. Marlett, and W. E. Hanson, 1958, The quantitative determination of low molecular weight aromatic hydrocarbons in aquatic sediments: Presented at 134th Mtg. Am. Chem. Soc., Chicago.

Fischer, A. G., and R. E. Garrison, 1967, Carbonate lithification on the sea floor: Jour. Geology, v. 75, no. 4, p. 488-496.

Fischer, E. K., 1950, Colloidal dispersions: New York, John Wiley and Sons, 387 p.

Forsman, J. P., 1963, Geochemistry of kerogen, Chap. 5 in Organic geochemistry: New York, Macmillan Co. (Internat. Ser. Mons. Earth Sci.), p. 148-182.

———and J. M. Hunt, 1958, Insoluble organic matter (kerogen) in sedimentary rocks of marine origin, in Habitat of oil: Am. Assoc. Petroleum Geologists, p. 747-778.

Franks, A. J., and B. D. Goodier, 1922, Preliminary study of the organic matter of Colorado oil shales: Colorado School Mines Quart., v. 17, no. 4, suppl. A, 16 p.

Fuechtbauer, H., 1963, Discussion, in W. Philipp et al., The history of migration in the Gifhorn trough (NW Germany): 6th World Petroleum Cong. Proc., Frankfurt, Germany, sec. I, p. 457-478.

Fuloria, R. C., 1967, Source rocks and criteria for their recognition: Am. Assoc. Petroleum Geologists Bull., v. 51, no. 6, p. 842-848.

Gehman, H. M., Jr., 1962, Organic matter in limestones: Geochim. et Cosmochim. Acta, v. 26, no. 8, p. 885-897.

Ginsburg, R. N., 1957, Early diagenesis and lithification of shallow-water carbonate sediments in south Florida, in Regional aspects of carbonate deposition: Soc. Econ. Paleontologists and Mineralogists Spec. Pub. 5, p. 80-99.

——— E. A. Shinn, and J. H. Schroeder, 1967, Submarine cementation and internal sedimentation within Bermuda reefs (abs.): Geol. Soc. America Spec. Paper 115, p. 78-79.

Giraud, A., 1970, Application of pyrolysis and gas chromatography to geochemical characterization of kerogen in sedimentary rock: Am. Assoc. Petroleum Geologists Bull., v. 54, no. 3, p. 439-455.

Gordon, J. E., and R. L. Thorne, 1967, Salt effects on non-electrolyte activity coefficients in mixed aqueous electrolyte solutions: II. Artificial and natural sea waters: Geochim. et Cosmochim. Acta, v. 31, no. 12, p. 2433-2443.

Grim, R. E., 1953, Clay mineralogy, 1st ed.: New York, McGraw-Hill, 384 p.

——— 1968, Clay mineralogy, 2d ed.: New York, McGraw-Hill, 596 p.

Gussow, W. C., 1954, Differential entrapment of oil and gas: a fundamental principle: Am. Assoc. Petroleum Geologists Bull., v. 38, no. 5, p. 816-853.

——— 1955, Time of migration of oil and gas: Am. Assoc. Petroleum Geologists Bull., v. 39, no. 5, p. 547-574.

Haeberle, F. R., 1951, Relationship of hydrocarbon gravities to facies in Gulf Coast: Am. Assoc. Petroleum Geologists Bull., v. 35, no. 10, p. 2238-2248.

Hedberg, H. D., 1936, Gravitational compaction of clays and shales: Am. Jour. Sci., 5th ser., v. 31, no. 184, p. 241-287.

——— 1964, Geological aspects of origin of petroleum: Am. Assoc. Petroleum Geologists Bull., v. 48, no. 11, p. 1755-1803.

——— 1967, Geologic controls on petroleum genesis: 7th World Petroleum Cong. Proc., Mexico, v. 2, p. 73-77.

——— 1968, Significance of high-wax oils with respect to genesis of petroleum: Am. Assoc. Petroleum Geologists Bull., v. 52, no. 5, p. 736-750.

Heling, D., 1970, Microfabrics of shales and their rearrangement by compaction: Sedimentology, v. 15, nos. 3-4, p. 247-260.

Hill, G., 1959, Oil migration (abs.): Am. Assoc. Petroleum Geologists Bull., v. 43, no. 7, p. 1770-1771.

Hiltabrand, R. R., 1970, Experimental diagenesis of argillaceous sediment (abs.): Dissert. Abs., no. 70-18,534, v. 31, no. 4, sec. B, p. 2063-B.

Hlauschek, H., 1950. Roumanian crude oils: Am. Assoc. Petroleum Geologists Bull., v. 34, no. 4, p. 755-781.

Hobson, G. D., 1954, Some fundamentals of petroleum geology: London, Oxford Univ. Press, 139 p.

——— 1967, Oil and gas accumulations, and some allied deposits, in Fundamental aspects of petroleum geochemistry: New York, Elsevier, p. 1-36.

Hodgson, G. W., B. L. Baker, and E. Peake, 1967, Geochemistry of porphyrins, in Fundamental aspects of petroleum geochemistry: New York, Elsevier, p. 177-259.

——— B. Hitchon, and K. Taguchi, 1964, The water and hydrocarbon cycles in the formation of oil accumulations, in Recent researches in the fields of hydrosphere, atmosphere, and nuclear geochemistry: Tokyo, Maruzen, p. 217-242.

——— N. Ushijima, K. Taguchi, and I. Shimada, 1963, The origin of petroleum porphyrins. Pigments in some crude oils, marine sediments and plant material of Japan: Tohoku Univ. Sci. Rept., ser. 3, no. 8, p. 483-513.

Hoering, T. C., and R. M. Mitterer, 1967, Thermal reactions of the organic matter in a recent sediment (abs.): Geol. Soc. America Abs. with Programs, p. 99.

Holmquest, H. J., 1966, Stratigraphic analysis of source-bed occurrences and reservoir oil gravities: Am. Assoc. Petroleum Geologists Bull., v. 50, no. 7, p. 1478-1486.

Hoshkiw, M. E., 1970, Reliability of saturation-pressure method of dating time of oil accumulation: Am. Assoc. Petroleum Geologists Bull., v. 54, no. 5, p. 699-708.

Hubbard, A. B., and W. E. Robinson, 1950, A thermal decomposition study of Colorado oil shale: U.S. Bur. Mines Rept. Inv. 4744, 24 p.

Hunt, J. M., 1961, Distribution of hydrocarbons in sedimentary rocks: Geochem. et Cosmochim. Acta, v. 22, no. 1, p. 37-49.

—— 1967, The origin of petroleum in carbonate rocks, *in* Carbonate rocks, physical and chemical aspects: New York, Elsevier, Developments in Sedimentology, v. 9B, p. 225-251.

—— 1968, How gas and oil form and migrate: World Oil, v. 167, no. 4, p. 140, 145, 148-150.

—— and G. W. Jamieson, 1956, Oil and organic matter in source rocks: Am. Assoc. Petroleum Geologists Bull., v. 40, no. 3, p. 477-488.

Illing, L. V., 1959, Deposition and diagenesis of some upper Paleozoic carbonate sediments in western Canada: 5th World Petroleum Cong. Proc., New York, sec. 1, p. 23-52.

Illing, V. C., 1938, The migration of oil, *in* The science of petroleum, v. 1: London, Oxford Univ. Press, p. 209-215.

Jirgensons, B., and M. E. Straumanis, 1962, A short textbook of colloidal chemistry: New York, Macmillan Co., 500 p.

Judson, S. S., Jr., and R. C. Murray, 1956, Modern hydrocarbons in two Wisconsin lakes: Am. Assoc. Petroleum Geologists Bull., v. 40, no. 4, p. 747-750.

Jurg, J. W., and E. Eisma, 1964, Petroleum hydrocarbons: generation from fatty acid: Science, v. 144, no. 3625, p. 1451-1452.

Karimov, A. K., 1963, Primary migration of hydrocarbons of the oil series: Geologiya Nefti i Gaza, v. 7, no. 8, p. 414-418 (1968, *Engl. trans.* from Russian).

Kartsev, A. A., 1964, Geochemical transformation of petroleum, *in* Advances in organic geochemistry: Earth Sci. Ser. Mono. 15, New York, Pergamon Press, p. 11-14.

—— O. I. Abramova, and M. Ya. Dudova, 1970, Organic substance in mineralized waters: Internat. Geology Rev., v. 12, no. 3, p. 270-271.

—— Z. A. Tabasaranskii, M. I. Subbota, and G. A. Mogilevski, 1954, Geochemical methods of prospecting and exploration for petroleum and natural gas: Los Angeles, Univ. California Press, 347 p. (1959, Engl. trans. from Russian).

—— N. B. Vassoyevich, A. A. Geodekian, S. G. Neruchev, and V. A. Sokolov, 1971, The principal stage in the formation of petroleum: 8th World Petroleum Cong. Preprints, Moscow, PD 1, paper 1, 17 p.

Kasatochkin, V. I., and O. I. Zilberbrand, 1956, Chemical structure of shale kerogen: Akad. Nauk SSSR Doklady, v. 3, p. 1031-1034 (in Russian).

Khitarov, N. I., and V. A. Pugin, 1966, Behavior of montmorillonite under elevated temperatures and pressures: Geochem. Internat., v. 3, no. 4, p. 621-626.

Kidwell, A. L., and P. A. Dickey, 1959, Origin of oil (abs.): Am. Assoc. Petroleum Geologists Bull., v. 43, no. 7, p. 1769-1770.

—— and J. M. Hunt, 1958, Migration of oil in recent sediments of Pedernales, Venezuela, *in* Habitat of oil: Am. Assoc. Petroleum Geologists, p. 790-817.

Kogerman, P. N., 1931, On the chemistry of the Esthonian oil shale kukersite: Archiv Naturkunde Estlands, v. 10, p. 6-27, 37-47, 66-68, 73-75, 82-85; Abs., 1960, *in* W. E. Robinson and K. E. Stanfield: U.S. Bur. Mines Inf. Circ. 7968, p. 50-52.

Kollerov, D. K., 1956, The rate of the thermal decomposition of the organic substances of oil shales: Khim. Tekhnol. Topliva, no. 10, p. 55-62 (in Russian); 1957, Abs., *in* Chem. Abs., v. 51, p. 5392.

Kontorovich, A. E., L. I. Bogorodskaya, L. F. Lipnitskaya, V. M. Mel'nikova, and O. F. Stasova, 1965, Dispersed hydrocarbons in Jurassic deposits of the West Siberian lowland: Akad. Nauk. SSSR Doklady, v. 162, no. 2, p. 428-431; 1965, Abs., *in* Petroleum Abs., v. 5, no. 33, abs. 57,721, p. 1827.

Konyukhov, A. I., and G. I. Teodorovich, 1969, Optimum depths of oil formation in the terrigenous Jurassic in the eastern Caucasus region: Akad. Nauk. SSSR Doklady, v. 188, no. 2, p. 408-411 (in Russian); 1970, Abs., *in* Petroleum Abs., v. 10, no. 17, abs. 127,124, p. 1149.

Kovacheva, I. S., 1968, The change of syngenetic bitumoids

with depth as a probable criterion of identification of oil source terrigenous rocks of the Volga-Ural area: Geologiya Nefti i Gaza, no. 8, p. 43-46 (in Russian); 1969, Abs., *in* Petroleum Abs., v. 9, no. 28, abs. 114,816, p. 1910.

Krebs, W., 1969, Early void-filling cementation in Devonian fore-reef limestones (Germany): Sedimentology, v. 12, nos. 3-4, p. 279-299.

Kroepelin, H., 1967, Panel discussion, origin of oil and gas: 7th World Petroleum Cong. Proc., Mexico, v. 2, p. 73-77.

Kvenvolden, K. A., 1962, Normal paraffin hydrocarbons in sediments from San Francisco Bay, California: Am. Assoc. Petroleum Geologists Bull., v. 46, no. 9, p. 1643-1652.

—— 1965, Saturated fatty acids and normal paraffin hydrocarbons in Lower Cretaceous sediments (abs.): Geol. Soc. America Spec. Paper 87, p. 91.

—— 1966a, Molecular distributions of normal fatty acids and paraffins in some Lower Cretaceous sediments: Nature, v. 209, no. 5023, p. 573-577.

—— 1966b, Evidence for transformation of normal fatty acids in sediments (abs.): 3d Internat. Geochem. Soc., European Branch Div. Mtg., London; Petroleum Abs., v. 6, no. 45, abs. 74,187, p. 2662.

—— and D. Weiser, 1967, A mathematical model of a geochemical process; normal paraffin formation from normal fatty acids: Geochim. et Cosmochim. Acta, v. 31, no. 8, p. 1281-1309.

Landes, K. K., 1959, Petroleum geology, 2d ed.: New York, John Wiley and Sons, 443 p.

—— 1960, Ubiquity of petroleum: Am. Assoc. Petroleum Geologists Bull., v. 44, no. 8, p. 1416-1419.

Link, T. A., 1950, Theory of transgressive and regressive reef (bioherm) development and origin of oil: Am. Assoc. Petroleum Geologists Bull., v. 34, no. 2, p. 263-294.

Link, W. K., 1949, Approach to origin of oil: Am. Assoc. Petroleum Geologists Bull., v. 33, no. 10, p. 1767-1769.

Lopatin, N. V., 1969, The principal phase of oil formation: Akad. Nauk SSSR Izv. Ser. Geol., no. 5, p. 69-76 (in Russian); 1970, Abs., *in* Petroleum Abs., v. 10, no. 17, abs. 172,125, p. 1149.

Louis, M., 1966, Geochemical study of the "schistes cartons" of the Toarcian of the Paris basin: 2d Internat. Cong. Org. Geochem. Proc., Rueill-Malmaison, France, 1964, p. 85-94.

—— amd B. Tissot. 1967. Influence de la température et de la pression sur la formation des hydrocarbures dans les argiles à kérogène: 7th World Petroleum Cong. Proc., Mexico, v. 2, p. 47-60.

Mabesoone, J. M., 1971, Recent marine limestones from the shelf off tropical Brazil: Geologie en Mijnbouw, v. 50, no. 3, p. 451-459.

Martin, G. B., 1969, The subsurface Frio of South Texas: stratigraphy and depositional environments as related to the occurrence of hydrocarbons: Gulf Coast Assoc. Geol. Socs. Trans., v. 19, p. 489-501.

McAuliffe, C., 1963, Solubility in water of C_1-C_9 hydrocarbons: Nature, v. 200, no. 4911, p. 1092-1093.

McIver, R. D., 1967, Composition of kerogen—clue to its role in the origin of petroleum: 7th World Petroleum Cong. Proc., Mexico, v. 2, p. 25-36.

—— C. B. Koons, M. O. Denekas, and G. W. Jamieson, 1963, Maturation of oil, an important natural process (abs.): Geol. Soc. America Ann. Mtg. Prog., New York, p. 113A-114A; 1964, Geol. Soc. America Spec. Paper 76, p. 113A-114A.

McKee, R. H., and E. E. Lyder, 1921, The thermal decomposition of shales. heat effects: Ind. Eng. Chem., v. 13, p. 613-618.

—— and P. D. V. Manning, 1927, Shale oil, 2, Gases from oil shale: Oil Bull., v. 13, p. 489-493, 729-737, 837-843.

Meinhold, R., 1969, Research and progress in geochemistry of hydrocarbons: Zeitschr. Angew. Geologie. Bd. 15, Hft. 4, p. 168-178.

Meinschein, W. G., 1959, Origin of petroleum: Am. Assoc. Petroleum Geologists Bull., v. 43, no. 5, p. 925-943.

—— 1961, Significance of hydrocarbons in sediments and petroleum: Geochim. et Cosmochim. Acta, v. 22, no. 1, p. 58-64.

—— Y. M. Sternberg, and R. W. Klusman, 1968, Origins of natural gas and petroleum: Nature, v. 220, no. 5173, p. 1185-1189.

Milliman, J. D., 1966, Submarine lithification of carbonate sediments: Science, v. 153, no. 3739, p. 994-997.

Mirchink, K., et al., 1971, Main concepts of the theory of oil and gas origin and their accumulation in the light of the most recent investigations: 8th World Petroleum Cong. Preprints, Moscow, PD 1, paper 3, 13 p.

Morandi, J. R., and H. B. Jensen, 1966, Comparison of porphyrins from shale oil, oil shale, and petroleum by adsorption and mass spectroscopy: Jour. Chem. and Eng. Data, v. 11, p. 81-88.

Mysels, K. J., 1959, Introduction to colloidal chemistry, 1st ed.: New York, Interscience Pub., 475 p.

Neruchev, S. G., 1967, Regularities of transformation of dispersed organic matter in buried deposits as basis of diagnosis of oil-producing deposits and of quantitative estimation of processes of migration from source rocks: USSR All-Union Oil and Gas Genesis Symposium Proc., Moscow, p. 71-77; 1968, Abs., in Petroleum Abs., v. 8, abs. 90,929, p. 3248.

—— 1970, Katagenesis of dispersed organic material of rocks and generation of oil and gas during the process of burial: Akad. Nauk SSSR Doklady, v. 194, no. 5, p. 1186-1189.

Newell, N. D., J. K. Rigby, A. G. Fischer, A. J. Whiteman, J. E. Hickox, and J. S. Bradley, 1953, The Permian reef complex of the Guadalupe Mountains region, Texas and New Mexico: San Francisco, W. H. Freeman and Co., 236 p.

Orr, W. L., and K. O. Emery, 1956, Composition of organic matter in marine sediments—preliminary data on hydrocarbon distribution in basins off southern California: Geol. Soc. America Bull., v. 67, no. 9, p. 1247-1257.

Perry, E., and J. Hower, 1970a, Burial diagenesis in Gulf Coast pelitic sediments: Clays and Clay Minerals, v. 18, no. 3, p. 165-177.

—— and —— 1970b, Diagenesis and dehydration of Gulf Coast pelitic sediments (abs.): Geol. Soc. America Abs. with Programs, v. 2, no. 7, p. 648-649.

Peterson, D. H., 1967, Fatty acid composition of certain shallow-water marine sediments (abs.): Dissert. Abs., v. 28, no. 5, p. 1997B-1998B.

Philipp, W., H. J. Drong, H. Fuechtbauer, H. G. Haddenhorst, and W. Jankowsky, 1963, The history of migration in the Gifhorn trough (NW Germany): 6th World Petroleum Cong. Proc., Frankfurt, Germany, sec. 1, p. 457-481.

Philippi, G. T., 1957, Identification of oil source beds by chemical means, in Geología del petróleo: 20th Internat. Geol. Cong. Proc., Mexico, sec. 3, p. 25-38.

—— 1965, On the depth, time and mechanism of petroleum generation: Geochim. et Cosmochim. Acta, v. 29, no. 9, p. 1021-1049.

—— 1969, Essentials of the petroleum formation process are organic source material and a subsurface temperature controlled chemical reaction mechanism, in Advances in organic geochemistry, 1968: Oxford, Pergamon Press, p. 25-46; 1970, Abs., in Abs. North Am. Geology, abs. 07208, p. 1094.

Pirson, S. J., 1964, Projective well log interpretation, pt. 3: World Oil, v. 159, no. 5, p. 180-182.

Pow, J. R., G. H. Fairbanks, and W. J. Zamora, 1963, Descriptions and reserve estimates of the oil sands in Alberta: Research Council Alberta, Inf. Ser. 45, p. 1-14.

Powers, M. C., 1959, Adjustment of clays to chemical change and the concept of the equivalent level, in Clays and clay minerals: New York, Pergamon Press, Internat. Ser. Mons. Earth Sci., v. 2, p. 309-326.

—— 1967, Fluid-release mechanisms in compacting marine mudrocks and their importance in oil exploration: Am. Assoc. Petroleum Geologists Bull., v. 51, no. 7, p. 1240-1254.

Powers, R. W., 1962, Arabian Upper Jurassic carbonate reservoir rocks, in Classification of carbonate rocks: Am. Assoc. Petroleum Geologists Mem. 1, p. 122-192.

Pratt, W. E., 1947, Petroleum on continental shelves: Am. Assoc. Petroleum Geologists Bull., v. 31, no. 4, p. 657-672.

Pray, L. C., 1960, Compaction in calcilutites (abs.): Geol. Soc. America Bull., v. 71, no. 12, p. 1946.

Purser, B. H., 1969, Syn-sedimentary marine lithification of Middle Jurassic limestones in the Paris basin: Sedimentology, v. 12, nos. 3-4, p. 205-230.

Rainwater, E. H., 1963, The environmental control of oil and gas occurrence in terrigenous clastic rocks: Gulf Coast Assoc. Geol. Socs. Trans., v. 13, p. 79-94.

—— 1966, Exploration for natural gas, in Natural gas: Inst. Petroleum, Exploration and Production Group Symposium Proc., London, June 17, 1966, p. 1-31.

Ramsden, R. M., 1952, Stylolites and oil migrations: Am. Assoc. Petroleum Geologists Bull., v. 36, no. 11, p. 2185-2186.

Renz, H. H., H. Alberding, K. F. Dallmus, J. M. Patterson, R. H. Robie, N. E. Weisbord, and J. MasVall, 1958, The Eastern Venezuelan basin, in Habitat of oil: Am. Assoc. Petroleum Geologists, p. 551-600.

Robinson, W. E., and K. E. Stanfield, 1960, Constitution of oil-shale kerogen: bibliography and notes on Bureau of Mines research: U.S. Bur. Mines Inf. Circ. 7968, 79 p.

—— J. J. Cummins, and G. U. Dinneen, 1963, Alteration of paraffin compounds in Green River shale after deposition (abs.): Geol. Soc. America Ann. Mtg. Prog., New York, p. 138-139; 1964, Geol. Soc. America Spec. Paper 76, p. 138-139.

—— —— and —— 1965, Changes in Green River oil-shale paraffins with depth: Geochim. et Cosmochim. Acta, v. 29, no. 4, p. 249-258.

Rodionova, K. F., Ye. P. Shishenina, and Y. M. Korolev, 1963, Study of the composition of asphaltenes of disseminated bituminous matter: Geologiya Nefti i Gaza (in Russian); 1968, Engl. trans., McLean, Virginia, Petroleum Geology, v. 7, no. 8, p. 419-425.

Rogers, J. K., 1969, Possible ground water influence on the habitat of oil in the Gulf Coast: Gulf Coast Assoc. Geol. Socs. Trans., v. 19, p. 119-130.

Schmidt, G. W., 1971, Interstitial water composition and geochemistry of Gulf Coast deep shales and sandstones (abs.): Am. Assoc. Petroleum Geologists Bull., v. 55, no. 2, p. 363.

Schnackenberg, W. D., and C. H. Prien, 1953, The effect of solvent properties in thermal decomposition of oil shale kerogen: Ind. Eng. Chem., v. 45, p. 313-322.

Shinn, E. A., 1969, Submarine lithification of Holocene carbonate sediments in the Persian Gulf: Sedimentology, v. 12, nos. 1-2, p. 109-144.

—— R. M. Lloyd, and R. N. Ginsburg, 1968, Tidal-flat sedimentation and dolomite distribution on Andros Island tidal flats, Bahamas: Geol. Soc. America Ann. Mtg., Mexico City, p. 276; 1969, Geol. Soc. America Spec. Paper 121, p. 276.

Silverman, S. R., 1964, Investigations of petroleum origin and evolution mechanisms by carbon isotope studies, in Isotopic and cosmic chemistry: Amsterdam, North Holland Pub., p. 92-102.

Smith, J. E., J. G. Erdman, and D. A. Morris, 1971, Migration, accumulation and retention of petroleum in the earth: 8th World Petroleum Cong. Preprints, Moscow, PD 1, paper 2, 28 p.

Smith, P. V., Jr., 1952, Preliminary note on origin of petroleum: Am. Assoc. Petroleum Geologists Bull., v. 36, no. 2, p. 411-413.

—— 1954, Studies on origin of petroleum—occurrence of

hydrocarbons in recent sediments: Am. Assoc. Petroleum Geologists Bull., v. 38, no. 3, p. 377-404.

—— 1955, Status of our present information on the origin and accumulation of oil: 4th World Petroleum Cong. Proc., Rome, sec. 1, p. 359-376.

Snarsky, A. N., 1961, Relationship between primary migration and compaction of rocks: Geologiya Nefti i Gaza (in Russian); 1964, Engl. trans., McLean, Virginia, Petroleum Geology, v. 5, no. 7, p. 362-365.

—— 1970, Nature of primary oil migration: Vyssh. Ucheb. Zavedeniy Izv., Nefti Gaza, no. 8, p. 11-15 (in Russian); 1971, Abs., in Petroleum Abs., v. 11, no. 16, abs. 143,313, p. 1135.

Sokolov, V. A., 1957, Possibilities of formation and migration of oil in young sedimentary deposits: Lvov Conf. Proc., May 8-12; 1959, Gostoptekhizdat, Moscow, p. 59-63 (in Russian).

—— 1967, The organic and inorganic formation of hydrocarbons: USSR All-Union Oil and Genesis Symposium Proc., Moscow, p. 113-133 (in Russian); 1967, Abs., in Petroleum Abs., v. 7, no. 49, abs. 90,916, p. 3245.

—— and S. I. Mironov, 1962, On the primary migration of hydrocarbons and other oil components under the action of compressed gases, in The chemistry of oil and oil deposits: Acad. Sci. USSR, Inst. Geol. and Exploit. Min. Fuels, p. 38-91 (in Russian); 1964, Engl. trans. by Israel Prog. Sci. Trans., Jerusalem.

—— T. P. Zhuse, N. B. Vassoyevich, P. L. Antonov, G. G. Grigoriyev, and V. P. Kozlov, 1963, Migration processes of gas and oil, their intensity and directionality: 6th World Petroleum Cong. Proc., Frankfurt, Germany, sec. 1, p. 493-505.

Spencer, D. W., and G. R. Harvey, 1971, Significance of water-soluble organic matter in the origin of crude oil (abs.): Am. Chem. Soc., 15th Ann. Petroleum Research Fund Rept., abs. 3416-A2, p. 180-181.

Starikova, N. D., 1959, Organic matter of the liquid phase of marine muds (abs.): Washington, D.C., Am. Assoc. Adv. Sci., Internat. Oceanographic Cong., Preprints of Abs., p. 980-981.

—— 1970, Vertical distribution patterns of dissolved organic carbon in sea water and interstitial solutions: Oceanology (Acad. Sci. USSR), v. 10, no. 6, p. 796-807 (in Russian); Engl. trans. by Am. Geophys. Union.

Stevens, N. P., E. E. Bray, and E. D. Evans, 1956, Hydrocarbons in sediments of Gulf of Mexico: Am. Assoc. Petroleum Geologists Bull., v. 40, no. 5, p. 975-983.

Stockdale, P. B., 1926, The stratigraphic significance of solution in rocks: Jour. Geology, v. 34, no. 5, p. 399-414.

Sujkowski, Z. L., 1958, Diagenesis: Am. Assoc. Petroleum Geologists Bull., v. 42, no. 11, p. 2692-2717.

Swanson, V. E., and J. G. Palacas, 1965, Humate in coastal sands of northwest Florida: U.S. Geol. Survey Prof. Paper 1214-B, p. B1-B29.

—— A. H. Love, T. G. Ging, Jr., and P. M. Gerrild, 1968, Hydrocarbons and organic fractions in recent tidal-flat and estuarine sediments, northeastern Gulf of Mexico (abs.): Am. Chem. Soc. Petroleum Chem. Div. Preprints, v. 13, no. 4, Petr. 33, p. B35-B37.

Taft, W. H., 1967, Modern carbonate sediments, Chap. 2, in Carbonate rocks—origin, occurrence and classification: New York, Elsevier, p. 29-50.

Teas, L. P., and C. R. Miller, 1933, Raccoon Bend oil field, Austin County, Texas: Am. Assoc. Petroleum Geologists Bull., v. 17, no. 12, p. 1459-1491.

Terzaghi, K., 1925, Principles of soil mechanics, II, compressive strength of clays: Eng. News Record, v. 95, p. 796-800.

Thompson, G., V. T. Bowen, W. G. Melson, and R. Cifelli, 1968, Lithified carbonates from the deep-sea of the equatorial Atlantic: Jour. Sed. Petrology, v. 38, no. 4, 1305-1312.

Timofeev, G. E., 1970, Influence of primary migration on bituminological characteristics of sandy-aleuritic rocks: Geolo-

giya Nefti i Gaza, no. 7, p. 39-43 (in Russian); 1971, Abs., in Petroleum Abs., v. 11, no. 3, abs. 138,944, p. 135.

Tissot, B., and R. Pelet, 1971, Nouvelles données sur les mécanismes de genèse et de migration du pétrole simulation mathématique et application à la prospection: 8th World Petroleum Cong. Preprints, Moscow, PD1, paper 4, 23 p.

—— Y. Califet-Debyser, G. Deroo, and J. L. Oudin, 1971, Origin and evolution of hydrocarbons in early Toarcian shales, Paris basin, France: Am. Assoc. Petroleum Geologists, v. 55, no. 12, p. 2177-2193.

Trask, P. D., 1932, Origin and environment of source sediments of petroleum: Houston, Gulf Pub. Co., 323 p.

Treibs, A., 1936, Chlorophyll and hemin derivatives in organic materials: Angew. Chem., Bd. 49, p. 682-686.

Uspensky, V. A., 1962, The geochemistry of processes of primary migration: Geokhimiya, no. 12, p. 1027-1045 (in Russian); 1962, Engl. trans., Geochemistry, no. 12, p. 1157-1178.

Van der Weide, B., 1967, Evolution of n-paraffins through heat treatments of marine recent sediments: Bull. Centre Rech. Pau-SNPA, v. 1, no. 1, p. 161-164.

Van Moort, J. C., 1971, A comparative study of the diagenetic alteration of clay minerals in Mesozoic shales from Papua, New Guinea and in Tertiary shales from Louisiana, U.S.A.: Clays and Clay Minerals, v. 19, p. 1-20.

Van Olphen, H., 1963, Clay-organic complexes and the retention of hydrocarbons by source rocks: New York, Pergamon Press, Internat. Clay Conf. Proc., v. 1, p. 307-317.

Van Tuyl, F. M., and B. H. Parker, 1941, The time of origin and accumulation of petroleum: Colorado School Mines Quart., v. 36, no. 2, 180 p.

Vassoyevich, N. B., Yu. I. Korchagina, N. V. Lopatin, E. G. Gladkova et al., 1971, Manifestations of the main phase of formation of oil in Maykop clay deposits of western Kuban depression: Neftegazovaya Geologiya i Geofizika, no. 2, p. 12-15 (in Russian).

—— I. V. Visotski, A. N. Guseva, and V. B. Olenin, 1967, Hydrocarbons in the sedimentary mantle of the earth: 7th World Petroleum Cong. Proc., Mexico, v. 2, p. 37-45.

Veber, V. V., and N. M. Turkel'taub, 1958, Gaseous hydrocarbons in recent sediments: Geologiya Nefti i Gaza, no. 8, p. 39-44 (in Russian).

von Engelhardt, W., 1960, Der Porenraum der Sedimente: Berlin, Springer-Verlag, 207 p.

—— and K. H. Gaida, 1963, Concentration changes of pore solution during compaction of clay sediments: Jour. Sed. Petrology, v. 33, no. 4, p. 919-930.

—— and W. L. M. Tunn, 1954, Über das Strömen von Flüssigkeiten durch Sandsteine (The flow of fluids through sandstones): Beitr. Mineralogie u. Petrographie, v. 4, nos. 1-2, p. 12-25; 1955, Engl. trans. by P. A. Witherspoon, Jr., Illinois Geol. Survey Circ. 194, 17 p.

Vyshemirskii, V. S., A. A. Gontsov, V. N. Krymova, G. D. Ushakov, and L. S. Yamkovaya, 1971, Experimental study of the newly formed bitumens and their migration in clay compaction: Geol. Geofiz., no. 1, p. 17-19 (in Russian); 1971, Abs., in Petroleum Abs., v. 11, no. 41, abs. 151,334, p. 2924.

Weaver, C. E., 1960, Possible uses of clay minerals in search for oil: Am. Assoc. Petroleum Geologists Bull., v. 44, no. 9, p. 1505-1518.

—— 1967, The significance of clay minerals in sediments, in Fundamental aspects of petroleum geochemistry: New York, Elsevier, p. 37-75.

—— and K. C. Beck, 1969, Changes in the clay-water system with depth, temperature and time (abs.): Geol. Soc. America Abs. with Programs, pt. 7, p. 233-234.

—— and —— 1971, Clay water diagenesis during burial: how mud becomes gneiss: Geol. Soc. America Spec. Paper 134, 78 p.

—— and J. M. Wampler, 1970, K, Ar, illite burial: Geol. Soc. America Bull., v. 81, no. 11, p. 3423-3430.

Weeks, L. G., 1952, Factors of sedimentary basin development that control oil occurrence: Am. Assoc, Petroleum Geologists Bull., v. 36, no. 11, p. 2071-2124.

—— 1958a, Habitat of oil and factors that control it, *in* Habitat of oil: Am. Assoc. Petroleum Geologists, p. 1-61.

—— ed., 1958b, Habitat of oil: Am. Assoc. Petroleum Geologists, 1384 p.

—— 1961, Origin, migration, and occurrence of petroleum, *in* Petroleum exploration handbook: New York, McGraw-Hill, p. 5-11-5-50.

Weller, J. M., 1959, Compaction of sediments: Am. Assoc. Petroleum Geologists Bull., v. 43, no. 2, p. 273-310.

Welte, D. H., 1965, Relation between petroleum and source rock: Am. Assoc. Petroleum Geologists Bull., v. 49, no. 12, p. 2246-2268.

—— 1966, Correlation problems among crude oils (abs.): 3d Internat. Geochem. Soc., European Branch Div. Mtg., London, Program Abs.

—— 1967, Evolutionary history of hydrocarbon oils in the light of geochemical and geological research: Erdöl Kohle-Erdgas-Petrochemie., v. 20, no. 2, p. 65-67.

White, D. E., 1965, Saline waters of sedimentary rocks, *in* Fluids in subsurface environments: Am. Assoc. Petroleum Geologists Mem. 4, p. 342-366.

Woodward, H. P., 1958, Emplacement of oil and gas in Appalachian basin, *in* Habitat of oil: Am. Assoc. Petroleum Geologists, p. 494-510.

Yurkevich, I. A., 1962, The study of geochemical facies as a means towards characterizing the sources of petroleum deposits, *in* The geochemistry of oil and oil deposits: Acad. Sci. USSR, Inst. Geol. and Exploit. Min. Fuels, p. 129-149 (in Russian); 1964, *Engl. trans.* by Israel Prog. Sci. Trans., Jerusalem.

Zankl, H., 1969, Structural and textural evidence of early lithification in fine-grained carbonate rocks, *in* Lithification of carbonate sediments, 2: Sedimentology, v. 12, nos. 3-4, p. 241-256.

Zhuze, T. P., V. I. Sergeevich, V. F. Burmistrova, and E. A. Esakov, 1971, On solubility of hydrocarbons in water in reservoir conditions: Akad. Nauk SSSR Doklady, v. 198, no. 1, p. 206-209 (in Russian); 1971, Abs., *in* Petroleum Abs., v. 11, no. 35, abs. 149,822, p. 2600.

ZoBell, C. E., 1945, The role of bacteria in the formation and transformation of petroleum hydrocarbons: Science, v. 102, no. 2650, p. 364-369.

—— 1946, Studies on redox potential of marine sediments: Am. Assoc. Petroleum Geologists Bull., v. 30, no. 4, p. 477-513.

—— 1947, Microbial transformation of molecular hydrogen in marine sediments, with particular reference to petroleum: Am. Assoc. Petroleum Geologists Bull., v. 31, no. 10, p. 1709-1751.

—— 1951, Contributions of bacteria to the origin of oil: 3d World Petroleum Cong. Proc., The Hague, 1951, sec. 1, p. 414-420.

ADDENDUM

The following references were noted recently and should be added to the references.

Cahen, R. M., J. E. Marechal, M. P. della Faille, and J. J. Fripiat, 1965, Pore-size distribution by a rapid, continuous flow method: Anal. Chemistry, v. 37, no. 1, p. 133-137.

Slabaugh, W. H., and A. D. Stump, 1964, Surface areas and porosity of marine sediments: Jour. Geophys. Research, v. 69, no. 22, p. 4773-4778.

Price, L. P., 1972, Solubility of petroleum forming hydrocarbons in aqueous solutions as applied to primary petroleum migration (abs.): Geol. Soc. America, Abs. with Programs, v. 4, no. 3, p. 220.

The two reports on pores (Cahen *et al.*, 1965; Slabaugh and Stump, 1964) deal with pore sizes of fine-grained samples. Analysis of these samples, previously prepared by artificial crushing and grinding, showed pore diameters predominantly less than 50-100 Å. As these pore sizes are much smaller than Welte (1965) computed for buried fine-grained sediments, their significance relative to modes of primary migration needs further investigation.

Price's (1972) report bears on the problem of hydrocarbon release from true solution. He concludes from laboratory experiments that hydrocarbon solubility in water at elevated temperatures is sufficiently effective to explain primary migration, and that hydrocarbon release by increasing salinity can account for oil accumulation.

Reprinted from:
BULLETIN OF THE AMERICAN ASSOCIATION OF PETROLEUM GEOLOGISTS
VOL. 58, NO. 1 (JANUARY, 1974), PP. 149-154

DISCUSSION AND REPLY

Depth of Oil Origin and Primary Migration: a Review and Critique: Discussion[1]

G. T. PHILIPPI[2]

Estes Park, Colorado 80517

Robert J. Cordell (1972) published, with other subjects, a critique of the research on depth, time, and mechanism of petroleum generation which I had done for Shell Development Company (Philippi, 1965). Following is a reply to his critique.

On the basis of the depth range where most of the oil is formed, and on the basis of the depth range where shale hydrocarbons and crude oil hydrocarbons become nearly equal in composition, I concluded in my 1965 paper that in the Los Angeles and Ventura basins of California most petroleum is formed in deep upper Miocene D and E shales, which are considerably below many of the producing oil zones.

According to Cordell (p. 2055), "difficulties arise in explaining the great distance of vertical migration through shales. First, at the depths of origin specified, the supply of water as a migration medium in shale would be severely limited. Migration of oil as a separate phase, without benefit of water, appears implausible because of the disseminated nature of the hydrocarbons in the shale and the lack of a suitable mechanism for hydrocarbon collection and concentration." Summarizing, Cordell endorsed primary migration of petroleum in colloidal solution, and flushing of the colloidal solution out of the source rock by water of compaction. Furthermore, hydrocarbon release from the colloid in reservoirs was postulated (p. 2047).

To circumvent the difficulty of lack of water in deep shales, Cordell assumed a much shallower origin of the Los Angeles and Ventura basin oils than I did on the basis of my analytical data. Cordell postulated that most petroleum is released from the source beds in a depth range where the subsurface temperature is relatively low, not much oil has been generated as yet, and crude oil and shale hydrocarbons definitely are unlike in composition.

Cordell's assumptions and criticism are based on his firm conviction that water is the essential driving agent in the primary migration of petroleum from a source rock. On this point we disagree as follows. In 1959 Hill (1959, in Nagy and Colombo, 1967, p. 359) suggested, "that the residual organic matter in shales might be the vehicle for the migration of hydrocarbons. If the volume of residual organic matter is referred to the pore volume, it can be seen that, when the shale porosity reaches values as low as 15–20 percent, the fraction of such pore volume, which is actually occupied by the organic matter, is rather important. Eventually, according to Hill, the organic matter may form a continuous network through the water-wet shale."

The study of thin sections of source rocks by Heacock (1972) is not incompatible with the idea of a continuous organic network. Heacock observed that in many moderately rich and rich oil-source beds the organic matter is deposited in thin continuous bands parallel with the bedding plane, probably as a seasonal phenomenon. Hydrocarbon fluids originating in these individual organic bands may be interconnected by microfissures originating from differential compaction and tectonics, and, possibly, by direct contact of the organic bands at various points.

As a working hypothesis, I postulate that much of the organic matter in moderately rich and rich oil-source beds forms an interconnected hydrophobic organic network, as described previously or otherwise. The oil is formed within the organic matter and initially sorbed there. Only in the later stages of the oil-formation process, when enough oil is formed to exceed the sorption capacity of the organic matter, can oil be expelled as a separate phase (Philippi, 1965). The oil formed will move through the hydrophobic organic network in

[1] Discussion received, February 12, 1973; accepted June 4, 1973.

[2] P.O. Box 986.

the direction of the pressure gradient. Oil will be expelled where the organic matter ends if the pressure is relatively low. This occurs in sufficiently open fractures in the shale, and in porous sands and limestones above and below the source rock. When compaction is increased, the hydrophobic organic network of a source rock is squeezed like a sponge, greatly aiding oil expulsion, but only after the sorption capacity of the organic network is exceeded because of sufficient oil generation. Once the oil is in porous and permeable territory, it will move along because of the hydrodynamic gradient and gravity.

The mechanism of primary migration outlined does not require appreciable water movement through source rocks. Thus, Cordell's main objection against my conclusion of deep source rocks in the Los Angeles and Ventura basins—that not enough compaction water is available for oil expulsion—cannot be maintained. Moreover, the hypothesis that oil moves as a colloidal solution during primary migration becomes superfluous. This is fortunate, because the colloidal-solution hypothesis is questionable. Even more questionable and vague is the "solution mechanism" which has been suggested to fabricate "mature" hydrocarbons in petroleum out of "immature" hydrocarbons in intermediate depth shales. Neither a satisfactory theory nor experimental proof of the "solution mechanism" has been presented. In summary, no adequate explanation has been given by Cordell of the significant difference in n-paraffin composition and in naphthene index of crude oils and shale hydrocarbons of intermediate depth (3,900–9,600 ft in the Los Angeles basin and 7,000–10,700 ft in the Ventura basin).

I do not share Cordell's pessimism concerning vertical migration of petroleum through fractures. Oil seeps predominantly are near fault zones, which usually are full of fractures. Commonly, fracture planes contain deposits such as quartz, carbonates, metal compounds, or asphalt-like material, which indicate that solutions have passed through the fractures over long periods of time.

Finally I should like to reply to Cordell's opinion of my being a proponent of ultradeep petroleum origin. In the Miocene upper Telissa source beds of south Sumatra (Soengei Taham field) the shale hydrocarbons begin to mature at 7,000–7,500 ft. In West Texas some mature oil-source beds of Paleozoic age are only 5,000–8,000 ft deep, very much shallower than the upper Miocene source beds of the Los Angeles and Ventura basins. A fixed depth of oil generation does not exist. The depth interval of flush-oil generation and hydrocarbon maturity depends on the local temperature gradient and on the temperature history and age of the source sediments (Philippi, 1965). In the Los Angeles and Ventura basins the oil source beds happen to be ultra young (upper Miocene D, E); therefore, very deep burial was required for flush oil generation and hydrocarbon maturity.

REFERENCES CITED

Cordell, R. J., 1972, Depths of oil origin and primary migration: a review and critique: Am. Assoc. Petroleum Geologists Bull., v. 56, p. 2029–2067.

Heacock, R. L., 1972, Pyrolysis and thin-section examination of petroleum source rocks (abs.): Geol. Soc. America Abs. with Programs, v. 4, no. 7, p. 532.

Hill, G. A., 1959, Oil migration: Paper presented at Am. Assoc. Petroleum Geologists Ann. Mtg., Dallas (unpub.); quoted in B. Nagy and U. Colombo, eds., 1967, Fundamental aspects of petroleum geochemistry: New York, Elsevier, p. 359.

Philippi, G. T., 1965, On the depth, time, and mechanism of petroleum generation: Geochim. et Cosmochim. Acta, v. 29, p. 1021–1049.

Reply to G. T. Philippi[1]

ROBERT J. CORDELL[2]

Richardson, Texas 75080

The discussion by G. T. Philippi focuses on differences between our interpretations of his organic geochemical results (Philippi, 1965) in the Los Angeles and Ventura basins, California. This problem was considered in the latter part of my recent paper in the *Bulletin* (Cordell, 1972, p. 2054–2058, 2061).

Philippi's study of the organic components in southern California source beds is a key contribution. His conclusion that a temperature higher than 115°C was necessary for origin and primary migration of the bulk of the oil from upper Miocene shales has been noted by several authors (Kartsev *et al.*, 1971; Tissot and Pelet, 1971; Nixon, 1973). Indeed, Philippi's hydrocarbon-maturation temperatures, which he associated with major oil origin, were *much* higher than 115°C, as he listed 150°C for the Los Angeles basin and over 143°C for the Ventura. Significantly, major hydrocarbon occurrence in shales of similar (upper Miocene) age in the Gulf Coast begins at a temperature of only 96°C (LaPlante, 1972).

Application of Philippi's hydrocarbon-maturation temperatures to a general age-temperature model of oil origin could distort our understanding of origin–primary migration concepts. As shown in the following paragraphs, he makes several questionable assumptions in attempting to bridge the vertical gap between his depth range for major oil origin (beginning at 11,000 ft in the Los Angeles basin, and 15,000 ft in the Ventura) and the actual positions of most of the oil accumulations (several thousand feet shallower).

Philippi is incorrect in stating that my main objection to his deeper origin thesis is the lack of water at depth. Findings by clay mineralogists in the Gulf Coast (Powers, 1967; Burst, 1969; Weaver and Beck, 1971; Perry and Hower, 1972) suggest that some pore water might be present at 11,000–15,000 ft or deeper in the California basins. The important question is whether the small percentage of water at such depths would be adequate to transport vast amounts of hydrocarbons upward for long distances. Moreover, the following problems are just as important: (1) there is no plausible explanation of how oil can become sufficiently concentrated to begin migration *as oil* within the source bed (regardless of whether or not pore water is present); and (2) the vertical migration of enormous amounts of oil through several thousand feet of stratigraphic section which is largely shale seems insupportable, regardless of the presence of faults and fractures.

In attributing my main concern to the water problem, Philippi emphasized his own belief that oil migrates through source beds as a separate phase, without the benefit of appreciable water movement. Let us examine several aspects of this hypothesis in the light of pertinent factors.

Philippi suggests a continuous fabric of organic matter in source beds to serve as a pathway along which the oil initially would migrate. To support this idea he refers to a paper presented by Hill (1959a) during the 1959 national AAPG meeting in Dallas, in which Hill allegedly suggested this possible mechanism.

Actually, in his abstract, Hill (1959b) did not mention this mechanism at all, but he did emphasize that 15–25 percent of the pore space in a source bed would have to be occupied by oil for it to migrate by continuous flow. Such a concentration seems impossible unless the oil already has been concentrated by prior migration. Obviously one of the critical problems is to explain the beginning of primary migration before any concentration takes place. In a typical source bed with just a few percent of total organic matter (mostly kerogen), only a minor proportion of which can be expected to become liquid hydrocarbons, how could an *initial* 15–25 percent pore-space concentration of liquid hydrocarbons ever be achieved? And even if it were, how could the oil migrate through the extremely small pore spaces?

The great difficulty of hydrocarbon migration in the form of oil through intergranular pore spaces of source beds has been emphasized by many authors. Recently, Welte (1972)

[1] Reply received, April 10, 1973; accepted, June 5 1973.

[2] Sun Oil Company Production Research Laboratory

has shown that these pore spaces are much smaller, even, than generally has been assumed. A suggestion of primary migration through microfractures is probably feasible, but it leaves unanswered the means whereby oil in source-bed pores could reach the nearest microfracture.

In postulating a continuous framework of organic matter for migration pathways, one immediately encounters the problem not only of relatively small percentages of organic matter, but also of its distribution. Philippi appeals to a statement attributed by Nagy and Colombo (1967) to Hill (1959a), in which the organic matter is inferred to be concentrated in the linings of pore spaces, with the pore spaces constituting 15–20 percent of a compacting shale source rock. If this distribution were true, the organic matter might be relatively continuous.

However, organic matter in source rocks apparently is distributed in several ways: (1) as disseminated particles; (2) as isolated carbonized fossils; (3) as thin deposits along or inclined to bedding surfaces; (4) in interlayer positions of clay-mineral structure; (5) in the fluids; and (6) as adsorbed coatings (pore linings) on the grains of clays and other minerals. Hence only a part of the organic matter would be distributed as the adsorbed coatings or pore linings.

Philippi mentions a thin-section study by R. L. Heacock (1972) presented before the 1972 GSA meeting. Apparently, Heacock observed organic-rich material in thin, continuous depositional bands in various source beds. It was suggested that these bands may contain a continuous organic network, and that interconnecting microfissures may assist in the migration process.

Even if the continuous organic network were present, we have the additional problem of water-wet versus oil-wet character—that is, the more continuous the organic matter, the more likely that the sediment would be oleophilic (oil-attracting) and therefore not conducive to primary migration.

Philippi (1965) disposed of the oleophilic problem by stating that a limit exists to the hydrocarbon-adsorptive capacity of organic matter. He proposed that hydrocarbons generated after this limit is reached are free to migrate as oil along a decreasing pressure gradient, with fractures as conduits. But he presents no substantiating experimental evidence of an adsorptive limit which can be applied dependably to source-bed conditions. In an earlier report

(Philippi, 1957), he concluded that migrated oil in source beds could be identified by its deviation from the straight-line ratio of hydrocarbons to noncarbonate carbon which is characteristic of indigenous hydrocarbons. However, other factors probably contribute to deviations in this ratio; for example, variations in the original composition of the kerogen. Hence, caution should be observed in attempting to apply this ratio to migration. With the restriction on percentage of hydrocarbons that can be anticipated from a given amount of kerogen under normal source-bed conditions, it is questionable whether the adsorptive limit in Philippi's model would be reached.

As to the long interval of postulated vertical migration, two questions are critical: (1) Do faults and fractures at considerable depth in source beds have a significant capacity as oil conduits? And (2) could sufficient oil have been delivered upward to the main oil reservoirs to account for the very large reserves?

Philippi points to oil seeps near fault zones as evidence for sizable migration along faults. However, these are surface and near-surface phenomena; and fault and fracture zones near the surface, where there is relatively little confining pressure, should be sufficiently open to allow the passage and seepage of oil. Conditions in shales at depths of 11,000–15,000 ft are drastically different. It is doubtful that oil could migrate in significant quantities through the essentially closed faults and fractures of shales at these depths, even with the decreasing viscosity of oil in the higher temperature environment. Unless they are cemented strongly by silica, shales and mudstones should be sufficiently plastic under the high confining pressures to close the fractures to oil migration. Water, on the other hand, could seep slowly along such fractures, carrying its load of hydrocarbons in colloidal or true solution. However, the heavier, less soluble molecules would drop out of solution during the early stages of the long upward migration, leaving unexplained the rather large content of heavy hydrocarbons and nonhydrocarbon constituents in the shallow crude oil accumulations of the Los Angeles and Ventura basins.

Philippi mentions deposits of asphalt along fracture zones, and he could have added that commercial oil has been produced from shale fracture zones in many places. However, these occurrences can be related to rather shallow-depth environments or highly silicified shales. He also lists deposits of quartz, carbonates, and metal compounds along fracture zones. These minerals, of course, normally are deposited from water, which can move along fracture zones with much greater facility than oil.

In summary, the writer is convinced that both the preceding questions regarding the capacity of faults and fractures for deep-to-shallow oil migration must be answered in the negative.

One advantage of a water-migration medium is that it permeates throughout the source rock. Water is far more mobile than oil, hence it can migrate not only along nearly closed fractures but also between the grains of the sediment. Because of its distribution throughout the source bed and its mobility, water is in contact with source-bed organic matter everywhere and is susceptible to compaction influences on a large scale. These compaction forces ultimately can move the water and its acquired hydrocarbon load (dissolved and solubilized by colloids) out of the source bed and into a reservoir. The dehydration of clay minerals with increasing depth (increasing temperature and pressure) provides a replenishment of pore water for primary migration in the general depth interval for primary migration suggested by other lines of evidence. Moreover, as montmorillonite is changed progressively to illite with depth (increasing temperature), part of the adsorbed organic matter should be released to the pore spaces. This clay-mineral transformation is now well documented (Powers, 1967; Burst, 1969; Perry and Hower, 1972); and illite has less adsorptive capacity for organic matter than does montmorillonite (Grim, 1968).

With water as a medium, we have a basis for initiating primary migration, which is lacking in Philippi's hypothesis. Initiation of primary migration requires interstitial pore-to-pore movement which is apparently impossible for oil, either as an oil phase or as oil globules in water. Conversely, the universal contact of water with the organic fraction sets the stage for initiation of colloidal and true solution of hydrocarbons. As temperature increases with depth, water should incorporate colloids and liquid hydrocarbons in increasing amounts. The higher temperature not only increases the generation of potential colloids and hydrocarbons; it also reduces the attraction of these constituents to associated kerogen. During deposition and subsidence of the sedimentary section and accompanying compaction, a depth point eventually is reached where permeability is so low and pore water so reduced in percentage

that functional primary migration ceases.

Philippi stresses his view that hydrocarbon maturation in the source bed must develop before primary migration takes place. This maturation is defined as a smooth distribution of molecular sizes among liquid hydrocarbons, with general size-range percentages as found in crude oil. The fact that this maturation is achieved in shales of the Los Angeles and Ventura basins only at relatively great depths (11,000 and 15,000 ft or more, respectively) constitutes the crux of the migration problem.

Advocates of the concept of full maturation for hydrocarbons prior to their migration seem tacitly to assume that the *present* degree of organic maturation is representative of the time when the primary migration actually occurred. Realistically, the degree of maturation now existing must be more advanced in any given subsiding stratigraphic interval than it was during migration. After all, up to the present time, a stratigraphic interval in basins such as Los Angeles and Ventura has continued to be subjected to progressively higher temperatures as additional sediments were deposited. Hence, even if primary migration still were going on, most of the migration history must have been spent under conditions of lesser maturation. The only alternative is to postulate the absurdity that all primary migration is taking place now.

Both the colloidal and true solution concepts can offer tentative explanations for the general distribution of hydrocarbons in crude oil, in terms of migration from source beds in which the hydrocarbon composition is still immature. Solution processes in the source beds would incorporate similar amounts of normal paraffin molecules with even and odd numbers of carbons to give a smooth petroleumlike molecular distribution. Also, the lighter liquid hydrocarbons would be solubilized or dissolved preferentially. The resulting hydrocarbon accumulation would have a relatively large percentage of the lighter liquid hydrocarbons as do most crude oils, and the molecular distribution would be similarly smooth. To account for nonhydrocarbon constituents of crude oil, and parts of its heavier to intermediate hydrocarbon components, colloidal micelles could be postulated as a part of the progenitor material. A paper assessing the colloidal(soap)-migration idea recently has been published (Cordell, 1973).

There is a possible inconsistency with solution concepts in the ratio of odd-to-even-carbon-numbered *n*-paraffin molecules in the organically immature part of the shale section of the Los Angeles and Ventura basins. A high odd-to-even molecular ratio characterizes the immature n-paraffins. Hence, if we postulate that smooth distributions of hydrocarbons are solubilized or dissolved in the source beds, the odd-to-even molecular ratio in n-paraffins remaining in the source bed should increase progressively with depth. As this is not the case in the migration intervals postulated by the present writer, the following possible explanation is offered. In the source bed, the heavy end of the n-paraffin distribution, which mainly is involved in the odd-to-even molecule problem, may simply be relict; that is, it may be held over from early diagenesis. The heavy end of crude oil n-paraffins, on the other hand, may have been derived from long-chain fatty acids or waxes formed during thermocatalysis of kerogen in the source beds. These products should have had the smooth distribution of carbon atoms characteristic of thermocatalytic effects. Their solution in pore waters (quite likely as a colloidal soap) would have been followed by migration to a reservoir. In the reservoir, decarboxylation eventually would occur with continuation of relatively high-temperature conditions, and the resulting heavy normal paraffins should have the smooth distribution which is found in crude oil.

The writer does not imply that all the problems are solved with solution concepts of migration. Much additional research is needed, but it seems preferable to forge ahead on such research rather than accept an hypothesis incorporating mechanisms which physical and chemical principles apparently do not support.

REFERENCES CITED

Burst, J. F., 1969, Diagenesis of Gulf Coast clayey sediments and its possible relation to petroleum migration: Am. Assoc. Petroleum Geologists Bull., v. 53, no. 1, p. 73–93.

Cordell, R. J., 1972, Depths of oil origin and primary migration: a review and critique: Am. Assoc. Petroleum Geologists Bull., v. 56, no. 10, p. 2029–2067.

——— 1973, Colloidal soap as a proposed primary migration medium for hydrocarbons: Am. Assoc. Petroleum Geologists Bull., v. 57, no. 9, p. 1618–1643.

Grim, R. E., 1968, Clay mineralogy 2d ed.: New York, McGraw-Hill, 596 p.

Heacock, R. L., 1972, Pyrolysis and thin-section examination of petroleum source rocks (abs.): Geol. Soc. America Abs. with Programs, v. 4, no. 7, p. 532.

Hill, G., 1959a, Oil migration: Paper presented at Am. Assoc. Petroleum Geologists Ann. Mtg., Dallas (unpub.).

——— 1959b, Oil migration (abs.): Am. Assoc. Petroleum Geologists Bull., v. 43, no. 7, p. 1770.

Kartsev, A. A., N. B. Vassoevich, A. A. Geodekian, S. G. Neruchev, and V. A. Sokolov, 1971, The principal stage in the formation of petroleum: 8th World Petroleum Cong., Moscow, Preprints PD1 (1), 17 p.

LaPlante, R. E., 1972, Petroleum generation in Gulf Coast Tertiary sediments (abs.): Am. Assoc. Petroleum Geologists Bull., v. 56, no. 3, p. 635.

Nagy, B., and U. Colombo, 1967, Origin and evolution of petroleum, in Fundamental aspects of petroleum geochemistry: New York, Elsevier, p. 331–369.

Nixon, R. P., 1973, Oil source beds in Cretaceous Mowry Shale of northwestern interior United States: Am. Assoc. Petroleum Geologists Bull., v. 57, no. 1, p. 136–161.

Perry, E. A., and J. Hower, 1972, Late-stage dehydration in deeply buried pelitic sediments: Am. Assoc. Petroleum Geologists Bull., v. 56, no. 10, p. 2013–2021.

Philippi, G. T., 1957, Identification of oil-source beds by chemical means, in Geologia del petroleo: 20th Internat. Geol. Cong., Mexico, (1956), sec. 3, p. 25–38.

——— 1965, On the depth, time, and mechanism of petroleum generation: Geochim. et Cosmochim. Acta, v. 29, no. 9, p. 1021–1049.

Powers, M. C., 1967, Fluid-release mechanisms in compacting marine mudrocks and their importance in oil exploration: Am. Assoc. Petroleum Geologists Bull., v. 51, no. 7, p. 1240–1254.

Tissot, B., and R. Pelet, 1971, Nouvelles donnees sur les mechanismes de genese et de migration du petrole simulation mathematique et application a la prospection: 8th World Petroleum Cong., Moscow, Preprints PD1 (4), 23 p.

Weaver, C. E., and K. C. Beck, 1971, Clay-water diagenesis during burial; how mud becomes gneiss: Geol. Soc. America Spec. Paper 134, 96 p.

Welte, D. H., 1972, Petroleum exploration and organic geochemistry: Jour. Geochem. Exploration, v. 1, p. 117–136.